Encyclopedia of

ELECTRONIC CIRCUITS

Volume 6

To Russell
May you always be as self-confident as you are today.

Patent notice

Encyclopedia of

ELECTRONIC

CIRCUITS

Volume 6

Rudolf F. Graf
&
William Sheets

McGraw-Hill

New York San Francisco Washington, D.C. Auckland Bogotá
Caracas Lisbon London Madrid Mexico City Milan
Montreal New Delhi San Juan Singapore
Sydney Tokyo Toronto

McGraw-Hill

A Division of The McGraw·Hill Companies

pbk 1 2 3 4 5 6 7 8 9 0 FGR/FGR 9 0 0 9 8 7 6
hc   1 2 3 4 5 6 7 8 9 0 FGR/FGR 9 0 0 9 8 7 6

Library of Congress Cataloging-in-Publication Data
(Revised for vol. 6)
Graf, Rudolf F.
      The encyclopedia of electronics circuits

      Authors for v. 6–   : Ruldolf F. Graf & William
Sheets.
      Includes bibliographical references and indexes.
      1. Electronic circuits—Encyclopedias. I. Sheets,
William.   II. Title
TK7867G66   1985   621.3815   84-26772
ISBN 0-8306-0938-5 (v. 1)
ISBN 0-8306-1938-0 (pbk. : v. 1)
ISBN 0-8306-3138-0 (pbk. : v 2)
ISBN 0-8306-3138-0 (v. 2)
ISBN 0-8306-3348-0 (pbk. : v 3)
ISBN 0-8306-7348-2 (v. 3)
ISBN 0-8306-3895-4 (pbk. : v 4)
ISBN 0-8306-3896-2 (v. 4)
ISBN 0-07-011077-8 (pbk. : v 5)
ISBN 0-07-011076-X (v. 5)
ISBN 0-07-011275-4 (v. 6)
ISBN 0-07-011276-2 (pbk. : v 6)

McGraw-Hill books are available at special quantity discounts to use as premiums and sales promotions, or for use in corporate training programs. For more information, please write to the Director of Special Sales, McGraw-Hill, 11 West 19th Street, New York, NY 10011. Or contact your local bookstore.

Acquisitions editor: Roland S. Phelps
Editorial team: Lori Flaherty, Executive Editor
                Andrew Yoder, Book Editor
                Joann Woy, Indexer
Production team: Katherine G. Brown, Director
                Rose McFarland, Desktop Operator
                Nancy Mickley, Proofreading
Design team: Jaclyn J. Boone, Designer
                Katherine Lukaszewicz, Associate Designer

EL1
0112762

Contents

Introduction

The enthusiastic reception of the first five volumes of *The Encyclopedia of Electronic Circuits* prompted the authors to produce this volume—the sixth in the popular series.

Taken together, the six volumes contain approximately 6000 circuits—by far the largest and broadest collection of practical electronic circuits available anywhere.

As in the other volumes, the 1000+ circuits presented here are arranged alphabetically, by category. All circuits in this volume, as well as those from the previous five volumes, are included in the index, which now has approximately 6000 entries.

We express sincere appreciation to the many electronic industry sources and publishers who graciously allowed us to utilize some of their materials. Their cooperation is gratefully acknowledged.

Once again, it gives us great pleasure to extend our sincerest thanks to Loretta Gonsalves-Battiste, a fine lady whose skill at the computer and willingness to work long and hard made on-time delivery of the manuscript for this book possible.

Rudolf F. Graf & William Sheets
September 1995

1

AGC and ALC Circuits

The sources of the following circuits are contained in the Sources section, which begins on page 707. The figure number in the box of each circuit correlates to the entry in the Sources section.

AGC AUDIO PREAMP

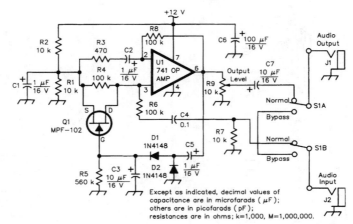

Except as indicated, decimal values of
capacitance are in microfarads (μF);
others are in picofarads (pF);
resistances are in ohms; k=1,000, M=1,000,000.

QST

Fig. 1-1

The circuit uses an easily obtained 741 op amp set for an internal gain of about 200. A portion of the op amp's output signal is rectified by the 1N4148 diodes, then filtered and fed to the gate of the FET input shunting circuit. As the output rises, more and more input shunting takes place. That is, more of the input signal is bypassed, effectively keeping the output level constant.

The circuit offers a 100:1 limiting action. The input level can change over a 100:1 ratio with little or no effect on the output level. The output level itself can be set from less than unity all the way up to nearly the gain of the amplifier, making the circuit usable in other applications as well.

3-MHz LOW-NOISE AGC SYSTEM

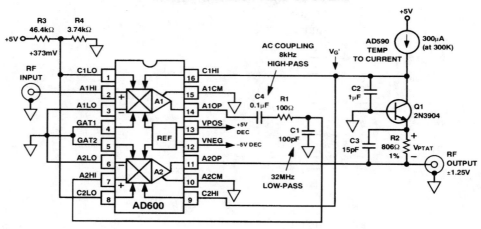

ANALOG DIALOG

Fig. 1-2

The AD600 dual voltage-controlled amplifier in this circuit provides a 3-MHz AGC system with 80-dB range.

IF AGC NETWORK

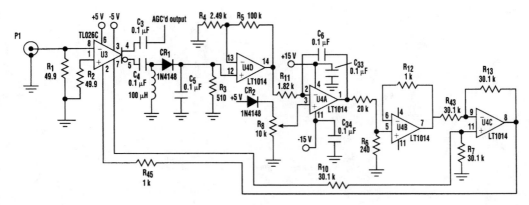

ELECTRONIC DESIGN

Fig. 1-3

A simple IF AGC circuit that features wide dynamic range and excellent linearity can be achieved with two chips: TI's TL026C voltage-controlled amplifier IC and Linear Technology's LT1014 (or any other similar basic quad op amp).

AUDIO LEVELER

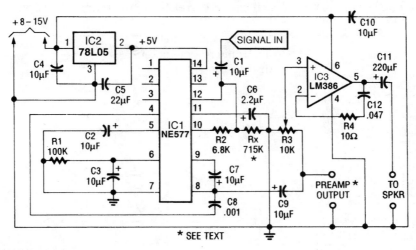

1994 EXPERIMENTERS HANDBOOK

Fig. 1-4

A low power programmable compandor chip, the Signetics NE577 IC is used. Incoming audio is compressed, rectified and conditioned so that the input signal level always remains about the noise level. The compressor is an ALC circuit that outputs a constant level and the expander part of the IC is not used.

3

2

Air-Flow Circuits

The sources of the following circuits are contained in the Sources section, which begins on page 707. The figure number in the box of each circuit correlates to the entry in the Sources section.

Hot-Wire Anemometer
Electronic Anemometer
Air Flow Detector

HOT-WIRE ANEMOMETER

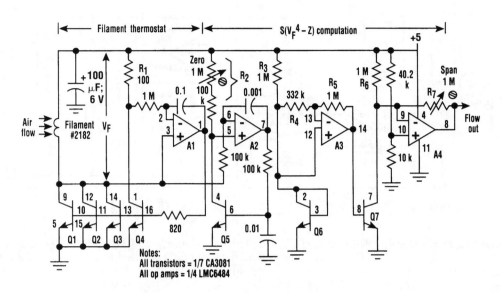

ELECTRONIC DESIGN

Fig. 2-1

An anemometer can be realized by utilizing the inherent transconductance match of transistors in the array, instead of passive series resistances, to control filament current. As a result, as A1 serves the collector current of Q4 and thereby the voltage across R1, it simultaneously adjusts the filament (a 2182-type incandescent lamp denuded of its glass envelope) voltage, V_f. The ratio of the filament to R1 current is stably maintained by the identical temperature and operating points of Q1 through Q4. The net result is that A1 drives the filament temperature to the value that causes filament resistance to equal $R1/3 = 33\ \Omega$. This is about double the cold resistance of the filament and therefore, assuming tungsten wire with a 0.0045/degree coefficient of resistance, represents a filament operating temperature of around 230°C. This is hot enough that moderate changes in ambient temperature are unimportant factors in filament power demand, but not so hot as to cause the filament to burn.

Rail-to-rail input amplifier A2 continuously serves the collector current of Q5 to V_f/R_2, making the V_{be} of LQ5 a logarithmic function of V_f. A3 multiplies this log by 4 and applies the product of Q7. Q7 does the antilog function so that its collector current is proportional to the fourth power of V_f. Thus, by King's law, it's proportional to air speed in the vicinity of the filament. This current is offset and scaled by A4 to produce a voltage output that, thanks to the rail-to-rail output capability of the LMC6484, can range from 0.01 to 4.99 V. Full-scale air speed can be adjusted, using R7, to any value in the range of 1 to 10 meters/s.

ELECTRONIC ANEMOMETER

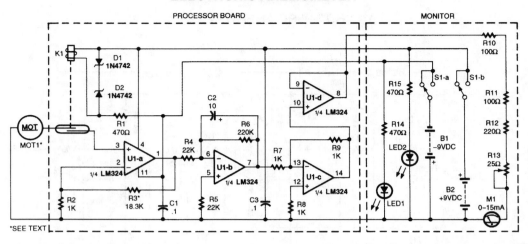

POPULAR ELECTRONICS

Fig. 2-2

A motor used as a generator is used as a transducer to generate a dc voltage that is proportional to wind speed. K1 prevents the transducer voltages from being applied to the circuit if no dc power is present. U1A through U1D is a dc amplifier, integrator, and buffer. This circuit drives the meter M1. The processor board is mounted in a housing along with the generator M1.

AIR FLOW DETECTOR

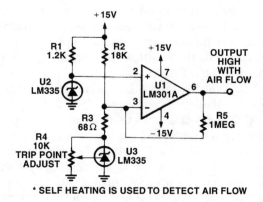

* SELF HEATING IS USED TO DETECT AIR FLOW

POPULAR ELECTRONICS

Fig. 2-3

The self heating of a semiconductor that is cooled by airflow is used as a sensing method.

3

Alarm and Security Circuits

The sources of the following circuits are contained in the Sources section, which begins on page 707. The figure number in the box of each circuit correlates to the entry in the Sources section.

Burglar Alarm Circuit
Auto Security System Transmitter
Home Security System
Auto Security System Receiver
Flashing Brake Light for Motorcycles
Car Alarm Decoy
Motorcycle Alarm
Simple Bike Horn
Door Ajar Indicator
Motorcycle Burglar Alarm
Horn Circuit for Motorcycle Use

BURGLAR ALARM CIRCUIT

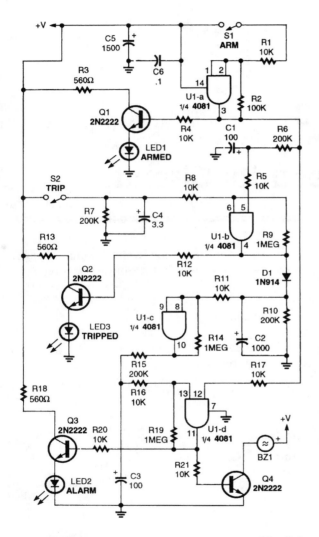

This alarm circuit is built around a single 4081 (CMOS) quad AND gate. It offers an exit and entry delay (around automatically reset two minutes after tripping, provided that the trip input is not left high).

The arming switch must go high to arm or low to disarm. After arming, U1-a begins to charge C1 via R6. Around 20 seconds later (after the exit delay), C1 has a sufficient charge to produce a high at the pin-5 input of U1-b. Also, when the circuit is armed, Q1 is turned on to indicate arming, and one input of U1-d is brought high.

After the exit delay times out, if the trip input opens, it causes an output on gate U1-b. Transistor Q1 is turned on, lighting the trip indicator (LED3), C2 instantly charges, and the output of U1-c goes high. At that point, C3 begins charging to provide the entry delay.

After 20 seconds, C3 has sufficient charge to produce a high at pin 13 of U1-d. That forces U1-d's output high, tuning Q3 and Q4 on, which activates the alarm indicator (LED2) and sounder (BZ1), respectively. If disarmed after a trip pulse, but before the 20-second, entry delay time out, pin 12 of U1-d goes low, so the gate's output does not go high and the alarm does not sound.

Components C2 and R10 hold U1-c on for around 2 minutes and 20 seconds to provide the two-minute alarm. After C2's charge drops below half of the supply voltage, U1-c's output goes low, awaiting another trip pulse to set it off again.

AUTO SECURITY SYSTEM TRANSMITTER

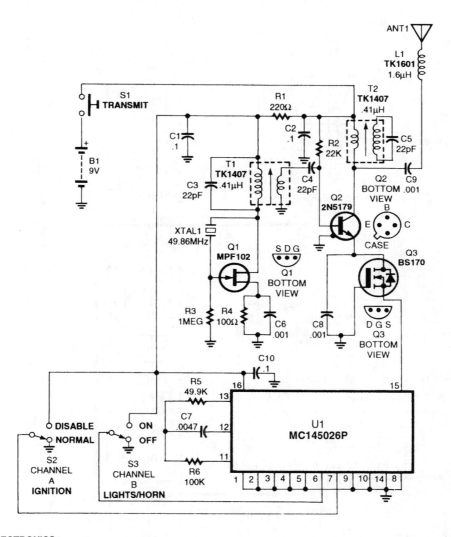

Fig. 3-2

This transmitter operates at 49 MHz and uses an M145026 programmable digital encoder to generate a unique digital code, depending on the positions of S2 and S3, to control ignition and lights or horn. Q1 is the oscillator, Q2 the power amplifier. The antenna is a 36-inch whip or wire antenna.

HOME SECURITY SYSTEM

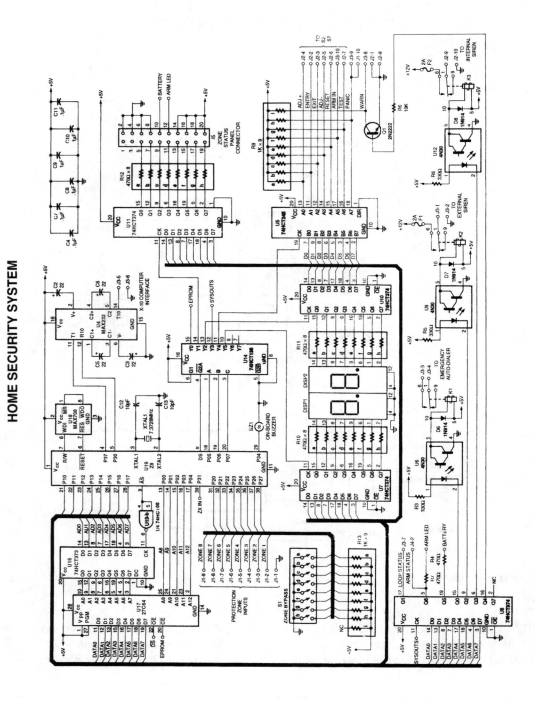

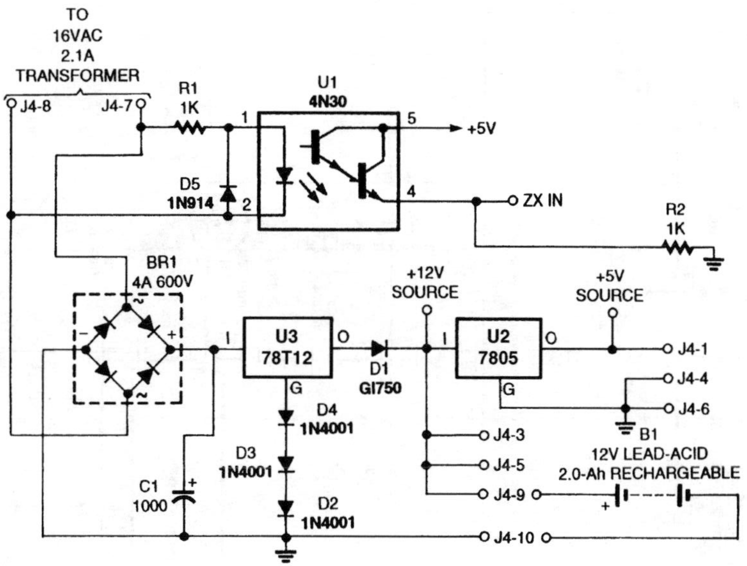

Fig. 3-3

TO 16VAC 2.1A TRANSFORMER

J4-8   J4-7

R1 1K

D5 1N914

U1 4N30

BR1 4A 600V

C1 1000

U3 78T12

D3 1N4001   D4 1N4001   D2 1N4001

D1 GI750

+12V SOURCE

U2 7805

R2 1K

ZX IN

+5V

+5V SOURCE

J4-1   J4-4   J4-6

B1 12V LEAD-ACID 2.0-Ah RECHARGEABLE

J4-3   J4-5   J4-9   J4-10

At the heart of the main-control unit is U16, Zilog's Z8-8-bit microcontroller, which receives its program instruction from U17, a 27C64 8K x 8 EPROM. The home security system in its most basic form, features eight individual protection zones, adjustable entry and exit delays, a panic switch (for emergency situations), automatic system reset, support for an auto-dialer (which, in case of emergency, dials pre-programmed telephone numbers), it's X-10 compatible (allowing it to control house lights and appliances), has a backup battery (to keep the system on-line during a power failure), and there is also an optional zone-status panel that is used to individually show the condition of each protection zone.

This system's power supply provides 12 Vdc for the sirens and digital keypad, and 5 Vdc for the on-board electronics, while also providing a constant 12-V output that's used to charge the backup-battery.

11

AUTO SECURITY SYSTEM RECEIVER

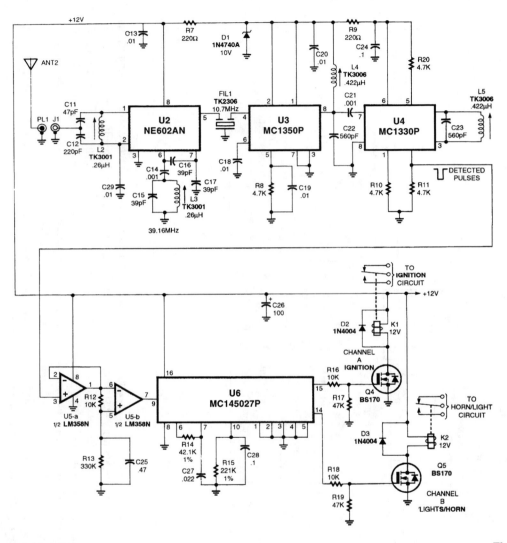

Fig. 3-4

This receiver is a superheterodyne type tuned to 49 MHz. U2 is a mixer, followed by a 10.7-MHz filter and two-stage IF (U3, U4) and detector. The encoded received RF pulse signal from the antenna produces detected pulse from the MC1330P. These pulses are amplified by U5 and fed to decoder IC U6, and MC1450278. Two channels are available at the output, which drives K1 and K2.

FLASHING BRAKE LIGHT FOR MOTORCYCLES

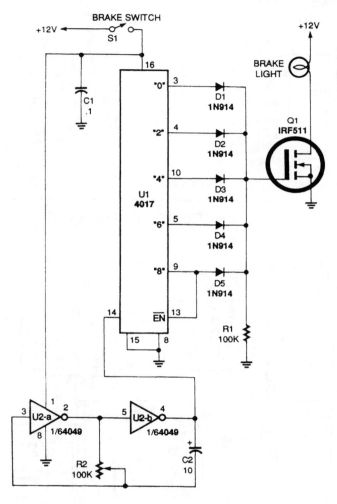

POPULAR ELECTRONICS

Fig. 3-5

When brake-light switch S1 is closed, power is applied to U1 and U2. Two inverters of U2, a 4049 hex inverting buffer, are connected in a low-frequency oscillator circuit that feeds clock pulses into U1, a 4017 decade counter/divider. Outputs 0, 2, 4, 6, and 8 of U1 are coupled to the gate of Q1 through a 1N914 diode. As the 4017 counts down, it turns the brake light on and off four times and then leaves it on until the brake switch is released. The on/off rate can be set by potentiometer R2; for best results, the on/off rate should be set so that it is rapid.

CAR ALARM DECOY

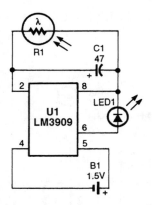

The device will simulate the presence of a burglar alarm in automobiles or homes. Mount R1 where daylight can fall on it. During darkness, LED1 flashes, making potential intruders think an alarm system is installed.

POPULAR ELECTRONICS

Fig. 3-6

MOTORCYCLE ALARM

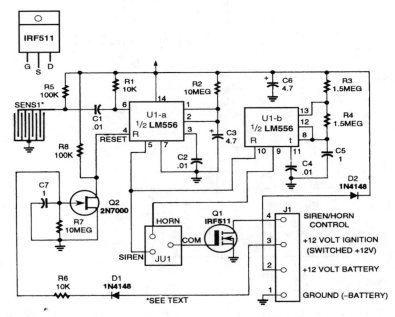

POPULAR ELECTRONICS

Fig. 3-7

A dual timer is used to generate a long pulse, which gates a second timer, producing a square wave (nonsymmetrical) and controls the on/off time of the horn. Siren operation can be selected with a jumper. In this case, the output of Q1 will be continuously on and not cycled. Sensor S1 is a row of adjacent circuit board traces with a stainless steel ball bearing laying on them. Any movement causes momentary shorting and opening of the circuit, triggering U1-a.

SIMPLE BIKE HORN

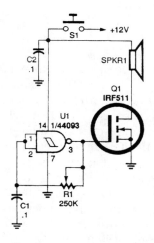

The horn circuit uses only one gate of a 4093 quad 2-input NAND Schmitt trigger, U1, connected in a simple, low-frequency, square-wave oscillator circuit. The oscillator's output, at pin 3, drives the gate of Q1. The drain of that FET drives a small horn speaker.

Potentiometer R1 can be adjusted to set the horn's output frequency. Some horn speakers are frequency sensitive, so play with the oscillator's frequency control for the best or loudest sound.

Fig. 3-8

DOOR AJAR INDICATOR

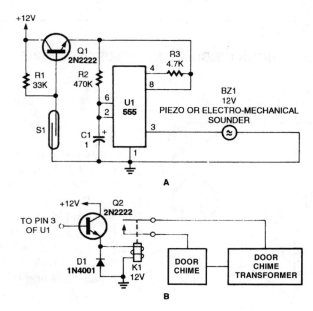

A

B

Fig. 3-9

This simple sounder (A) makes a good door annunciator. If the buzzer is replaced with the circuit in B, the annunciator can be made more pleasant to the ear.

MOTORCYCLE BURGLAR ALARM

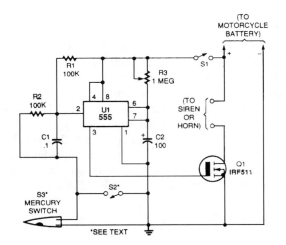

A 555 IC is connected in a one-shot timer circuit that turns on a FET transistor and either a siren or the bike's horn for a preset time period. Switch S1 is used as an on/off switch.

Closing either of two switches, S2 and S3, will trigger the IC. When either switch closes, pin 2 of U1 goes low. That triggers the IC to produce a positive output at pin 3 and sounds the alarm for the time period set by R3. The mercury switch, S3, is the switch that activates the alarm should anyone move your bike. Switch S2 can be used as a panic switch.

POPULAR ELECTRONICS

Fig. 3-10

HORN CIRCUIT FOR MOTORCYCLE USE

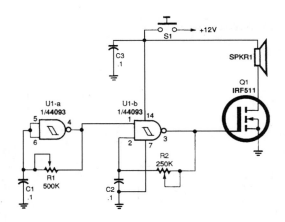

Gates U1-a and U1-b of the 4093 quad 2-input NAND Schmitt trigger are connected in variable, low-frequency, square-wave oscillator circuits. The output of gate U1-a is connected to one of the inputs of gate U1-b. The square-wave output of gate U1-a modulates oscillator U1-b, producing a two-tone output. A really interesting sound can be produced by carefully adjusting potentiometers R1 and R2.

POPULAR ELECTRONICS

Fig. 3-11

4

Amateur Circuits

The sources of the following circuits are contained in the Sources section, which begins on page 707. The figure number in the box of each circuit correlates to the entry in the Sources section.

1.2-kW 144-MHz Amplifier Power Supply
1.2-kW 144-MHz Amplifier Control Circuitry
1.2-kW 144-MHz Linear Amplifier
Four-Stage 75-Meter SSB Superhet Receiver
Improved CW Transmitter Keying Circuit
One-Chip AFSK Generator
Programmable CW Identifier
Audible SWR Detector Adapter
Audio Breakout Box
One-Watt CW Transmitter
PTT Control from Receiver Audio
Transceiver Memory Backup
80-Meter SSB Receiver
CW Audio Filter
RF Line Sampler/Coupler
Battery Pack and Reverse Polarity Protection
Simple Identifier
Transmit Keyer Interface Circuits
Mobile Radio On-Alarm Timer

1.2-kW 144-MHz AMPLIFIER POWER SUPPLY

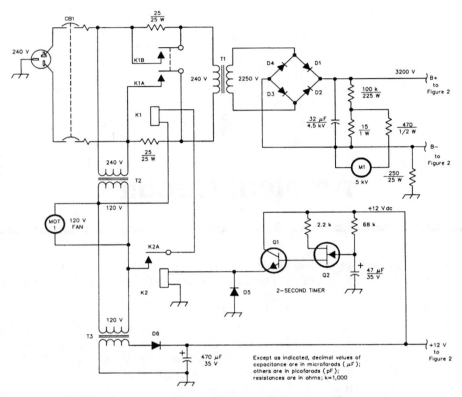

Schematic diagram of the high-voltage power supply recommended for use with the power amplifier.

D1-D4—Strings of 4 each, 1000-PIV, 3-A
   diodes, 1N5408 or equivalent)
K1—DPST relay, 120-V ac coil, 240-V-ac,
   20-A contacts (Midland Ross
   187-321200 or equivalent)
K2—SPDT miniature relay, 12-V dc coil
   (Radio Shack 275-248 or equivalent)
M1—High-voltage meter, 5 kV dc full scale

(1-mA meter movement used with series
   resistors shown in drawing)
MOT1—Cooling fan, Torin TA-300 or
   equivalent
Q1—2N2222A or equivalent
Q2—MPF102 or equivalent
S1—20-A hydraulic/magnetic circuit
   breaker (Potter and Brumfield
   W68X2Q12-20 or equivalent)

T1—High-voltage power transformer,
   240-V primary, 2250-V, 1.2-A secondary
   (Avatar AV-538 or equivalent)
T2—Stepdown transformer, Jameco
   112125, 240-V to 120-V, 100 VA
T3—Power transformer, Jameco 104379,
   120-V primary; 16.4 V, 1-A secondary
   (half used)

QST

Fig. 4-1

A schematic diagram of the high-voltage power supply recommended for use with the power transformer. This power supply can also be used for other equipment with similar requirements. CAUTION: hazardous high voltages.

1.2-kW 144-MHz AMPLIFIER CONTROL CIRCUITRY

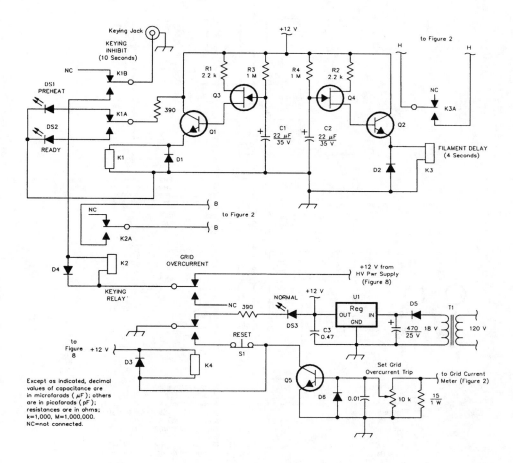

Fig. 4-2

Schematic diagram of the amplifier-control circuits.

1.2-kW 144-MHz LINEAR AMPLIFIER

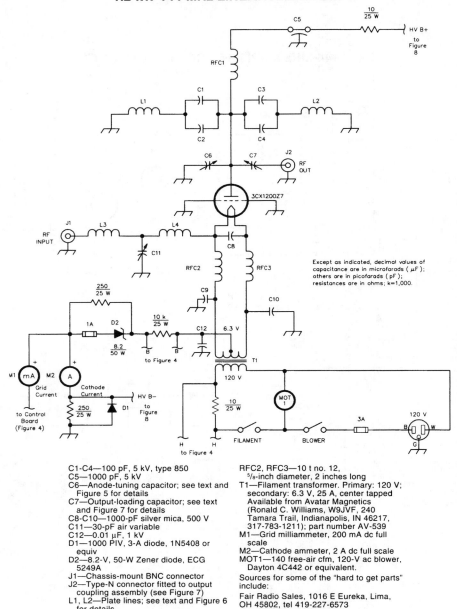

C1-C4—100 pF, 5 kV, type 850
C5—1000 pF, 5 kV
C6—Anode-tuning capacitor; see text and Figure 5 for details
C7—Output-loading capacitor; see text and Figure 7 for details
C8-C10—1000-pF silver mica, 500 V
C11—30-pF air variable
C12—0.01 µF, 1 kV
D1—1000 PIV, 3-A diode, 1N5408 or equiv
D2—8.2-V, 50-W Zener diode, ECG 5249A
J1—Chassis-mount BNC connector
J2—Type-N connector fitted to output coupling assembly (see Figure 7)
L1, L2—Plate lines; see text and Figure 6 for details
L3—5 t no. 14, ¹⁄₂-inch diameter, close wound
L4—3 t no. 14, ⁵⁄₈-inch diameter, ¹⁄₄-inch spacing
RFC1—7 t no.14, ⁵⁄₈-inch diameter, 1³⁄₈ inch long

RFC2, RFC3—10 t no. 12, ⁵⁄₈-inch diameter, 2 inches long
T1—Filament transformer. Primary: 120 V; secondary: 6.3 V, 25 A, center tapped Available from Avatar Magnetics (Ronald C. Williams, W9JVF, 240 Tamara Trail, Indianapolis, IN 46217, 317-783-1211); part number AV-539
M1—Grid milliammeter, 200 mA dc full scale
M2—Cathode ammeter, 2 A dc full scale
MOT1—140 free-air cfm, 120-V ac blower, Dayton 4C442 or equivalent.

Sources for some of the "hard to get parts" include:

Fair Radio Sales, 1016 E Eureka, Lima, OH 45802, tel 419-227-6573

Surplus Sales of Nebraska, 1502 Jones Street, Omaha, NE 68102, tel 402-346-4750.

Fig. 4-3

Schematic diagram of the 2-meter amplifier.

FOUR-STAGE 75-METER SSB SUPERHET RECEIVER

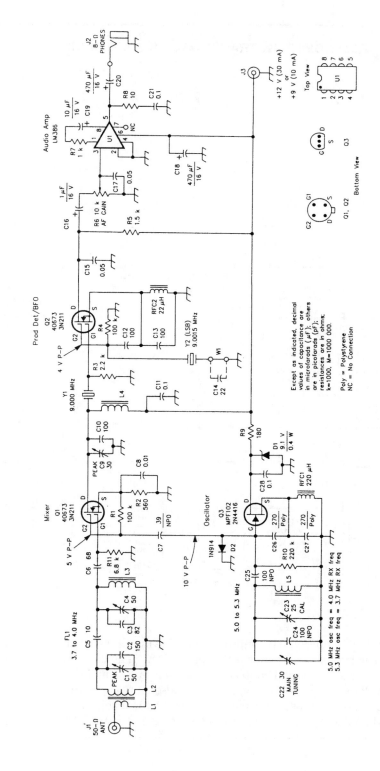

Fig. 4-4

A simple superhet receiver for SSB reception in the 75-meter amateur band is shown. Y1 acts as a crystal filter.

IMPROVED CW TRANSMITTER KEYING CIRCUIT

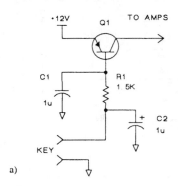

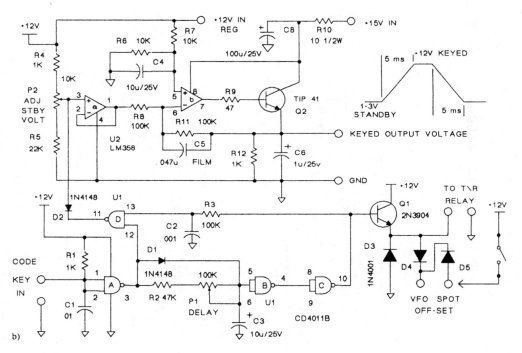

Fig. 4-5

Typical (A) QRP keying circuit; (B) Improved QRP keying circuit with CMOS T/R keying control. Op amp U2b is a basic inverting amplifier with a gain of one. The capacitor C5 across the feedback resistor R11 makes it an integrator. The RC time constant of R11 and C5 determine the ramp time. The values shown will produce a 5-ms ramp. Use a good-quality capacitor for C5, such as a mylar or polypropylene type. A power transistor is placed inside the feedback loop so that the circuit can supply several hundred milliamperes of current. Control P2 sets the stand-by output voltage as seen at

IMPROVED CW TRANSMITTER KEYING CIRCUIT (*Cont.*)

the emitter of Q1. U2a buffers the voltage from P2. This isolates the pot from the input of the integrator. With the key up, adjust the pot until you just start to see an output from your transmitter, then back it off a little. Typically, this will be between 2 and 4 V. Your output signal will now have the proper 5-ms leading and falling edges and there will be no delay between key closure and the start of the output signal.

You must supply the op amp and collector of Q2 with at least 15 V to produce a full 12-V output on the emitter.

Parts list

R1, R4, R12	1 kΩ ¼ W	C1	0.01 µF disk
R2	47 kΩ ¼ W	C2	0.001 µF disk
R3, R8, R11	100 kΩ	C3, C4	10 µF, 25 V electrolytic
R6, R7	22 kΩ	C5	0.047 µF poly-film type
R9	47 Ω	C6	1 µF, 25 V electrolytic
R10	10 Ω	C7	Skipped
P1	100 kΩ or 500 kΩ trimpot	C8	100 µF, 26 V electrolytic
P2	10 kΩ trimpot	D1, D2	1N4148 diode
		D3, D4, D5	1N4001 1-A diode
Q1	2N3904 NPN		
Q2	Tip 41-to-220 NPN		
U1	4011B CMOS NAND gates		
U2	LM358 dual op amp		

ONE-CHIP AFSK GENERATOR

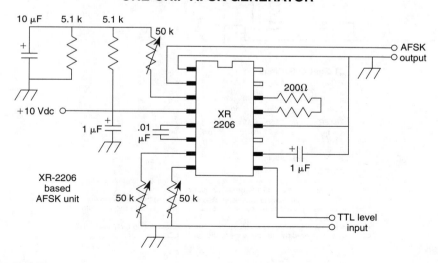

Fig. 4-6

Built around an XR2206 IC, this circuit will generate AFSK signals in the 1000- to 3000-Hz range.

PROGRAMMABLE CW IDENTIFIER

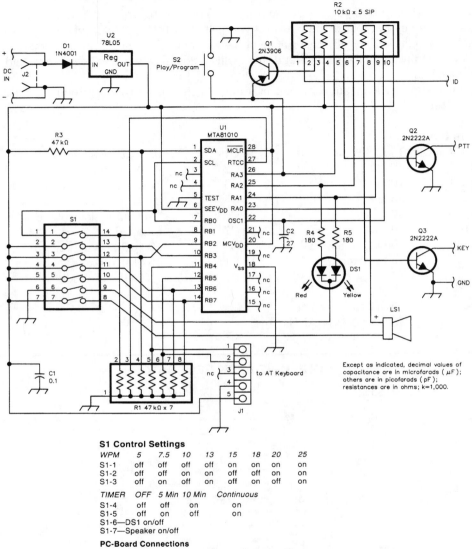

S1 Control Settings

WPM	5	7.5	10	13	15	18	20	25
S1-1	off	off	off	off	on	on	on	on
S1-2	off	off	on	on	off	off	on	on
S1-3	off	on	off	on	off	on	off	on

TIMER	OFF	5 Min	10 Min	Continuous
S1-4	off	off	on	on
S1-5	off	on	off	on

S1-6—DS1 on/off
S1-7—Speaker on/off

PC-Board Connections

ID—A momentary ground on this terminal causes the IDer to play its message; same as pressing the **PLAY/PROGRAM** pushbutton.

PTT—An open-collector output which goes to ground 250 ms before the CW output occurs. This output is used to place radio in transmit mode and is monitored by the red LED.

KEY—An open-collector output that goes to ground during CW keying. This output is monitored by the speaker and the yellow LED.

QST

Fig. 4-7

The identifier uses an MTA81010 microchip, containing a 1024-bit serial EEPROM and a micro-controller. It runs from a 9-V battery. A standard AT-type keyboard is used to program the desired message. Speed varies from 5 to 25 wpm.

AUDIBLE SWR DETECTOR ADAPTER

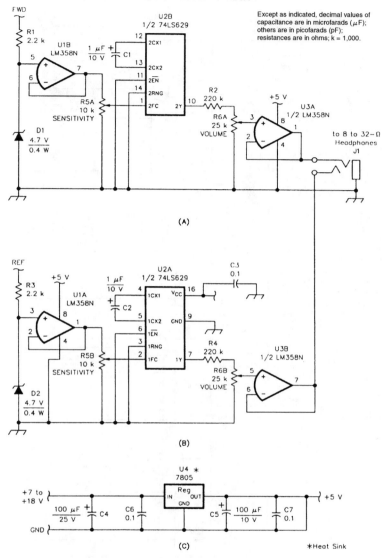

Except as indicated, decimal values of capacitance are in microfarads (μF); others are in picofarads (pF); resistances are in ohms; k = 1,000.

(A)

(B)

(C)

*Heat Sink

Fig. 4-8

This SWR detector audio adapter is designed specifically for blind or vision-impaired amateurs, but anyone can use it. Instead of using a meter (or meters) to indicate antenna system forward and reflected voltages, this adapter generates two tones with frequencies that are proportional to the respective voltages. The tones are fed to a pair of stereo headphones (the miniature types are ideal) so that one ear hears the forward-voltage tone and the other ear hears the reflected-voltage tone. Thus, tuning up a transmitter is simply a matter of tuning for the highest-pitched tone in the left ear and the lowest-pitched tone in the right ear.

AUDIO BREAKOUT BOX

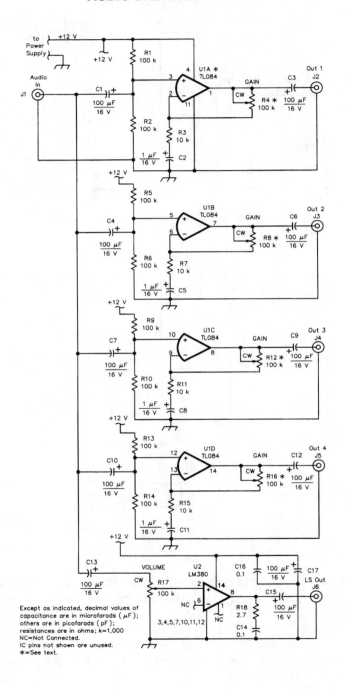

Except as indicated, decimal values of
capacitance are in microfarads (μF);
others are in picofarads (pF);
resistances are in ohms; k=1,000
NC=Not Connected.
IC pins not shown are unused.
✳=See text.

Fig. 4-9

AUDIO BREAKOUT BOX (*Cont.*)

In many radio shacks, one receiver audio-output line feeds a multitude of add-ons, such as one or more TNXs, SSTV modems, PC plug-in boards, and, perhaps, speakers. Having to manually plug the audio source from one accessory to another is inconvenient, if not frustrating as well. Overloading the sources by connecting the loads in parallel isn't satisfactory, either.

The audio breakout box takes the audio output from a receiver (or other audio source) and applies it to the inputs of four identical, independent, low-level AF buffer/amplifiers and one high-level (1-W output) AF channel. Each low-level output channel can provide up to 20 dB of gain that's independently adjustable.

ONE-WATT CW TRANSMITTER

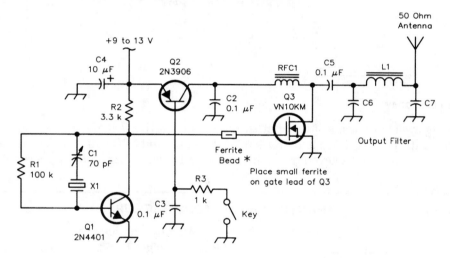

QST

Fig. 4-10

C6,C7
820 pF disc ceramic (160 meters)
470 pF disc ceramic (80 meters)
220 pF disc ceramic (40 meters)
150 pF disc ceramic (30 meters)
100 pF disc ceramic (20 meters)
82 pF disc ceramic (17 meters)

L1
33 turns, #30, T37-2 (160 meters)
23 turns, #30, T37-2 (80 meters)
17 turns, #26, T37-2 (40 meters)
14 turns, #26, T37-2 (30 meters)
12 turns, #26, T37-2 (20 meters)
10 turns, #26, T37-2 (17 meters)

PTT CONTROL FROM RECEIVER AUDIO

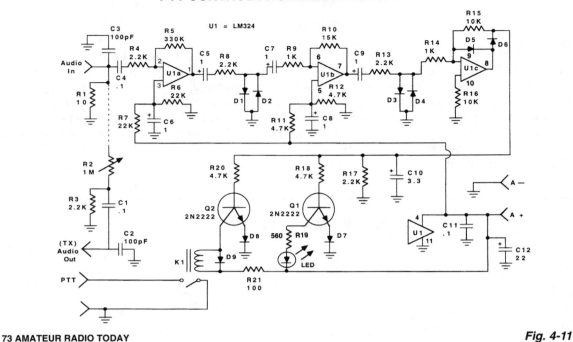

Fig. 4-11

This system will key a transmitter or other device that can be keyed by a relay closure. Audio is amplified, limited, and is rectified and drives relay driver Q2 and LED indicator. The transmitter audio output was used to feed a keyed transmitter and can be deleted or ignored where this feature is unnecessary.

TRANSCEIVER MEMORY BACKUP

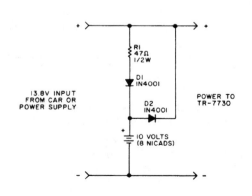

Although designed for a Kenwood TR7730, this idea might be adapted to other transceivers. This circuit will retain the frequencies in memory while moving the rig from car to house and vice versa. When connected to an external power source, battery B1 is charged through R1 and D1. D1 prevents B1 from discharging when connected to an external supply that is turned off. When external power is removed, D2 provides a current path to the TR-7730 to retain the memory's contents. However, the TR-7730 power switch should be turned off before external power is removed because B1 will not provide power for normal operation.

Fig. 4-12

80-METER SSB RECEIVER

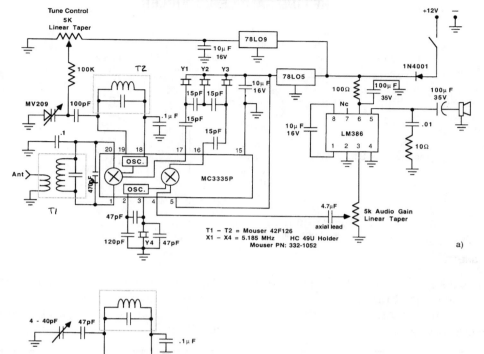

Fig. 4-13

This circuit uses an MC3335P IF chip and features a 3-pole crystal filter made from micro-processor crystals. Tuning is done either with a varactor diode or air-variable capacitor, as shown. Values are for 80 meters.

CW AUDIO FILTER

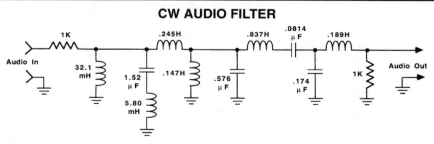

Fig. 4-14

A high-performance passive filter. The center frequency is 700 Hz; –3-dB bandwidth is 200 Hz.

RF LINE SAMPLER/COUPLER

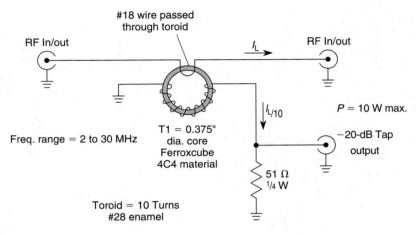

WILLIAM SHEETS

Fig. 4-15

Suitable for RF line sampling, this coupler is useful where an indirect measurement of line current is needed. A 10:1 turn ratio yields a secondary current about ⅒ (ideally) of the line current. A 51-Ω resistor terminates the secondary. Insertion loss in the main line is negligible, < 0.1 dB. For higher power levels, use proportionately larger core for T1.

BATTERY PACK AND REVERSE POLARITY PROTECTION

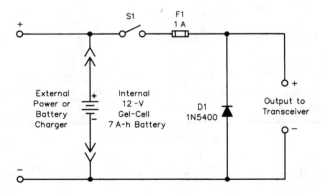

QST

Fig. 4-16

Schematic diagram and parts list for the reverse-polarity protection circuit (optional).

D1   1N5400 silicon diode
F1   1-A fast-acting fuse
S1   SPST rocker switch

SIMPLE IDENTIFIER

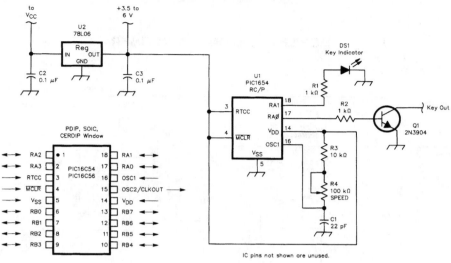

Fig. 4-17

This identifier uses a PIC 16C54 microcontroller which must be programmed for your desired identifier.

TRANSMIT KEYER INTERFACE CIRCUITS

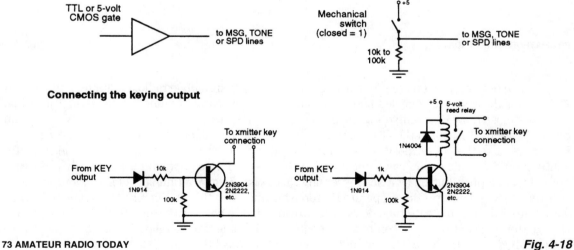

Fig. 4-18

These circuits are for use with Morse keyers and identifiers. They can be used to interface various devices with the identifier circuitry.

MOBILE RADIO ON-ALARM TIMER

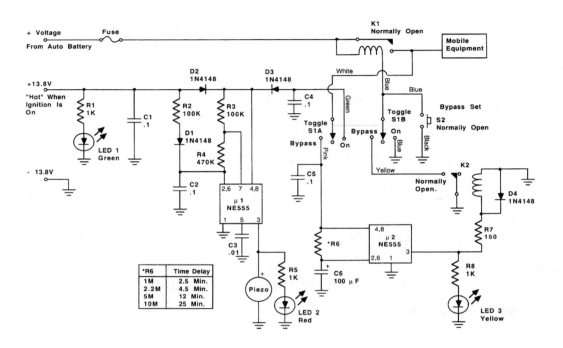

Fig. 4-19

This system will prevent you from accidentally leaving your mobile radio on, draining the battery. LED1 will light when the vehicle's ignition is on, or while the car is running. Switch S1 in the ON position will close relay K1, completing the power circuit to the equipment. If the ignition switch is shut off, and switch S1 is still in the ON position, an alarm (piezo) will begin to beep and LED2 will flash. Returning S1 to the center position will shut everything off. If equipment operation is desired after shutting off the vehicle, you can place switch S1 in the AUTO position and momentarily press S2, a normally open push-button switch. Depressing this switch begins a timing cycle. The length of time that the Mobile-ON alarm/timer operates before shutting everything off can be "programmed" by selecting R6. The approximate time delays are provided in the chart with the schematic. Or, you could change the value of C6. These components control the holding time of relay K1. LED3 will light while the circuit is in AUTO status. Incidentally, you can also cancel the time delay at any time during the delay period by simply switching it off.

5

Amateur Television (ATV) Circuits

The sources of the following circuits are contained in the Sources section, which begins on page 707. The figure number in the box of each circuit correlates to the entry in the Sources section.

5-W ATV Transmitter for 440 MHz
5-W ATV Transceiver
Mini ATV Transmitter
Dummy Load and Video Detector for Transmitter Tests
Mast-Mounted ATV Preamp
Three-Channel 902- to 928-MHz ATV Transmitter
ATV Downconverter for 902 to 928 MHz
Three-Channel 420- to 450-MHz ATV Transmitter
ATV Downconverter for 420 to 450 MHz

5-W ATV TRANSMITTER FOR 440 MHz

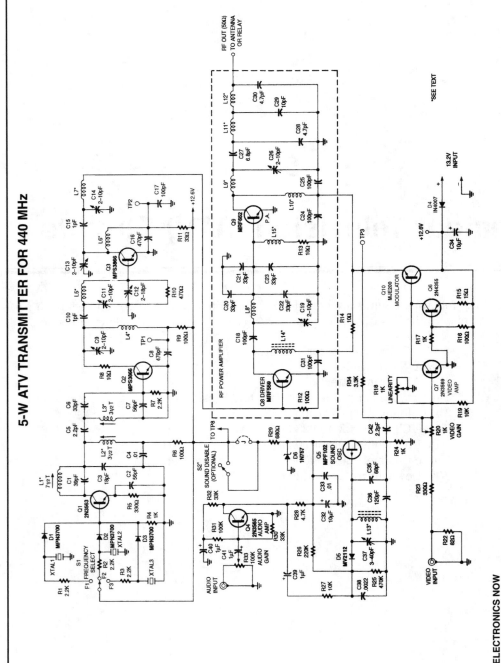

Fig. 5-1

ELECTRONICS NOW

The circuit will produce typically 6 W RF output on synch tips. A crystal oscillator drives a doubler to produce a 220-MHz output. Another doubler produces 440 MHz to drive the power amplifier. A high-level series modulator provides the video modulation capability. A sound subcarrier is generated using a VCO circuit and combined with the video information. A complete kit of parts, including the PC board, is available from North Country Radio, P.O. Box 53, Wykagyl Station, New Rochelle, NY 10804-0053A.

5-W ATV TRANSCEIVER

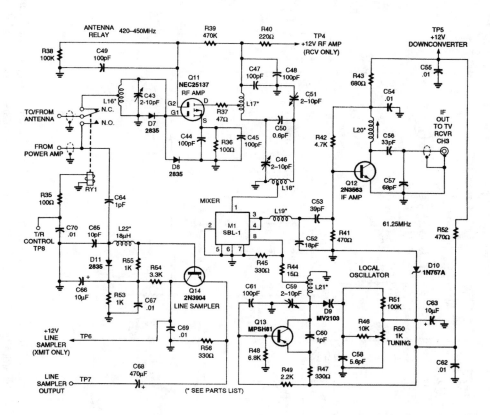

Fig. 5-2

For the transmitter schematic (part of this transceiver), see entry entitled "5-W ATV Transmitter for 440 MHz, Fig. 5-1." The downconverter portion is shown here.

This transmitter contains both a video and sound section. Five to six watts PEP on synch tips of NTSC video are produced. Three channels are available. Channel switching is via PIN diodes. Power supply voltage is 12 to 14 Vdc. The receiver function is provided with a downconverter circuit and is tunable. A relay is used for T-R switching. A complete kit of parts, including PC board, is available from North Country Radio, P.O. Box 53, Wykagyl Station, New Rochelle, NY 10804-0053A.

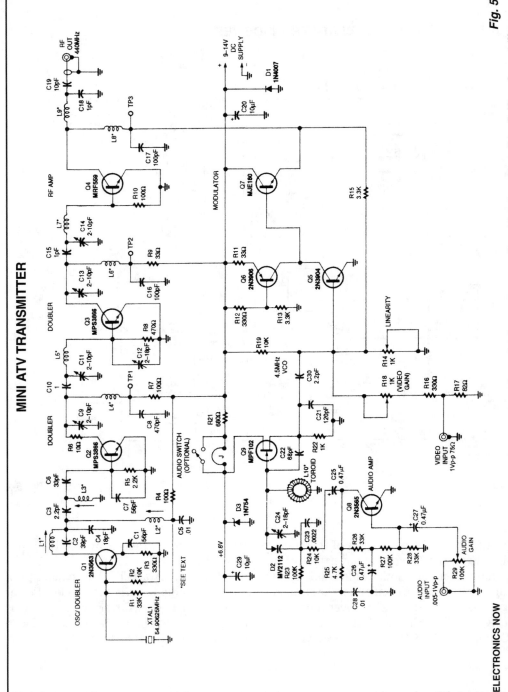

MINI ATV TRANSMITTER

Fig. 5-3

ELECTRONICS NOW

This low-power NTSC video and sound transmitter is useful for amateur radio, video handie-talkie, R/C and surveillance purposes. A crystal oscillator-multiplier RF power amplifier. Video modulation is via a three-transistor series modulator. The sound subcarrier is generated with a VCO circuit and is combined with the video information. The output is 0.4 to 1.2 W with supply voltages of 9 to 14 volts. A complete kit of parts, including PC board, is available from North Country Radio, P.O. Box 53, Wykagyl Station, New Rochelle, NY 10804-0053A.

DUMMY LOAD AND VIDEO DETECTOR FOR TRANSMITTER TESTS

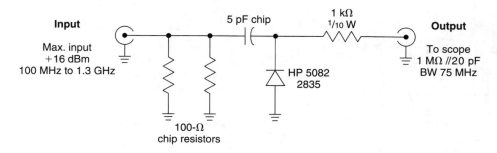

WILLIAM SHEETS

Fig. 5-4

This circuit is useful as a video modulation monitor for testing low-power video transmitters. For higher power inputs, use a suitable attenuator between the detector and the source. The detector should be connected to scope with as short a cable as possible to preserve video bandwidth.

MAST-MOUNTED ATV PREAMP

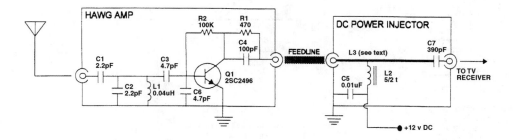

73 AMATEUR RADIO TODAY

Fig. 5-5

This simple ATV preamp covers the 427- to 439-MHz ATV frequencies and can be mast mounted and dc powered through the feedline.

THREE-CHANNEL 902- TO 928-MHz ATV TRANSMITTER

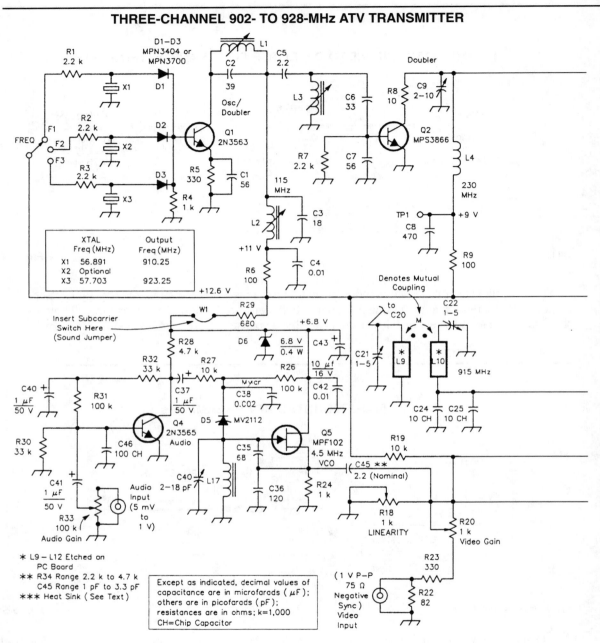

XTAL Freq (MHz)		Output Freq (MHz)
X1	56.891	910.25
X2	Optional	
X3	57.703	923.25

Insert Subcarrier Switch Here (Sound Jumper)

* L9 – L12 Etched on PC Board
** R34 Range 2.2 k to 4.7 k
   C45 Range 1 pF to 3.3 pF
*** Heat Sink (See Text)

Except as indicated, decimal values of capacitance are in microfarads (µF); others are in picofarads (pF); resistances are in ohms; k=1,000 CH=Chip Capacitor

This transmitter is for ATV applications in the 902- to 928-MHz band. It has three crystal-controlled channels, and will accept standard NTSC video input. It also has a 4.5-MHz sound subcarrier. Because this is an AM transmitter, audio can be transmitted as AM on the RF carrier. Simply use the

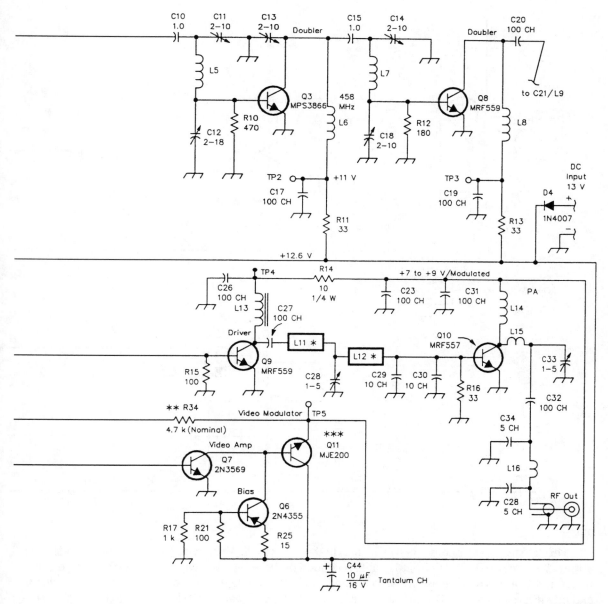

Fig. 5-6

video input. Bandwidth of audio can be restricted to 20 kHz by placing a capacitor with a value of about 0.002 µF across R34. The output is 1.5 to 2 watts PEP into a 50-Ω load. A complete kit of parts including PC board, is available from North Country Radio, P.O. Box 53, Wykagyl Station, New Rochelle, NY 10804-0053A.

ATV DOWNCONVERTER FOR 902 TO 928 MHz

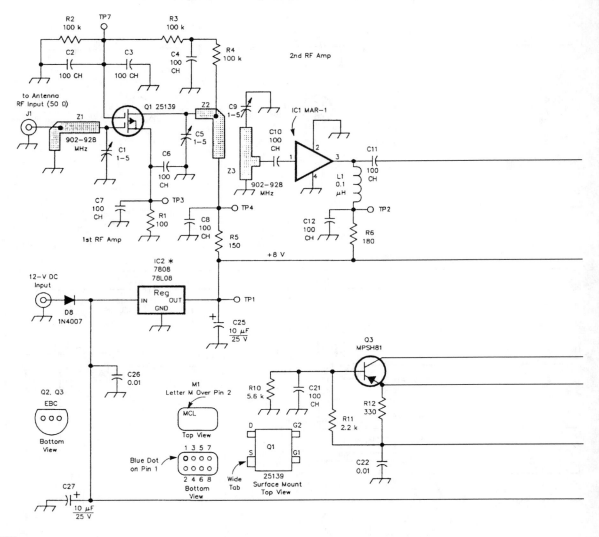

QST

This downconverter converts ATV signals in the 902- to 928-MHz range to a 61.25- or 67.25-MHz IF output frequency (CH 3 or CH 4) to enable reception of these signals on a standard VHF TV receiver or monitor. It features a low-noise RF amp feeding a Schottky diode double-balanced mixer, a tunable LO and one IF preamp stage. The RF amplifier is a low-noise dual-gate GASFET that is followed by a second RF stage using an MMIC. Five tuned circuits are used in the RF amplifier. This feeds a packaged Schottky diode mixer assembly for better dynamic range and reduced susceptibility to intermodulation and strong signal areas. The on-board local oscillator (LO) is voltage tuned and if desired can be set up for remote tuning. All necessary circuitry for remote tuning is on board for coax dc and IF feed. This en-

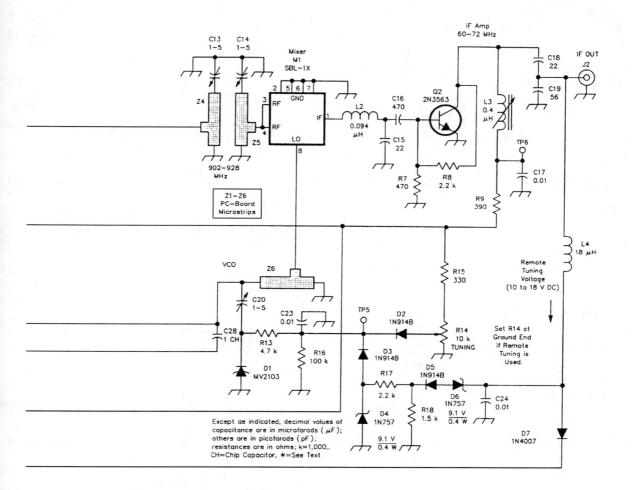

Fig. 5-7

ables the downconverter to be mast mounted to get around feedline losses generally associated with this frequency range. No separate dc feed is necessary because the coax (RG59/U recommended) carries dc power, tuning voltage, and IF signal. A dc block is used at the receiver for the purpose of separating dc voltage supply and the tuning voltage. This allows a cable run of several hundred feet, if needed.

By using this downconverter and transmitter, a physically small 915-MHz ATV station or even a video HT can be constructed because both units are each 2.50 × 4.00 inches × 1.00 high, and can be stacked together. A complete kit of parts, including PC board is available from North Country Radio, P.O. Box 53, Wykagyl Station, New Rochelle, NY 10804-0053A.

THREE-CHANNEL 420- TO 450-MHz ATV TRANSMITTER

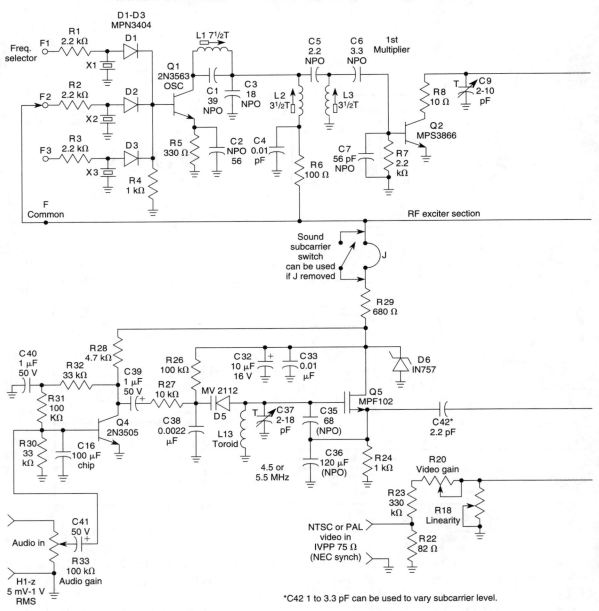

RUDOLF F. GRAF AND WILLIAM SHEETS

This transmitter is a 2-watt PEP output device for 420- to 450-MHz amateur TV operation. It has three crystal-controlled channels and will accept standard NTSC video input. It also has a 4.5-MHz sound subcarrier capability. Because this transmitter has AM modulation, audio can be transmitted in

42

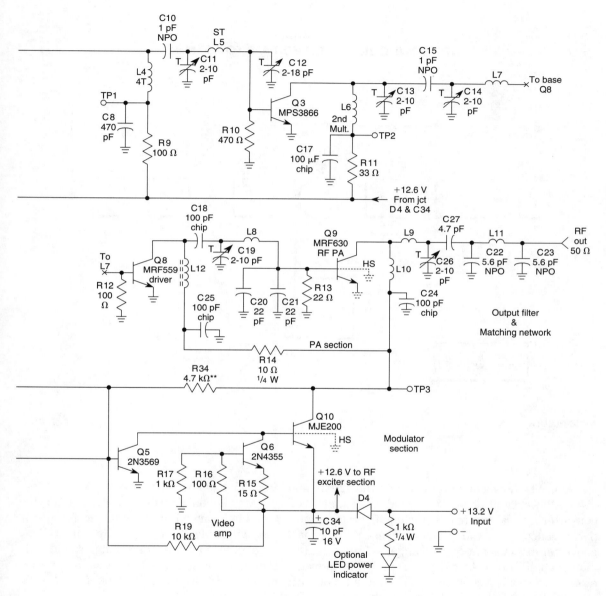

Fig. 5-8

AM form on the RF carrier by applying audio to the video input. Bandwidth of audio can be restricted to 20 kHz by placing a 0.002-μF capacitor across R34. A complete kit of parts, including PC board is available from North Country Radio, P.O. Box 53, Wykagyl Station, New Rochelle, NY 10804-0053A.

ATV DOWNCONVERTER FOR 420 TO 450 MHz

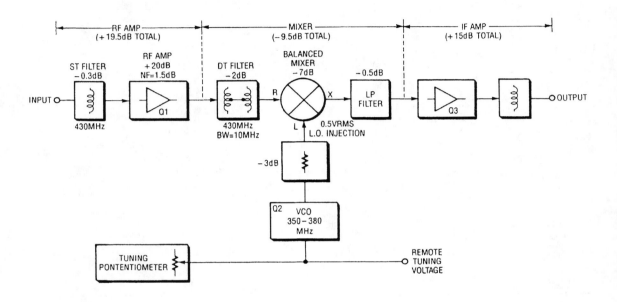

Fig. 5-9

This ATV downconverter converts the 420- to 450-MHz ATV band, which is several channels below the lower limit of the UHF band, to channel 3 or 4 for viewing on virtually any TV. The downconverter has a low-noise preamplifier stage and a double-balanced passive mixer for good performance and a wide dynamic range. That is necessary with today's crowded UHF bands. The converter draws about 27 milliamperes from a 13.2-volt dc source, so it can be used in portable and mobile applications. An extra IF stage gives an overall gain of about 25 dB. A block diagram of the downconverter is also shown. A complete kit of parts, including PC board, is available from North Country Radio, P.O. Box 53, Wykagyl Station, New Rochelle, NY 10804-0053A.

ATV DOWNCONVERTER FOR 420 TO 450 MHz (Cont.)

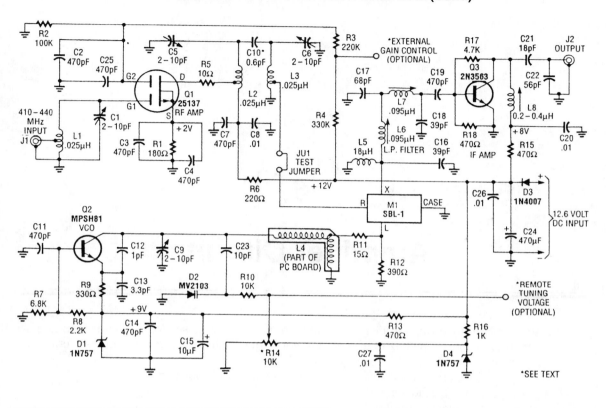

6

Amplifier Circuits

The sources of the following circuits are contained in the Sources section, which begins on page 707. The figure number in the box of each circuit correlates to the entry in the Sources section.

Operational Transconductance Amplifier with Booster
VCR Head Amplifier Tester
Lowpass Amplifier
Highpass Amplifier
ISD 1000A Record/Playback Circuit
Remote Amplifier
Programmable Gain Amplifier
Programmable Input Amplifier
Remotely Powered Sensor Amplifier
Tuned Amplifier
Difference Amplifier with Wide Input Common-Mode Range
Bandpass Amplifier
High-Side Current-Sensing Amplifier
High-Input Impedance ac Amplifier
MOSFET Push-Pull Amplifier
Low-Voltage Microphone Preamp
Basic Logarithmic Amplifier Using Op Amp
Crystal Tuned Amplifier

OPERATIONAL TRANSCONDUCTANCE AMPLIFIER WITH BOOSTER

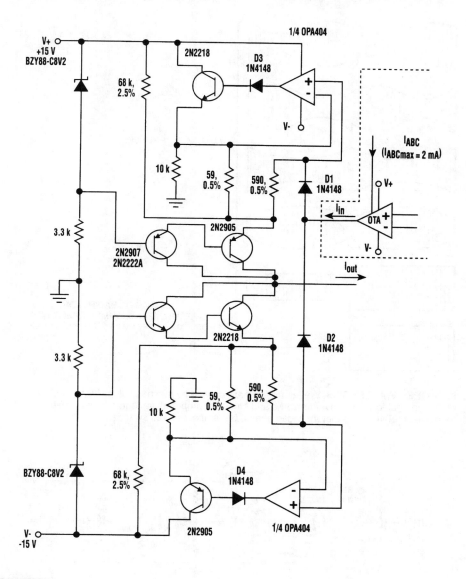

Fig. 6-1

Implementing a bidirectional precision current amplifier in an operational transconductance amplifier (OTA) can boost the OTA's output current. To accomplish this task, two diodes and a complementary stage are added to this otherwise simple design.

VCR HEAD AMPLIFIER TESTER

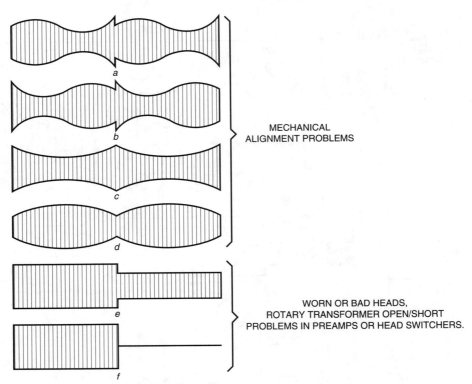

MECHANICAL
ALIGNMENT PROBLEMS

WORN OR BAD HEADS,
ROTARY TRANSFORMER OPEN/SHORT
PROBLEMS IN PREAMPS OR HEAD SWITCHERS.

IMPROPER WAVEFORMS. Waveforms *a*-*d* are caused by mechanical misalignment of the tape guides. The waveforms in *e* and *f* indicate proper alignment, but show that there's a problem with either the video heads, pre-amps, or head switcher.

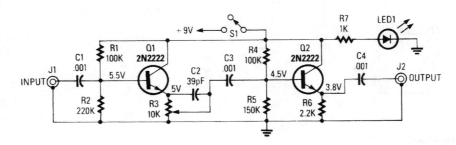

R-E EXPERIMENTERS HANDBOOK

Fig. 6-2

This amplifier enables you to use a signal from a working VCR to test the head amplifiers of a suspected defective VCR. The circuit is basically a video amplifier.

48

LOWPASS AMPLIFIER

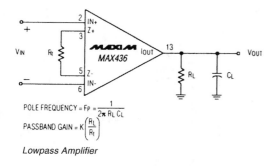

$$\text{POLE FREQUENCY} = F_P = \frac{1}{2\pi\, R_L\, C_L}$$

$$\text{PASSBAND GAIN} = K\left(\frac{R_L}{R_t}\right)$$

Lowpass Amplifier

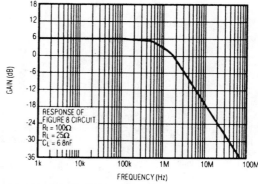

Lowpass Amplifier Gain vs. Frequency

MAXIM

Fig. 6-3

HIGHPASS AMPLIFIER

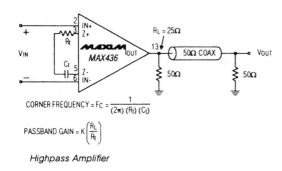

$$\text{CORNER FREQUENCY} = F_C = \frac{1}{(2\pi)\,(R_t)\,(C_t)}$$

$$\text{PASSBAND GAIN} = K\left(\frac{R_L}{R_t}\right)$$

Highpass Amplifier

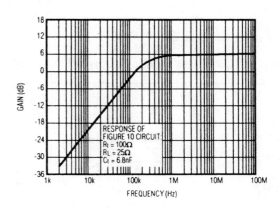

Highpass Amplifier Gain vs. Frequency

MAXIM

Fig. 6-4

49

ISD 1000A RECORD/PLAYBACK CIRCUIT

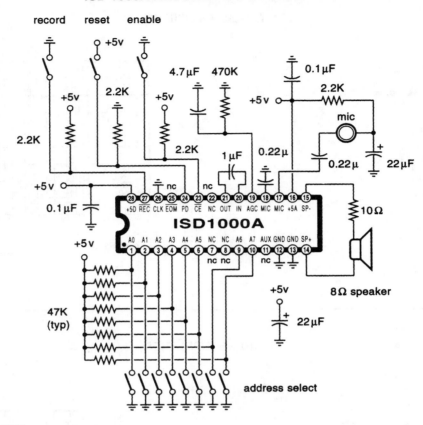

ELECTRONICS NOW

Fig. 6-5

This circuit uses the Information Storage Devices ISD1000A chip (Radio Shack P/N 276-1325).

REMOTE AMPLIFIER

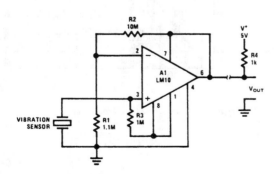

Useful for transducers and such where a single two-wire pair are the only leads available.

NATIONAL SEMICONDUCTOR

Fig. 6-6

PROGRAMMABLE GAIN AMPLIFIER

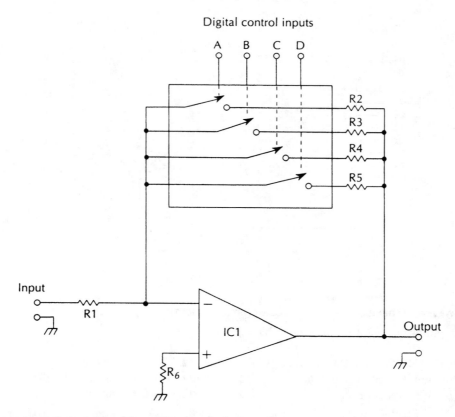

Fig. 6-7

The gain of this amplifier is $-R_f/R_1$ where R_f = effective value of resistance selected by the digital R1 inputs.

IC1   op amp
IC2   CD4066 quad bilateral switch
R1   1-kΩ, ¼-W 5% resistor
R2   10-kΩ, ¼-W 5% resistor
R3   4.7-kΩ, ¼-W 5% resistor
R4   2.2-kΩ, ¼-W 5% resistor
R5   1-kΩ, ¼-W 5% resistor
R6   2.2-kΩ, ¼-W 5% resistor

PROGRAMMABLE INPUT AMPLIFIER

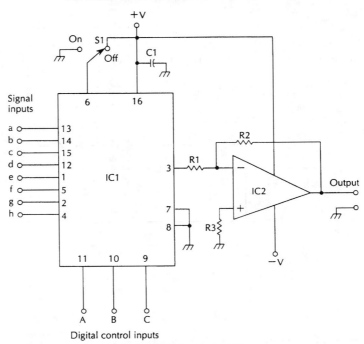

AMPLIFIERS, WAVEFORM GENERATORS & OTHER LOW-COST IC PROJECTS

Fig. 6-8

This amplifier has eight inputs selectable digitally.

IC1	CD4051 SP8T bilateral switch	R1	10-kΩ, ¼-W 5% resistor
IC2	op amp to suit application	R2	22-kΩ, ¼-W 5% resistor
C1	0.1-µF capacitor	R3	18-kΩ, ¼-W 5% resistor
S1	SPST switch		

REMOTELY POWERED SENSOR AMPLIFIER

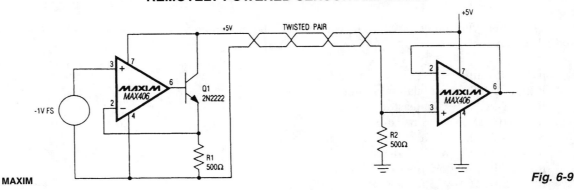

MAXIM

Fig. 6-9

For remote sensor applications, this circuit enables use of a single twisted pair.

TUNED AMPLIFIER

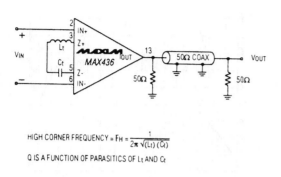

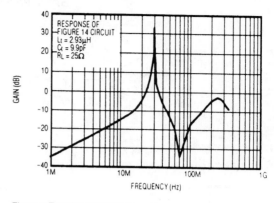

HIGH CORNER FREQUENCY = $F_H = \dfrac{1}{2\pi \sqrt{(L_t)(C_t)}}$

Q IS A FUNCTION OF PARASITICS OF L_t AND C_t

Figure A Tuned Amplifier

Figure B Tuned Amplifier Gain vs. Frequency

MAXIM

Fig. 6-10

This circuit is a tuned amplifier circuit, tuned to the resonant frequency of the LC transconductance network:

$$F_c = \frac{1}{2\pi \sqrt{L_t \, C_t}}$$

The impedance of the transconductance network is a minimum at the resonant frequency, providing maximum amplifier gain at that frequency. The Q of the amplifier is a function of the parasitic components associated with the LC network. The graph is the frequency response of the circuit, with $L_t = 2.93\ \mu\text{H}$ and $C_t = 9.9\ \text{pF}$.

DIFFERENCE AMPLIFIER WITH WIDE INPUT COMMON-MODE RANGE

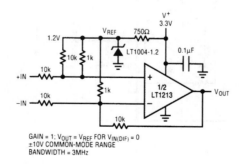

GAIN = 1; $V_{OUT} = V_{REF}$ FOR $V_{IN(DIF)} = 0$
±10V COMMON-MODE RANGE
BANDWIDTH = 3MHz

LINEAR TECHNOLOGY

Fig. 6-11

BANDPASS AMPLIFIER

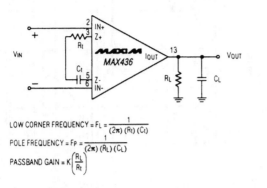

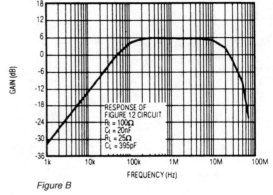

LOW CORNER FREQUENCY = $F_L = \dfrac{1}{(2\pi)(R_t)(C_t)}$

POLE FREQUENCY = $F_P = \dfrac{1}{(2\pi)(R_L)(C_L)}$

PASSBAND GAIN = $K\left(\dfrac{R_L}{R_t}\right)$

Figure A Bandpass Amplifier

Figure B

MAXIM

Fig. 6-12

The circuit A is a bandpass amplifier, with the low corner frequency set by the impedance of the transconductance network. The high corner frequency is set by the impedance of the RC network at the amplifier output. The passband gain is (k) x (R_1/R_t). Figure B is a plot of the circuit in Figure A, with $R_t = 100\ \Omega$, $C_t = 20$ nF, $R_1 = 25\ \Omega$, and $C_1 = 395$ pF.

HIGH-SIDE CURRENT-SENSING AMPLIFIER

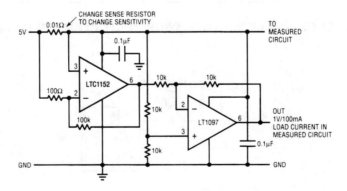

LINEAR TECHNOLOGY

Fig. 6-13

HIGH INPUT IMPEDANCE ac AMPLIFIER

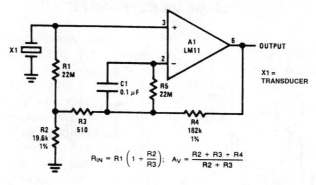

$$R_{IN} = R1 \left(1 + \frac{R2}{R3}\right); \quad A_V = \frac{R2 + R3 + R4}{R2 + R3}$$

NATIONAL SEMICONDUCTOR

Fig. 6-14

This figure shows an op amp used as an ac amplifier. It is unusual in that dc bootstrapping is used to obtain high input resistance without requiring high-value resistors. In theory, this increases the output offset because the op amp offset voltage is multiplied by the resistance boost.

But when conventional resistor values are used, it is practical to include R5 to eliminate bias-current error. This gives less output offset than if a single, large resistor were used. C1 is included to reduce noise.

MOSFET PUSH-PULL AMPLIFIER

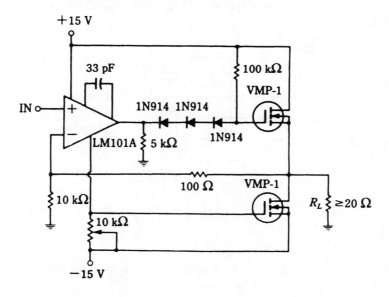

MCGRAW-HILL

Fig. 6-15

This amplifier can be used for audio or as a driver for inverter service.

LOW-VOLTAGE MICROPHONE PREAMP

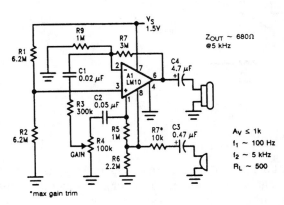

A microphone amplifier is shown. The reference, with a 500-kHz unity-gain bandwidth, is used as a preamplifier with a gain of 100. Its output is fed through a gain-control potentiometer to the op amp, which is connected for a gain of 10. The combination gives a 60-dB gain with a 10-kHz bandwidth, unloaded, and 5 kHz loaded at 500 Ω. Input impedance is 10 kΩ.

Potentially, using the reference as a preamplifier in this fashion can cause excess noise. However, because the reference voltage is low, the noise contribution, which adds root-mean-square, is likewise low. The input noise voltage in this connection is 440-500 nV Hz, about equal to that of the op amp.

NATIONAL SEMICONDUCTOR   *Fig. 6-16*

BASIC LOGARITHMIC AMPLIFIER USING OP AMP

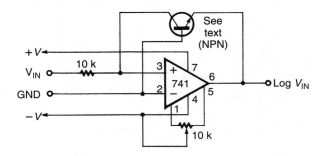

ELECTRONICS NOW!   *Fig. 6-17*

This logarithmic amplifier uses a single op amp. The current in the feedback loop of the op amp is equal to the current flow at the input of the op amp.

CRYSTAL TUNED AMPLIFIER

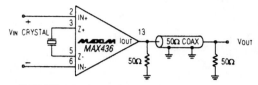

CENTER FREQUENCY = Fc = CRYSTAL FREQUENCY

Crystal Tuned Amplifier

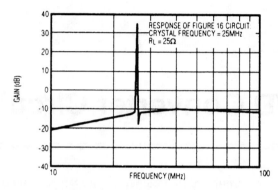

Crystal Tuned Amplifier Gain vs. Frequency

7

A/D Converter Circuits

The sources of the following circuits are contained in the Sources section, which begins on page 707. The figure number in the box of each circuit correlates to the entry in the Sources section.

High-Speed A/D Converter System
A/D Converter for PCs

HIGH-SPEED A/D CONVERTER SYSTEM

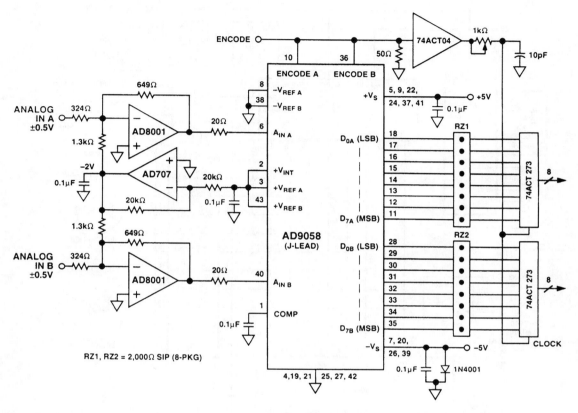

ANALOG DEVICES

Fig. 7-1

The AD8001 is well suited for driving high-speed analog-to-digital converters, such as the AD9058. The AD9058 is a dual 8-bit 50 Msps ADC. In the circuit shown, there are two AD8001s driving the inputs of the AD905f8 which are configured for 0- to +2-V ranges. Bipolar input signals are buffered, amplified (−2×), and offset (by +1.0 V) into the proper input range of the ADC. Using the AD9058's internal +2-V reference connected to both ADCs (as shown) reduces the number of external components required to create a complete data acquisition system. The 20-Ω resistors in series with ADC input are used to help the AD8001 drive the 10-pF ADC input capacitance. The two AD8001s only add 100 mW to the power consumption while not limiting the performance of the circuit.

A/D CONVERTER FOR PCs

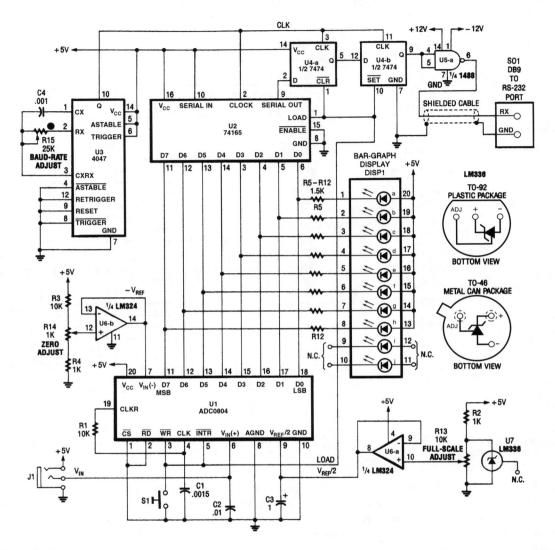

Fig. 7-2

An ADC0804 A/D converter converts analog data to digital. This is fed to a 74165 8-bit shift register and converted to serial data. U3 provides a baud-rate clock. U4A and U4B are used to generate start and stop bits needed at beginning and end of each data word.

8

Antenna Circuits

The sources of the following circuits are contained in the Sources section, which begins on page 707. The figure number in the box of each circuit correlates to the entry in the Sources section.

Remote Tuned Active HF Antenna
Miniature Broadband Antenna (3 to 30 MHz)
FM Auto Radio Diversity Antenna
Tunable FM Antenna Booster
Matchbox Antenna Tuner
Antenna Tuner
Active Antenna for UHF Scanners

REMOTE TUNED ACTIVE HF ANTENNA

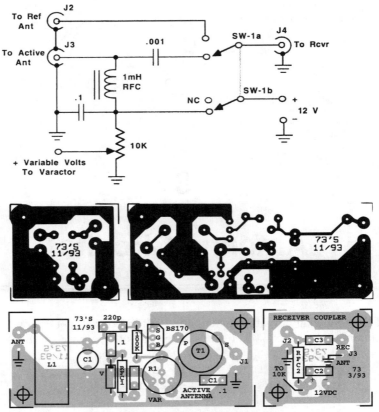

PC board pattern and parts placement diagram.

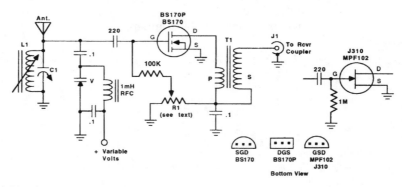

Fig. 8-1

An MV1662/S varactor diode tunes this active antenna/preamplifier. R1 varies gate bias on the BS170 FET. T1 is a 3:1 toroidal winding suitable for the frequencies of interest.

MINIATURE BROADBAND ANTENNA (3 TO 30 MHz)

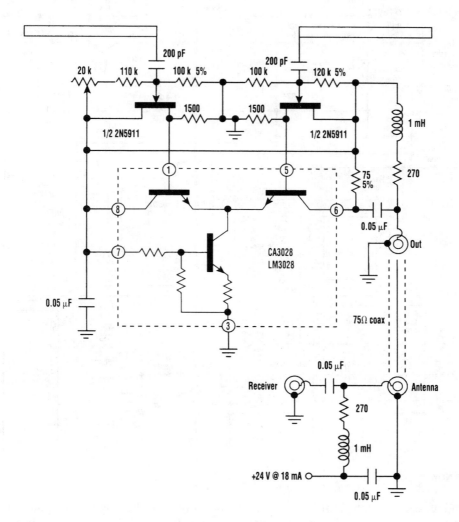

ELECTRONIC DESIGN

Fig. 8-2

A short dipole antenna and impedance converter combined together can be rotated to null out an interfering signal. The converter supplies a tremendous current gain so that the voltage appearing at the dipole's output eventually drives a 75-Ω load.

FM AUTO RADIO DIVERSITY ANTENNA

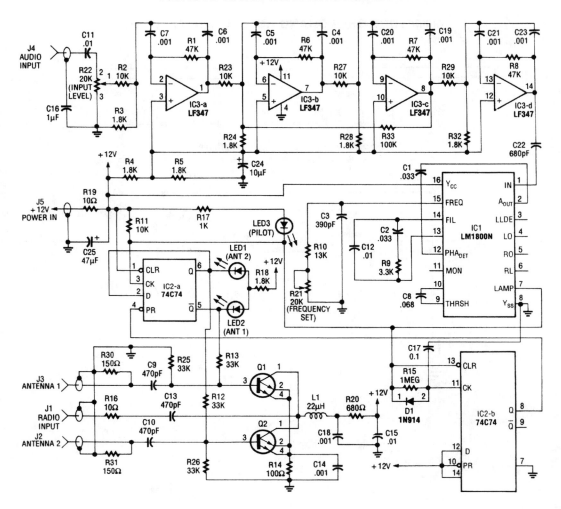

Fig. 8-3

A second antenna, installed on your vehicle as far away from the original equipment antenna as practical, provides the second FM signal. The figure is a simplified block diagram of the diversity system.

The cables from both antennas are connected to the electronic antenna switch. The 19-kHz pilot signal from the receiver's audio output is passed through a high-gain bandpass active filter, which attenuates audio programming that is much stronger than the pilot signal. After amplification, the pilot subcarrier becomes the reference frequency for a phase-locked loop (PLL) circuit. The output of the PLL locks to the 19-kHz pilot signal and functions as a subcarrier detector. When the reference frequency becomes noisy, the PLL will lose "lock" and trigger the flip-flop, whose output switches the

FM AUTO RADIO DIVERSITY ANTENNA (*Cont.*)

state of the electronic antenna switch. This action switches the alternate antenna into the system while disabling the original antenna.

If that second antenna is positioned for better reception, the received signal will clear, and the PLL will again lock to the subcarrier and hold the switch in that state until the pilot signal drops out again. If the second antenna does not restore the pilot signal reception after a 0.1-second delay, the primary antenna is switched back on.

When the radio is receiving AM, the absence of the 19-kHz subcarrier will also reactivate the primary antenna that is tuned to the receiver for the best AM reception.

TUNABLE FM ANTENNA BOOSTER

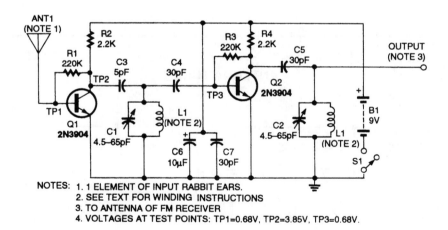

NOTES: 1. 1 ELEMENT OF INPUT RABBIT EARS.
2. SEE TEXT FOR WINDING INSTRUCTIONS
3. TO ANTENNA OF FM RECEIVER
4. VOLTAGES AT TEST POINTS: TP1=0.68V, TP2=3.85V, TP3=0.68V.

ELECTRONICS NOW   *Fig. 8-4*

This two-transistor amplifier circuit with tunable tank circuits boosts the distant FM signals. Coils L1 and L2 are 1½ turns #20 AWG bail tinned wire wound around a ⅜" diameter mandrel.

MATCHBOX ANTENNA TUNER

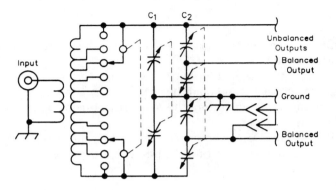

Fig. 8-5

C1 is a split stator capacitor and C2 is a dual differential capacitor. The top unbalanced output connection is used for high-impedance unbalanced loads, and the other is used for low-impedance unbalanced loads. In the latter case, the unused balanced load connection is grounded.

ANTENNA TUNER

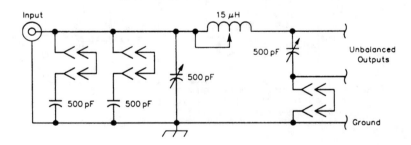

Fig. 8-6

This is a circuit diagram of the Collins Model 180S-1 antenna tuner. Three unbalanced configurations are available, two of which form an L-network and the other is a π-network. The tuning range is impressive.

ACTIVE ANTENNA FOR UHF SCANNERS

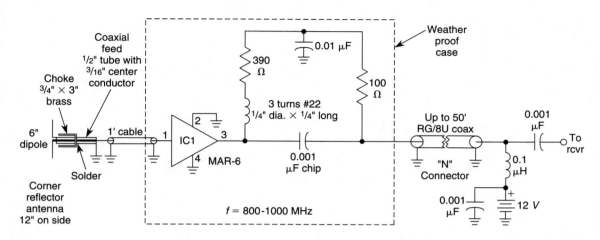

WILLIAM SHEETS

Fig. 8-7

This active antenna is a ½-wave dipole mounted in a 12" × 12" × 12" 90° corner reflector. A built-in active preamp IC1, fed dc through the RF coaxial line, provides 15 dB gain at 900 MHz to offset cable losses. This provides superior reception for scanners covering the 800- to 1000-MHz range.

9

Attenuator Circuits

The sources of the following circuits are contained in the Sources section, which begins on page 707. The figure number in the box of each circuit correlates to the entry in the Sources section.

Switchable Power Attenuator
Variable Voltage Attenuator

SWITCHABLE POWER ATTENUATOR

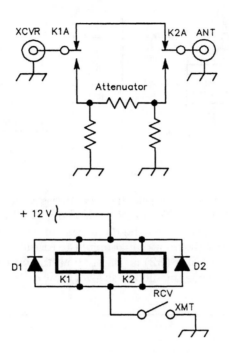

Fig. 9-1

Schematic diagram of a switchable power attenuator that can be used to reduce the power output of transmitters that don't have ALC lines.

Values for 10 and 20 dB:
10 dB: $R_1 = R_3 = 91\ \Omega$
$R_2 = 75\ \Omega$ nearest standard values

20 dB: $R_1 = R_3 = 62\ \Omega$
$R_2 = 240\ \Omega$ nearest standard values

Note: R1 must handle the largest share of the input power, and R2 somewhat less. This depends on attenuation selected.

VARIABLE VOLTAGE ATTENUATOR

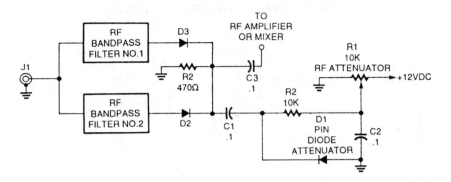

POPULAR ELECTRONICS

Fig. 9-2

The front-end of this circuit is a bank of selectable bandpass filters. The output of the filter banks are shunted to ground via capacitor (C1) and PIN diode (D1). The PIN diode acts like an electronically variable resistor. The resistance across the diode's terminals is a function of the applied bias voltage. This voltage, hence the degree of attenuation of the RF signal, is proportional to the setting of potentiometer R1. The series resistor (R2) is used to limit the current when the diode is forward biased. This step is necessary because the diode has a very low resistance when a certain rather low potential is exceeded.

10

Audio Signal Amplifier Circuits

The sources of the following circuits are contained in the Sources section, which begins on page 707. The figure number in the box of each circuit correlates to the entry in the Sources section.

Vacuum Tube Audio Amplifier
Micropower Linear Amplifier
NB FM Audio Amplifier
Two-Transistor Audio Amplifier
Personal Stereo Audio Amp
Transistor RIAA Preamp for Magnetic Phone Cartridges
Dynamic Microphone Preamp
Balanced Microphone Preamplifier
RIAA Line Amplifier/Driver
Single-Ended HI-*Z* Microphone Preamp
Low-Level Audio Amplifier
Simple 20-dB Gain Audio Amplifier
High-Gain Dynamic Microphone Preamplifier
FET Phono Cartridge Preamp
Simple High-Gain Audio Amplifier
RIAA Preamplifier
Basic Complementary Class-AB Single-Supply Amplifier
High-Impedance Microphone Input Circuit
Electronic-Ear Low-Noise Audio Amplifier (for Parabolic Dish Mikes)

VACUUM TUBE AUDIO AMPLIFIER

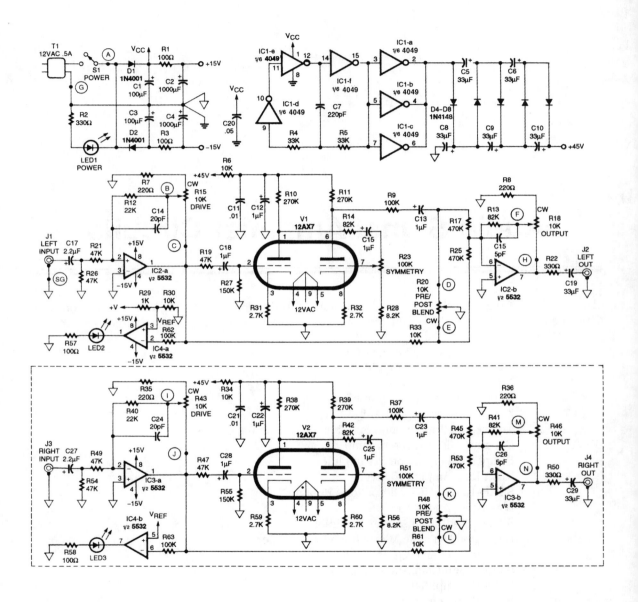

Fig. 10-1

VACUUM TUBE AUDIO AMPLIFIER (*Cont.*)

This schematic is for a tubehead amplifier. The output from transformer T1 is positive half-wave ac rectified by D1 and filtered by C1, C2, and R1 for a +15-V supply. A −15-V supply is available from D2, C3, C4, and R3. The plate supply for the 12AX7 tubes is produced by a voltage multiplier.

Some listeners prefer the sound of a vacuum tube audio system. Although this is rather subjective and a personal preference, this circuit can be used to simulate the "tube sound" preferred by these listeners.

MICROPOWER LINEAR AMPLIFIER

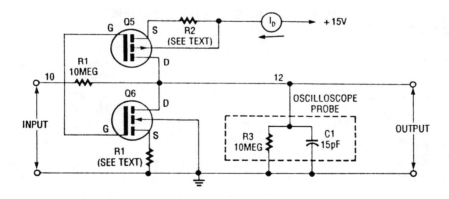

Fig. 10-2

This circuit, based on the inverter in the CD4007UB CMOS linear amplifier, shows a method for reducing drain current.

NB FM AUDIO AMPLIFIER

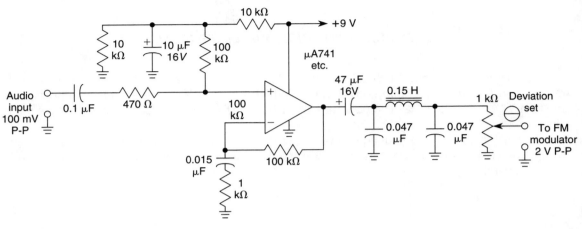

WILLIAM SHEETS

Fig. 10-3

This audio system amplifies, limits, and filters an audio voice signal for use with an FM modulator or VCO. It has pre-emphasis of 6-dB/octave 300–3000 Hz. Almost any suitable op amp can be used.

TWO-TRANSISTOR AUDIO AMPLIFIER

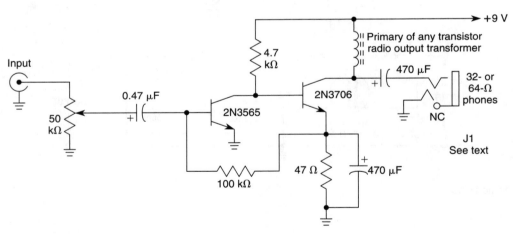

WILLIAM SHEETS

Fig. 10-4

This is a general-purpose audio amplifier for driving a pair of stereo earphones in monaural mode. Two can be used for stereo. In this case, ground the center top of the earphone (sleeve of J1).

PERSONAL STEREO AUDIO AMP

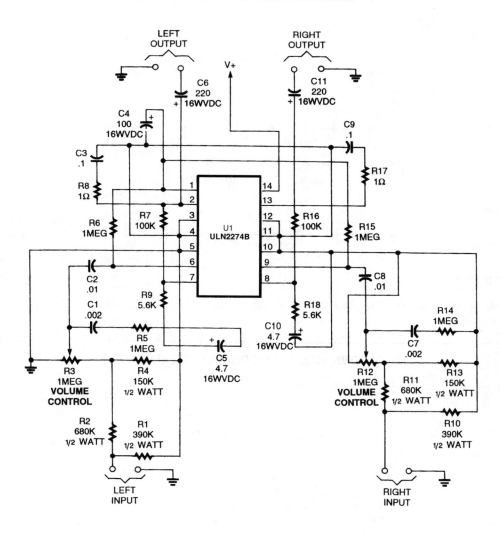

Fig. 10-5

You can make your personal stereo do double duty as a small room stereo by adding this 2-watt amplifier.

TRANSISTOR RIAA PREAMP FOR MAGNETIC PHONE CARTRIDGES

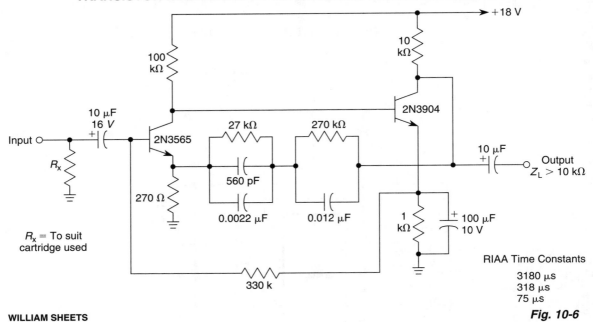

RIAA Time Constants
3180 μs
318 μs
75 μs

Fig. 10-6

This two-transistor circuit has around 40 dB (midband) gain at 1 kHz. A magnetic cartridge is used as a source.

DYNAMIC MICROPHONE PREAMP

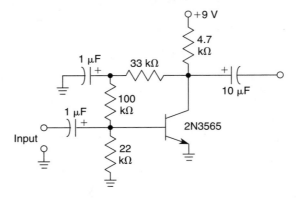

Fig. 10-7

This preamplifier provides 40- to 43-dB gain when used with a low-impedance (<1 kΩ) dynamic microphone.

BALANCED MICROPHONE PREAMPLIFIER

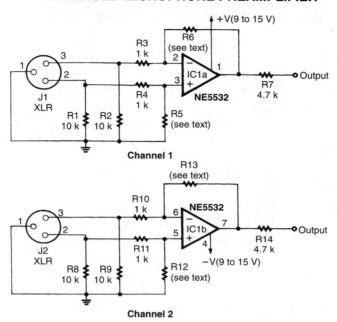

ELECTRONICS NOW

Fig. 10-8

A balanced input for microphones can solve hum and noise pickup problems. R6 and R13 should equal R5 and R12, respectively. Typical values would be 10 kΩ to 22 kΩ.

RIAA LINE AMPLIFIER/DRIVER

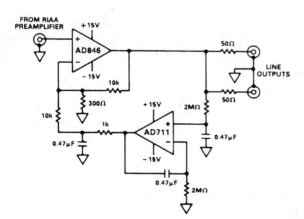

Two op amps by Analog Devices are used in this audio line amplifier, which is suitable for interfacing with an RIAA preamplifier.

ANALOG DEVICES   **Fig. 10-9**

SINGLE-ENDED HI-Z MICROPHONE PREAMP

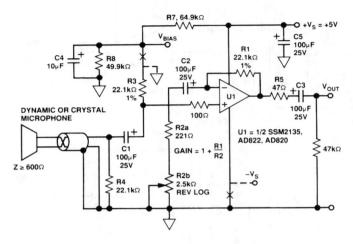

ANALOG DEVICES

Fig. 10-10

This low noise circuit works on a +5-V supply. Gain range is 20 to 40 dB and bandwidth is 20 kHz with the AD820. THD is 0.05% with 1 V RMS into a 2-kΩ load. Noise output with the input shorted is less than 200 μV.

LOW-LEVEL AUDIO AMPLIFIER

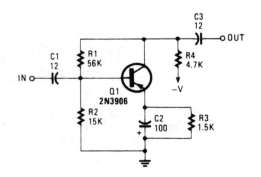

POPULAR ELECTRONICS

Fig. 10-11

SIMPLE 20-dB GAIN AUDIO AMPLIFIER

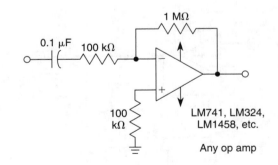

WILLIAM SHEETS

Fig. 10-12

HIGH-GAIN DYNAMIC MICROPHONE PREAMPLIFIER

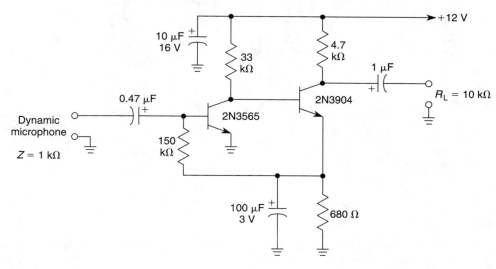

Fig. 10-13

This microphone preamplifier is capable of about 70 dB or more gain at audio frequencies. Its gain is approximately equal to the product of the h_{fe} of both transistors times the ratio of the load resistance to the input resistance of the preamp. As an approximation, these resistances are usually similar in value (≈ 2 to 5 kΩ) for most applications, so this ratio can be taken as unity.

FET PHONO CARTRIDGE PREAMP

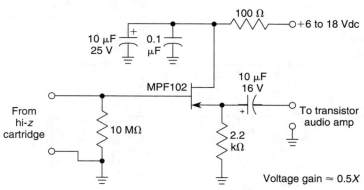

Fig. 10-14

A high-Z phono cartridge can be matched to a low-Z amplifier with this circuit. The FET provides a current gain of over 1000× and a voltage gain of about 0.5×.

SIMPLE HIGH-GAIN AUDIO AMPLIFIER

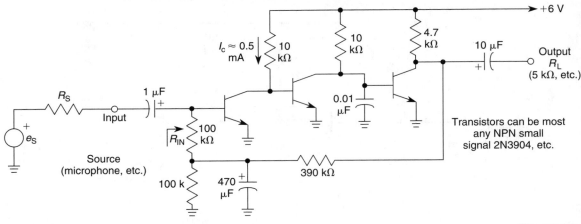

WILLIAM SHEETS

Fig. 10-15

This amplifier has a very high gain in the audio range and is approximately the product of the current gains of the three transistors multiplied by the ratio of R_L to $(R_{IN} + R_S)$. R_{IN} is approximately to:

$$\frac{(\beta_{Q1} + 1)\,(26)}{I_{EQ1}}$$

RIAA PREAMPLIFIER

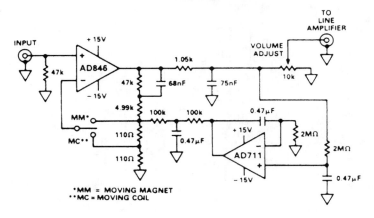

ANALOG DEVICES

Fig. 10-16

This preamp for RIAA phone use uses two op amps by Analog Devices. A switch selects compensation for moving magnet or moving coil pickups.

BASIC COMPLEMENTARY CLASS-AB SINGLE-SUPPLY AMPLIFIER

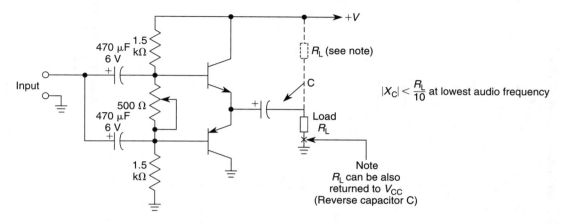

$|X_C| < \frac{R_L}{10}$ at lowest audio frequency

Note
R_L can be also
returned to V_{CC}
(Reverse capacitor C)

WILLIAM SHEETS

Fig. 10-17

HIGH-IMPEDANCE MICROPHONE INPUT CIRCUIT

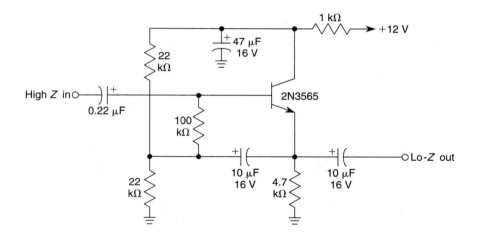

WILLIAM SHEETS

Fig. 10-18

This input circuit will enable use of a high-impedance microphone where a low-impedance microphone would be needed.

ELECTRONIC-EAR LOW-NOISE AUDIO AMPLIFIER (FOR PAROBOLIC DISH MIKES)

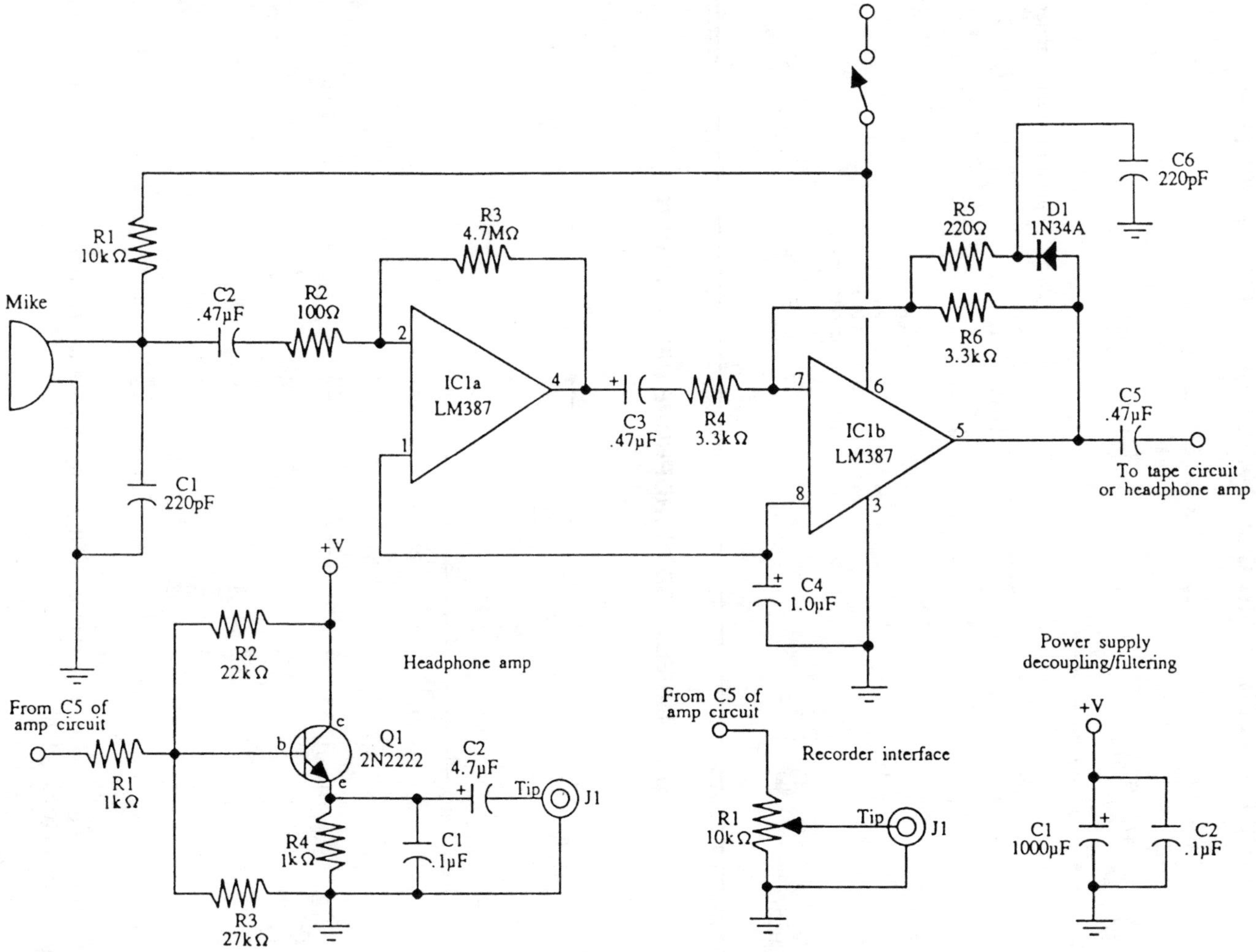

Fig. 10-19

MCGRAW-HILL

Use this circuit with a parabolic reflector microphone for eavesdropping on distant sounds.

11

Audio Power Amplifier Circuits

The sources of the following circuits are contained in the Sources section, which begins on page 707. The figure number in the box of each circuit correlates to the entry in the Sources section.

AUDIO POWER AMPLIFIER, 1.5 W, 12 V

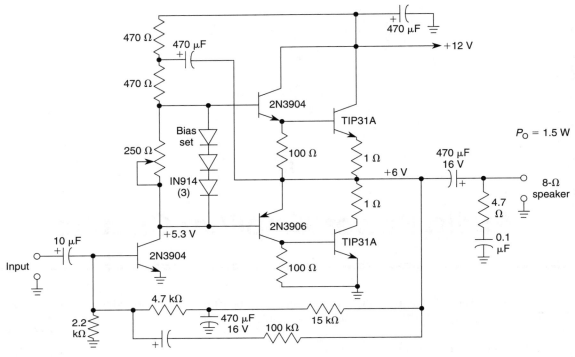

WILLIAM SHEETS

Fig. 11-1

Although ICs have largely replaced circuits such as this, this circuit still finds use where the flexibility of a discrete device design is desirable. Parts are easy to obtain and the problem of IC obsolescence is eliminated. The TIP31A can be heatsinked to a small metal heatsink, if desired.

PARALLEL POWER OP AMPS

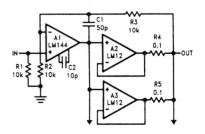

The power amplifiers, A2 and A3, are wired as followers and connected in parallel with the outputs coupled through equalization resistors.

NATIONAL SEMICONDUCTOR

Fig. 11-2

10-WATT AUDIO AMPLIFIER

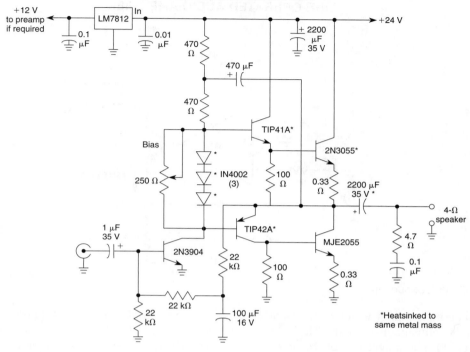

WILLIAM SHEETS

Fig. 11-3

This circuit is a general-purpose 10-W audio amplifier for moderate-power PA or modulator use in an AM transmitter. With higher voltages and a change in bias resistors, up to 30 W can be obtained.

POWER BRIDGE AMPLIFIER WITH SINGLE-ENDED OUTPUT

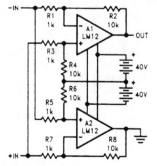

Bridge amplifier with a single-ended output uses floating supply. Either input can be grounded.

NATIONAL SEMICONDUCTOR

Fig. 11-4

LINE-OPERATED AUDIO AMPLIFIER

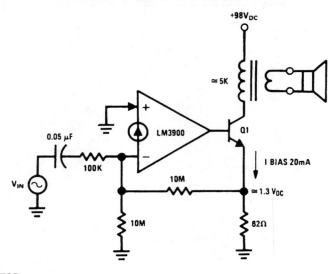

NATIONAL SEMICONDUCTOR

Fig. 11-5

An audio amplifier which operates off a +98-Vdc power supply (the rectified line voltage) is often used in consumer products. The external high-voltage transistor, Q1, is biased and controlled by the LM3900. The magnitude of the dc biasing voltage, which appears across the emitter resistor of Q1 is controlled by the resistor. The resistor is placed from the (–) input to ground.

BASIC COMPLEMENTARY CLASS-AB POWER AMPLIFIER

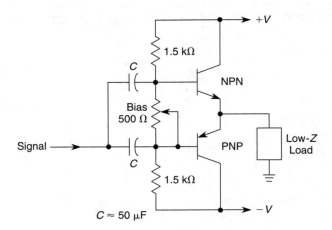

WILLIAM SHEETS

Fig. 11-6

SIMPLE VACUUM TUBE AMPLIFIER

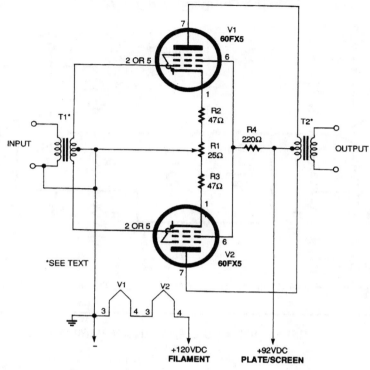

POPULAR ELECTRONICS

Fig. 11-7

Using a pair of 60 FX5 tubes, direct operation from 120 Vac is possible. However, the use of a power supply with an isolation transformer is recommended. R1 is adjusted for equal voltages at pin 1 of V1 and V2. The power output is about 2 to 3 watts.

POWER SUPPLY FOR VACUUM TUBE AMPLIFIER

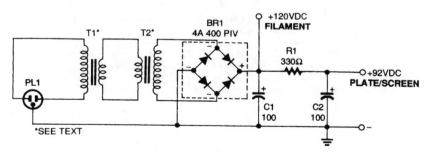

POPULAR ELECTRONICS

Fig. 11-8

The power supply for the amplifier uses two low-voltage transformers connected back-to-back. The full-wave bridge rectifier, BR1 provides dc for the filaments, plates, and screens.

16-W BRIDGE AMPLIFIER

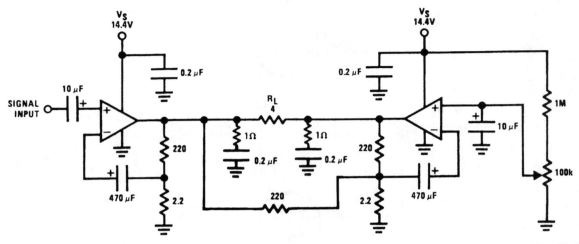

NATIONAL SEMICONDUCTOR

Fig. 11-9

This circuit delivers 16 W RMS audio into a 4-Ω load (R_L). The ICs are LM383s.

RFI-PROOF AUDIO POWER AMPLIFIER

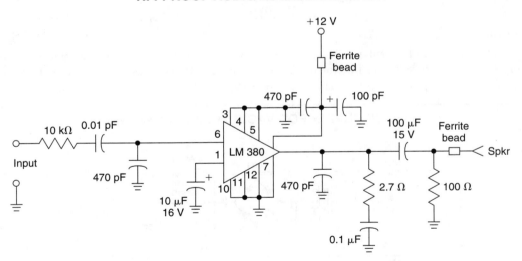

WILLIAM SHEETS

Fig. 11-10

This 1-watt audio amplifier was used in an FM repeater and proved to be immune to strong RF signal pickup. It functioned well in very strong RF fields.

BASIC QUASI-COMPLEMENTARY POWER AMPLIFIER WITH SPLIT POWER SUPPLIES

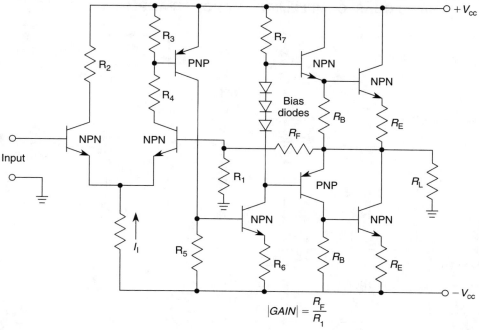

$$|GAIN| = \frac{R_F}{R_1}$$

WILLIAM SHEETS

Fig. 11-11

This is the basic circuit used in many audio power output stages where split supplies are used. This amplifier is inherently dc coupled and has high open loop gain and good dc stability if the feedback network is properly designed.

RIAA PHONO AMPLIFIER

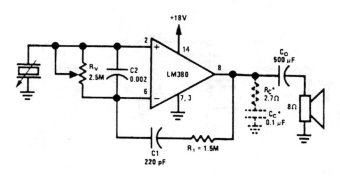

NATIONAL SEMICONDUCTOR

Fig. 11-12

$$Mid\text{-}band\ gain = \frac{R_1 + 150\ \text{k}\Omega}{150\ \text{k}\Omega}$$

BASIC QUASI-COMPLEMENTARY POWER AMPLIFIER CIRCUIT

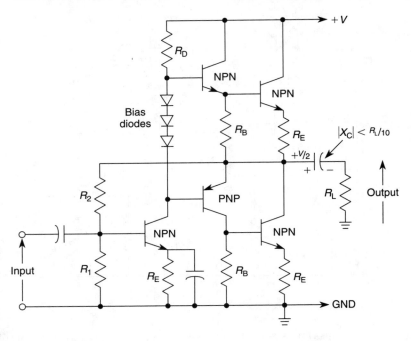

WILLIAM SHEETS

Fig. 11-13

PHONO AMP

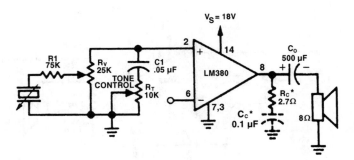

NATIONAL SEMICONDUCTOR

Fig. 11-14

   The figure shows the LM380 with a voltage-divider volume control and high-frequency roll-off tone control.

80-WATT IC AUDIO AMPLIFIER

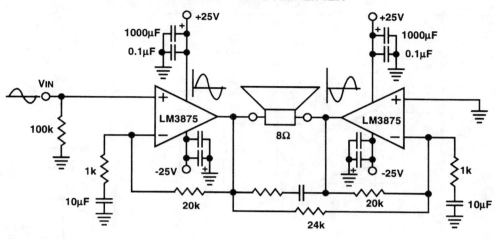

NATIONAL SEMICONDUCTOR

Fig. 11-15

This audio power amp will deliver 80 W of audio into an 8-Ω load. The LM3875 IC devices should be suitably heatsinked. Note that the amplifier is a bridged circuit, with both speaker leads "hot."

BASIC COMPLEMENTARY POWER AMPLIFIER CIRCUIT

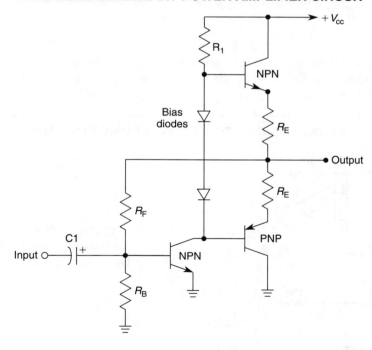

WILLIAM SHEETS

Fig. 11-16

GENERAL-PURPOSE AF AMPLIFIER

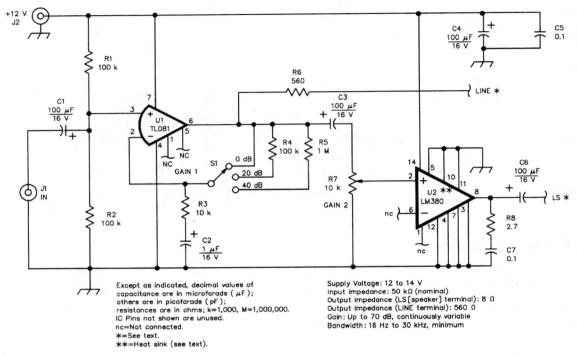

Except as indicated, decimal values of
capacitance are in microfarads (μF);
others are in picofarads (pF);
resistances are in ohms; k=1,000, M=1,000,000.
IC Pins not shown are unused.
nc=Not connected.
✳=See text.
✳✳=Heat sink (see text).

Supply Voltage: 12 to 14 V
Input impedance: 50 kΩ (nominal)
Output impedance (LS [speaker] terminal): 8 Ω
Output impedance (LINE terminal): 560 Ω
Gain: Up to 70 dB, continuously variable
Bandwidth: 16 Hz to 30 kHz, minimum

QST

Fig. 11-17

Schematic of the general-purpose AF amplifier. All resistors are ¼-W, 5%-tolerance carbon-composition or metal-film units. Equivalent parts can be substituted. General-purpose IC replacements are shown in parentheses.

BRIDGE CONNECTION OF TWO POWER OP AMPS

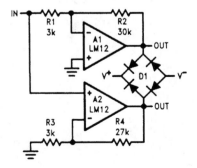

These bridge connections provide differential outputs that approach twice the total supply voltage. Diode bridge clamps output to the supplies.

NATIONAL SEMICONDUCTOR   **Fig. 11-18**

90-V 10-A HIGH-POWER AMPLIFIER

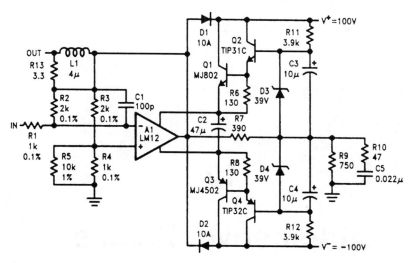

NATIONAL SEMICONDUCTOR

Fig. 11-19

This amplifier can drive ±90 V at 10 A, more than twice the output swing of the LM12. The IC provides current and power limiting for the discrete transistors.

MINI-MEGAPHONE

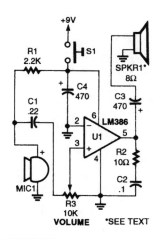

The Mini-Megaphone is comprised of an electret microphone (MIC1), and LM386 low-voltage audio-power amplifier (U1), a horn speaker (SPKR1), and a few other components.

POPULAR ELECTRONICS

Fig. 11-20

12

Automotive Circuits

The sources of the following circuits are contained in the Sources section, which begins on page 707. The figure number in the box of each circuit correlates to the entry in the Sources section.

Electronic Auto Stethoscope
Automotive Electrical Monitor
Car Alternator Monitor (Idiot Light)
Cigarette Lighter 9-V Adapter
Motorcycle Turn-Signal System
Tachometer Signal-Conditioning Circuit
Smart Turn Signal for Autos and Motorcycles
Turn-Signal Alarm
High-Power Audio Amp for Automotive Installation
High-Power 12-V IC Auto Amplifier
Capacitor Discharge Ignition System
Car Audio Power Supply
Motorcycle Headlight Monitor
Headlight-Off Indicator
Auto Battery Isolator Circuit
Automotive HI-Z Test Light

ELECTRONIC AUTO STETHOSCOPE

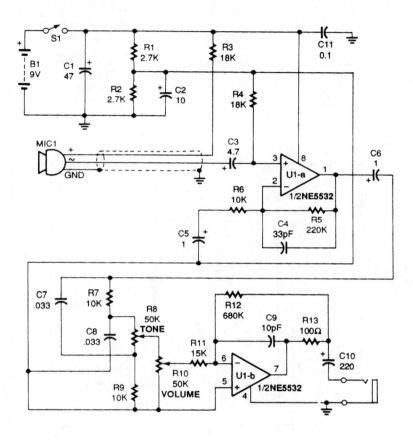

POPULAR ELECTRONICS

Fig. 12-1

The heart of the Stethoscope is the NE5532 audio op amp, U1. That component directly drives low impedances and allows the use of headphones without adding another amplifier.

AUTOMOTIVE ELECTRICAL MONITOR

TABLE 1—AUTOMOTIVE ELECTRICAL FAULTS

Condition	Normal Voltage	Possible Fault
Vehicle at rest	12.6 volts	<12.4 volts: bad cell or severely undercharged battery
Cranking	>9 volts	<9 volts: Weak battery
Idling	>12.8 volts	<12.8 volts: Not charging; bad alternator or wiring
Running minimum load	>13.4 volts	<13.4 volts: defective alternator or voltage regulator
Running minimum load	<15.2 volts	>15.2 volts: Overcharging; defective regulator
Running maximum load	>13.4 volts	<13.4 volts: alternator defective or belt slipping

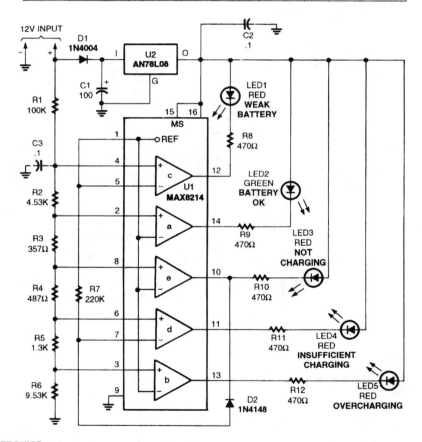

Fig. 12-2

The automotive electrical diagnostic system is built around a Maxim MAX8214ACPE five-stage voltage comparator, which contains a built-in 1.25-volt precision reference, and on-board logic that allows the outputs of two of the comparators to be inverted.

CAR ALTERNATOR MONITOR (IDIOT LIGHT)

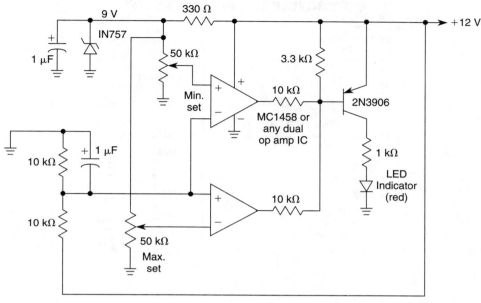

Fig. 12-3

A window comparator is used to detect a too-low or a too-high system voltage. The minimum and maximum settings are set with two 50-kΩ pots, as desired.

CIGARETTE LIGHTER 9-V ADAPTER

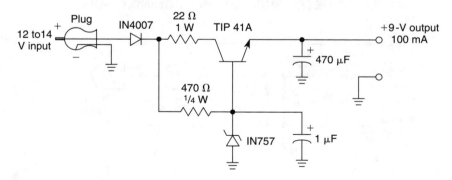

Fig. 12-4

A simple way to provide +9 V at 100 mA from a 12-V auto source. Applications include small radios, cassettes, etc.

MOTORCYCLE TURN-SIGNAL SYSTEM

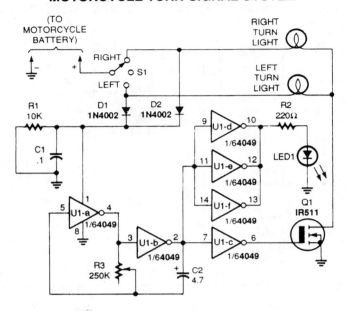

POPULAR ELECTRONICS

Fig. 12-5

Tired of making hand signals? Build this simple turn-signal system and keep your hands on the handlebars.

TACHOMETER SIGNAL-CONDITIONING CIRCUIT

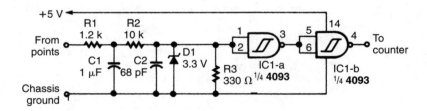

ELECTRONICS NOW

Fig. 12-6

This circuit, for use with auto tachometers, cleans up the ragged distribution waveform before it is sent to pulse counter circuits.

SMART TURN SIGNAL FOR AUTOS AND MOTORCYCLES

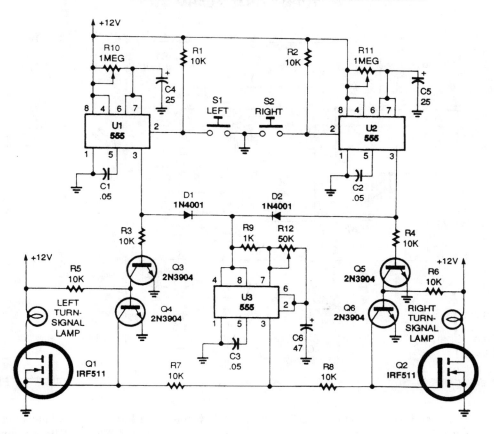

POPULAR ELECTRONICS

Fig. 12-7

Momentarily pressing S1 starts the left on-time timer and produces a positive output at pin 3 of U1. Power for the on/off signal timer, U3, is supplied through D1.

Also, a positive bias is supplied from U1's output to the base of Q3, turning it on and turning Q4 off. Unclamped Q1 turns the left turn-signal lamp on and off at that same low-frequency rate. Because U2 is not activated, its output at pin 3 is low, keeping Q5 off. With Q5 turned off, Q6 is on, clamping the gate of Q2 to ground and keeping it from responding and supplying an output for the right turn-signal lamp. The left turn signal continues to operate until the U1 timer circuit times out; the right turn signal operates in a similar manner, with U2 setting its operating time.

Potentiometer R10 sets the running time for the left turn signal and R11 sets that for the right turn signal.

TURN SIGNAL ALARM

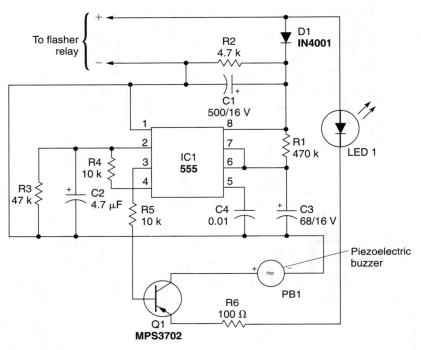

RADIO-ELECTRONICS

Fig. 12-8

This circuit can be used to tell the driver of a vehicle when his or her turn signal has been left on for too long. The circuit consists of IC1, a 555 timer; transistor Q1, and MPS3702 PNP preamp/driver; PB1, a piezoelectric buzzer; along with an assortment of resistors, capacitors, and diodes. The 555 is connected in the monostable mode, requiring only a momentary negative pulse at pin 2 to trigger the timing cycle.

Power for the circuit is picked off the flasher relay and applied to IC1, pin 8, provided by an initially discharged capacitor, C2. After the initial triggering, the voltage across C2 rises as it becomes charged through R4, a 10-kΩ resistor. This prevents subsequent interference with the delay function caused by false triggering.

Capacitor C3 and resistor R1 determine the delay. With the component values shown, a delay of about one minute will be provided before the intermittent tweet sound generated by the circuit begins. If higher values are used for C2 and R1, a longer delay time will result. The light-emitting diode, LED1, provides a voltage drop to assure complete transistor blocking during the off periods of the flasher. Alternatively, two diodes in series can be used.

HIGH-POWER AUDIO AMP FOR AUTOMOTIVE INSTALLATION

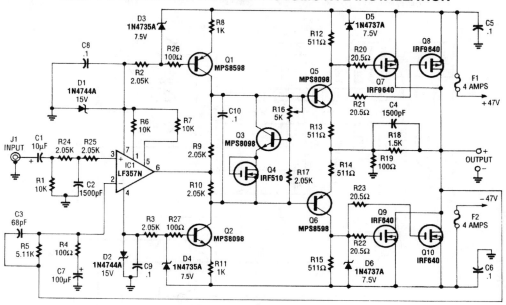

RADIO-ELECTRONICS

Fig. 12-9

Two of these audio amplifiers can be used to make a stereo amplifier 200 W per channel. IRF640 and IRF9640 power MOSFETs are used to drive the output load, which might be 4 or 8 Ω. Response is 12 Hz to 45 kHz (–3 dB), THD <0.1%. Power is supplied by a switching-type power supply, which is external to the amplifier (±47 V). About 600 W total power (peak) is needed.

HIGH-POWER 12-V IC AUTO AMPLIFIER

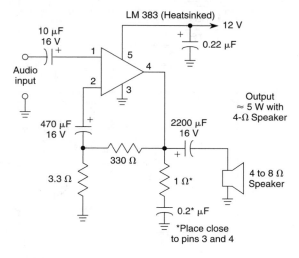

WILLIAM SHEETS

Fig. 12-10

CAPACITOR DISCHARGE IGNITION SYSTEM

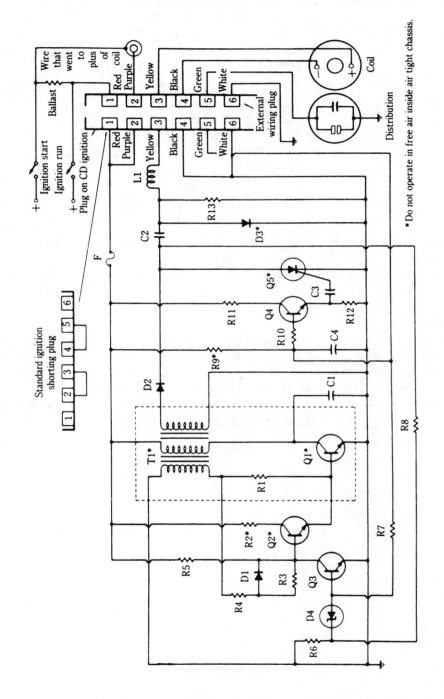

MCGRAW-HILL

Fig. 12-11

This ignition system charges a capacitor (C2) to 350 V and discharges it through the ignition coil.

*Do not operate in free air inside air tight chassis.

CAPACITOR DISCHARGE IGNITION SYSTEM (*Cont.*)

A Parts list

Q1—2N3055
Q2—2N3053
Q3—2N3241
Q4—2N3241
Q5—RCA 40657
D1—1N3193
D2—1N3195
D3—1N1763A
D4—12 V, ¼ W
C1—0.25 µF, 200 V
C2—1 µF, 400 V
C3—1 µF, 25 V
C4—0.25 µF, 25 V
F—5A
L1—10 µH, 100 Turns of No. 28 Wire Wound on a
 2-W Resistor (100 Ohms or More)
R1—1000 ohms, ½ W
R2—35 ohms, 5 W
R3—22,000 ohms, ½ W
R4—1000 ohms, ½ W
R5—18,000 ohms, ½ W
R6—15,000 ohms, ½ W
R7—8200 ohms, ½ W
R8—0.39 megohm, ½ W
R9—220 ohms, 1 W
R10—1000 ohms, ½ W
R11—68 ohms, ½ W
R12—4700 ohms, ½ W
R13—27,000 ohms, ½ W

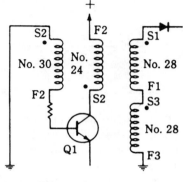

Details of inverter transformer

CAR AUDIO POWER SUPPLY

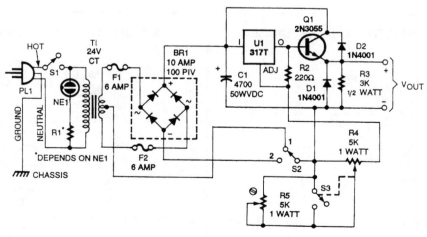

POPULAR ELECTRONICS

Fig. 12-12

This supply has a variable output voltage feature and a dual voltage switch, S2. Q1 should be adequately heatsinked.

103

MOTORCYCLE HEADLIGHT MONITOR

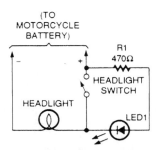

POPULAR ELECTRONICS   *Fig. 12-13*

The headlight on most newer bikes is keyed on with the ignition switch to guarantee that you are never underway without your headlight being on. However, many older bikes have a factory headlight switch, and a growing number of the newer bikes are owner-modified in the same way.

A simple headlight monitor circuit consists of just an LED and a current-limiting resistor wired across the headlight switch, as shown. When the ignition is on and the headlight switch is off, the LED will glow.

HEADLIGHT-OFF INDICATOR

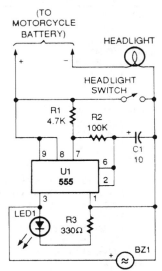

POPULAR ELECTRONICS   *Fig. 12-14*

Increasing the value of R_2 or C_1 will lower the oscillator's frequency and decreasing one of those values will increase the frequency. The IC's output at pin 3 drives the LED through R3 and sends power to the piezo sounder. Use a bright LED so that you will be able to see it in the daytime.

AUTO BATTERY ISOLATOR CIRCUIT

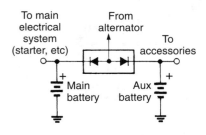

ELECTRONICS NOW   *Fig. 12-15*

The diodes ensure that current can flow in both batteries from the alternator, but the main battery can't feed the accessory system, nor vice versa.

AUTOMOTIVE HI-*Z* TEST LIGHT

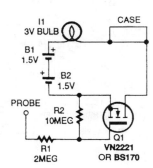

POPULAR ELECTRONICS   *Fig. 12-16*

This test light has a high-input impedance and draws only 1 mA at 12 V. Q1 switches dc to a battery and lamp circuit.

13

Battery Charger Circuits

The sources of the following circuits are contained in the Sources section, which begins on page 707. The figure number in the box of each circuit correlates to the entry in the Sources section.

Smart Battery Charger
Rechargeable LED Flashlight
Battery Charger Controller
Single-Cell Lithium Battery Charger
Battery-Charging Current Limiter
Three-Cell Lithium Charger
NiCad Battery Charger
Backup Battery Monitor/Charger/Alarm
NiCad Charger/Zapper
2- to 5-Cell Lithium Battery Charger
Lead-Acid Trickle Charger
NiCad Battery Charger

SMART BATTERY CHARGER

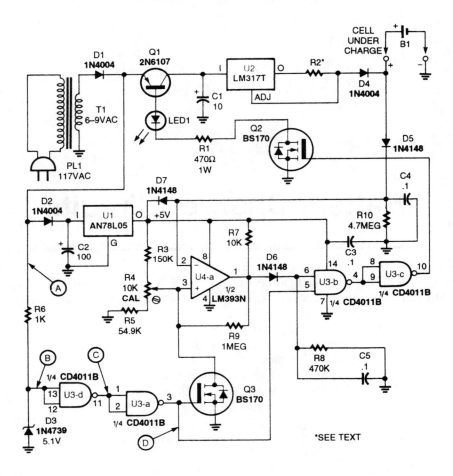

Fig. 13-1

This charger will work with NiCad or the new rechargeable alkaline batteries. The Smartcharger is comprised mainly of four chips—an AN78L05 5-V, 100-mA regulator (U1), an LM317T 1-A adjustable-voltage regulator (U2), a CD4011BE quad 2-input NAND gate (U3), and an LN393N dual-voltage comparator (U4). The value and rating of R2 is selected as described in the text. R2 is selected for a 1.2-V drop across it at the charging current (3 Ω for 400 mA, 6 Ω for 200 mA).

RECHARGEABLE LED FLASHLIGHT

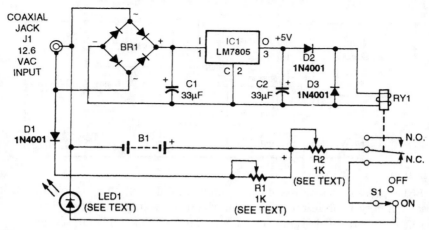

ELECTRONICS NOW

Fig. 13-2

This flashlight is useful for applications where night vision and/or darkness adaptation must be maintained. It uses an HLMP8150 T4 LED with a wavelength of 637 nm. This schematic is for the flashlight module. When the battery pack consisting of the four NiCad cells is fully charged (and there is no voltage at J1), 4.8 Vdc flows through trimmer potentiometer R2, the normally closed contact of relay RY1, and push-on/push-off power switch A1. Trimmer R2 limits the current flowing through LED1. Switch S1 can turn LED1 on and off when the battery is not being charged.

BATTERY CHARGER CONTROLLER

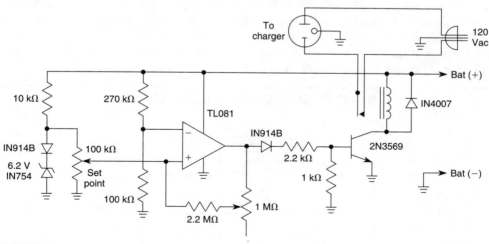

WILLIAM SHEETS

Fig. 13-3

When the battery voltage is low, the TL081 comparator produces a high output, turning on the 2N3569 relay driver. As the battery voltage approaches the set point, the relay driver is cut off, opening the 120-Vac supply.

SINGLE-CELL LITHIUM BATTERY CHARGER

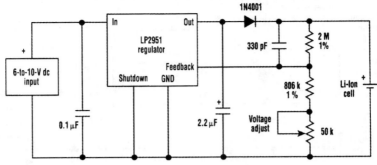

ELECTRONIC DESIGN

Fig. 13-4

An LP2951 regulator was chosen for this single lithium cell-charging circuit for its built-in current-limiting capability. In addition, the regulator's output voltage is extremely stable, which is a prerequisite for lithium battery charging. This figure details an example circuit designed to recharge a single cell. The required output set voltage was specified as 4.200 V (±0.025 V) with a maximum charging current of about 150 mA.

An LP2951 regulator was selected for two reasons. One is that its built-in current limiter holds the maximum current to 160 mA (typical). The other is because the output voltage can be very accurately set to 4.200 V, thanks to the regulator's stable internal bandgap reference.

The 1.23-V reference appears between the feedback pin and ground, which causes a precise current to flow in the output resistive-divider string. The amount of current flowing in these resistors determines (sets) the charger output voltage that appears across the battery terminals. Large-value resistors keep the battery drain below 2 µA when the dc input is removed (a customer requirement). A trimming potentiometer sets the output to 4.200 V. It must be adjusted when the battery isn't connected to the charger output. A blocking diode is required at the LP2951's output to prevent current from flowing out of the battery and back into the output when the dc-input source is removed. Because the diode is in series with the output, the minimum input-output voltage differential required for this circuit to operate is about 1.5 V.

BATTERY-CHARGING CURRENT LIMITER

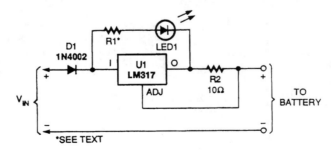

POPULAR ELECTRONICS

Fig. 13-5

This circuit uses an LM317 as a current regulator to limit charging current to a lead-acid battery. R2 should produce a 1.2-V drop at the desired limiting value of charging current.

THREE-CELL LITHIUM CHARGER

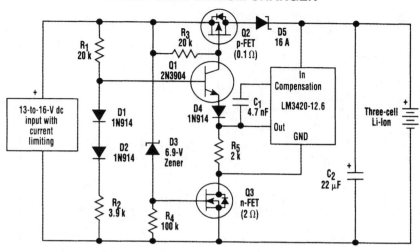

ELECTRONIC DESIGN

Fig. 13-6

This 3-A, three-cell charging circuit for lithium batteries includes a built-in on/off switch made up of Q3, R4, and D3. When a dc input is present, D3 turns on Q3, which allows current to flow through the LM3411 and Q1. If dc voltage is removed, Q3 turns off, cutting battery drain to zero.

NICAD BATTERY CHARGER

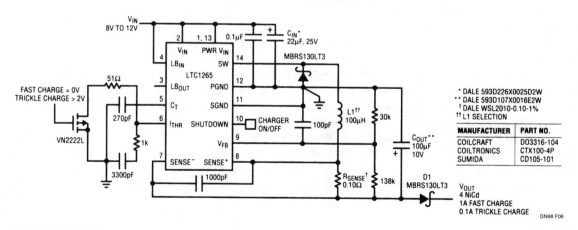

LINEAR TECHNOLOGY

Fig. 13-7

The LTC1265 is configured as a battery charger for a four-NiCad stack. It has the capability of performing a fast charge of 1 A, a trickle charge of 100 mA, or the charger can be shut off. In shut-off, diode D1 serves two purposes. First, it prevents the LTC1265 circuitry from drawing battery current and second, it eliminates "back powering" the LTC1265, which avoids a potential latch condition at power up.

BACKUP BATTERY MONITOR/CHARGER/ALARM

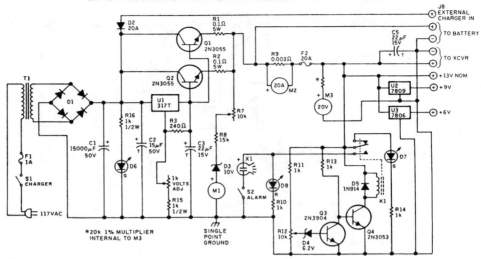

Fig. 13-8

Battery Condition
Meter Calibration

Lead-Acid Battery		Lead Calcium Battery	
Color	**Voltage**	**Color**	**Voltage**
Red	11.6 and below	Red	11.6 and below
Yellow	11.6 to 12.0	Yellow	11.6 to 12.0
Green	12.0 to 13.8	Green	12.0 to 13.5
Red	13.8 and higher	Red	13.5 and higher

Charging voltage is constant at the normal full-charge level, so the charging current drops as full charge is approached, and full charge is maintained with a trickle current. The charging voltage can be adjusted between approximately 10 and 15 Vdc to accommodate lead-acid (13.8 V) or lead-calcium (13.2 V, 13.5 V maximum) deep-cycle storage batteries.

A separate connection is provided so that an external charger can be used when greater than 3 A is needed to charge a partially discharged battery. Internal circuitry will maintain the charging voltage to the battery at the nominal full-charge voltage level, regardless of the voltage supplied by the external charger, which will be 2 V or more greater than that applied by the regulator to the storage battery. Warning: do not fast-charge deep-cycle storage batteries!

A pair of meters calibrated to indicate 20 Vdc and 20 Adc full-scale monitor voltage and current when battery power is used.

A separate, suppressed zero, expanded-scale meter calibrated over the range of about 10 to 15 Vdc allows immediate and constant indication of the state of charge of the station's backup battery. This meter scale is calibrated in bands of red, yellow, and green, as explained in the table. The narrow yellow segment is based on the assumption that solid-state transceivers might not operate properly below +12 Vdc. The internal power supply is used to calibrate this meter. A DMM should be used for greatest accuracy.

BACKUP BATTERY MONITOR/CHARGER/ALARM (*Cont.*)

An alarm circuit is included to indicate when the battery has been discharged by 60 percent to the 11.6-Vdc level. When battery voltage is above 11.6 V, the green LED will be illuminated; when voltage falls to 11.6 V, the green LED goes out and the red LED lights. A piezo audible alarm sounds at this low-voltage level unless silenced by the toggle switch controlling it.

A pair of fixed three-terminal regulators are included to provide +9 and +6 Vdc.

NICAD CHARGER/ZAPPER

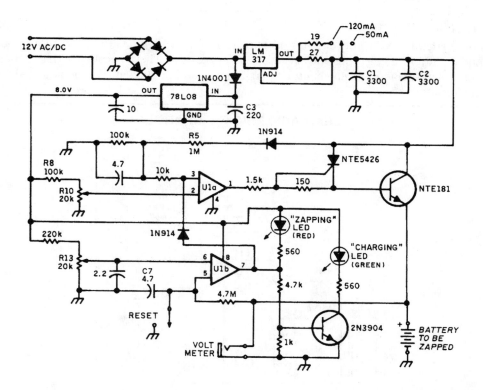

Fig. 13-9

The NiCad charger/zapper has a built-in charger and zapper circuit to clear shorted NiCads. This circuit delivers a high-current pulse to trim out internal shorts.

2- TO 5-CELL LITHIUM BATTERY CHARGER

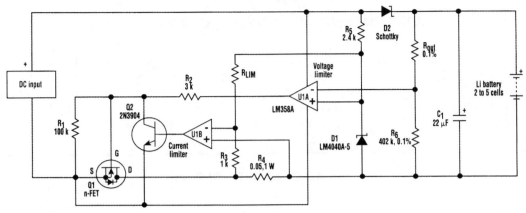

ELECTRONIC DESIGN

Fig. 13-10

A more generally applicable circuit-design concept for recharging lithium batteries could easily accommodate different cell types and various numbers of cells. That's because both the charger output-voltage set point and current limit, or maximum charging current, can be adjusted by simply changing a resistor.

LEAD-ACID TRICKLE CHARGER

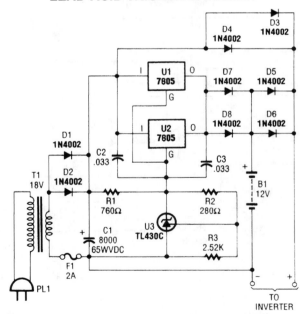

POPULAR ELECTRONICS

Fig. 13-11

This lead-acid battery trickle charger can be used as a stand-alone circuit (for alarm systems and such) or combined with the circuit in the figure to create an emergency lighting system.

NICAD BATTERY CHARGER

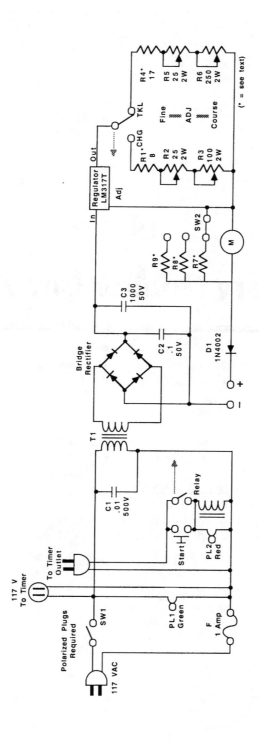

Fig. 13-12

73 AMATEUR RADIO TODAY

This circuit has a current regulator and uses an external timer to control the charging rate.

14

Battery Monitor Circuits

The sources of the following circuits are contained in the Sources section, which begins on page 707. The figure number in the box of each circuit correlates to the entry in the Sources section.

Battery Monitor
Battery Butler
Undervoltage Indicator for Single Cell
Battery Charger Probe
Low-Battery Circuit
Battery Charge Indicator
Battery Status Indicator
Lithium Memory Backup Replacement
Battery-Condition Indicator for 12-V Batteries

BATTERY MONITOR

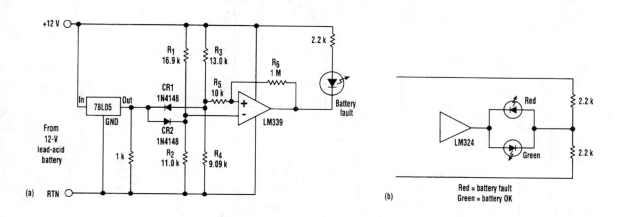

ELECTRONIC DESIGN

Fig. 14-1

One typical application for the detector involves monitoring a lead-acid battery. It indicates a fault when the battery voltage is outside an 11- to 14-V window. Because the circuit is powered by the battery, the input and reference were switched to keep the comparator inputs within its common-mode range.

The circuit's reference is 5.0 V. The resistor values in divider, R1/R2 were selected to produce 5.5 V at the inverting input when the battery voltage is 14.0 V. Divider R3/R4 is set to produce 4.5 V at the noninverting input when the battery voltage is equal to 11.0 V.

When the battery voltage is within the window, the noninverting input is more positive than the inverting input which is clamped at 4.5 V by CR2, the noninverting input continues below that, the comparator's output goes low, and the LED turns on. When the battery voltage rises above 14 V, the noninverting input is clamped at 5.5 V by CR1, the inverting input continues above that, the comparator output again goes low, and the LED turns on. Resistors R5 and R6 show that hysteresis might be added to this circuit in a conventional manner.

If an op amp, such as an LM324 is used as the comparator, two LEDs can be implemented. The green LED will turn on when the battery voltage is within the window, and the red LED turns on when the battery voltage is outside the window.

BATTERY BUTLER

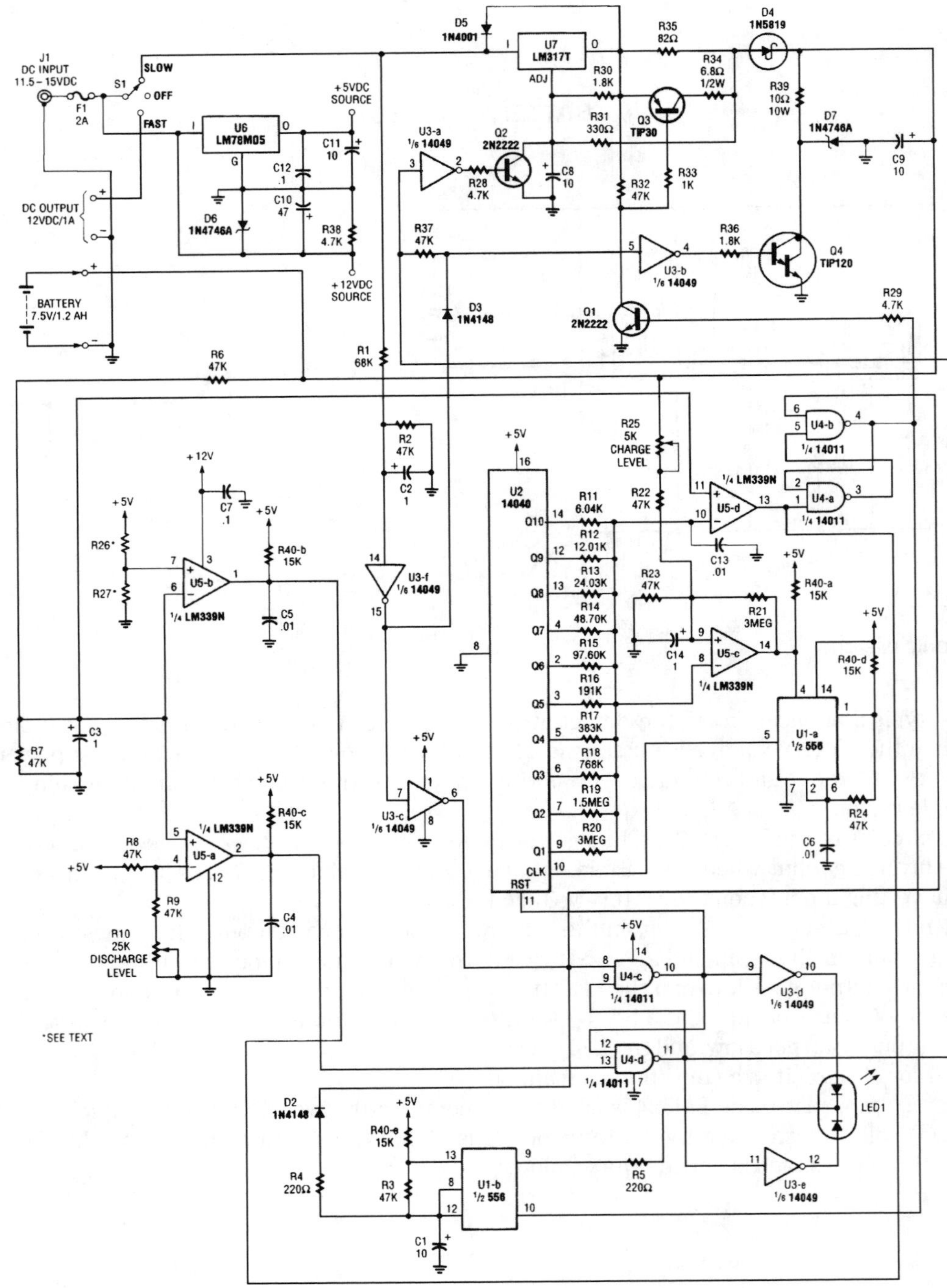

Fig. 14-2

BATTERY BUTLER (*Cont.*)

The battery butler solves the common problems associated with the maintenance and operation of NiCad batteries. The battery butler, by initially discharging a NiCad battery to a preset point, reduces the possibility of the "memory" effect occurring. Once discharged, a battery is then usually charged at 25% and reduce the internal cell pressure increase by 40% or more. Once the battery is fully charged, a trickle charge is provided to maintain the battery in a fully charged state. The battery butler circuit can be bypassed, and the existing fast-charger used, if needed.

UNDERVOLTAGE INDICATOR FOR SINGLE CELL

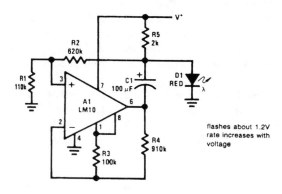

flashes about 1.2V
rate increases with
voltage

NATIONAL SEMICONDUCTOR

Fig. 14-3

When operating with a single cell, it is necessary to incorporate switching circuitry to develop sufficient voltage to drive the LED. A circuit that accomplishes this is drawn in the figure shown. Basically, it is a voltage-controlled asymmetrical multivibrator with a minimum operating threshold given by:

$$V_{TH} = \frac{R_4 \, (R_1 + R_2)}{R_1 \, (R_3 + R_4)} \, V_{REF}$$

Above this threshold, the flash frequency increases with voltage. This is a far more noticeable indication of a deteriorating battery than merely dimming the LED. In addition, the indicator can be made visible with considerably less power drain. With the values shown, the flash rate is 1.4 sec −1 at 1.2 V with 300-μA drain and 5.5 sec −1 at 1.55 V with 800-μA drain. Equivalent visibility for continuous operation would require more than 5-mA drain.

BATTERY CHARGER PROBE

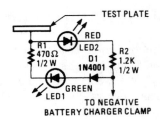

POPULAR ELECTRONICS

Fig. 14-4

This battery-charger probe can keep you from damaging batteries or yourself by testing to see if the charger is already on and/or connected improperly.

To use the probe, the positive cable clamp is first connected to the positive battery terminal. Then, the test plate is touched to the negative terminal of the battery. If the battery is connected properly, current will pass from the test plate through R1, LED1, D1, the negative charger, and into the positive side of the batteries. If LED1 (the green LED) lights, you can clamp on the negative lead and turn on the charger.

If the terminals are reversed, current will flow in the opposite direction, causing LED2 to light, warning you of danger. When the cable is reversed, D1 protects LED1 from excessive reverse voltage. If that happens, immediately turn the power off, and right the cable connections. Finally, if the battery charger is on, both LEDs will light because chargers actually produce pulsating dc and rely on the battery to act as a filter.

LOW-BATTERY CIRCUIT

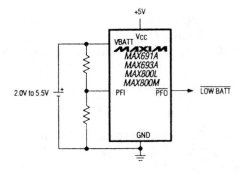

MAXIM

Fig. 14-5

A Maxim MAX691A series IC allows low-battery detection.

BATTERY CHARGE INDICATOR

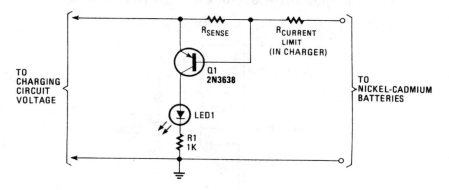

RADIO-ELECTRONICS   **Fig. 14-6**

When a battery is charging, a voltage drop across R_{SENSE} causes Q1 to conduct, and lights LED1. R_{SENSE} should be chosen as follows:

$$R_{\text{SENSE}} \text{ (ohms)} = \frac{0.65}{I_{\text{CHARGE}} \text{ (amps)}}$$

BATTERY STATUS INDICATOR

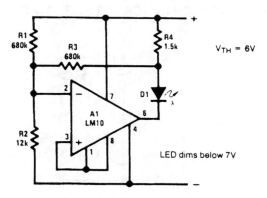

NATIONAL SEMICONDUCTOR   **Fig. 14-7**

In battery-powered circuitry, there are some advantages to having an indicator to show when the battery voltage is high enough for proper circuit operation. This is especially true for instruments that can produce erroneous data.

The battery status indicator is designed for a 9-V source. It begins dimming noticeably below 7 V and it extinguishes at 6 V. If the warning of incipient battery failure is not desired, R3 can be removed and the value of R_1 is halved.

LITHIUM MEMORY BACKUP BATTERY REPLACEMENT

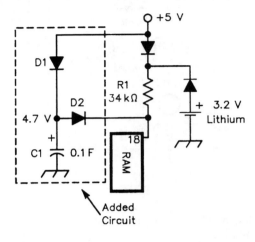

QST

Fig. 14-8

Physically very small high-capacitance capacitors are available for memory backup. Here, a 0.1-F (100,000 µF) capacitor and two diodes replace the lithium battery. The lithium battery can be retained as well, providing double backup.

BATTERY-CONDITION INDICATOR FOR 12-V BATTERIES

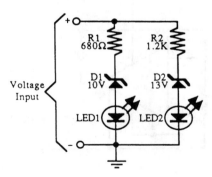

McGRAW-HILL

Fig. 14-9

A simple battery condition indicator. Choose the Zener diodes to provide a "window" for over/under voltage indication.

15

Bridge Circuits

The sources of the following circuits are contained in the Sources section, which begins on page 707. The figure number in the box of each circuit correlates to the entry in the Sources section.

Single-Supply Bridge Amplifier
Wheatstone Bridge
Bridge Amplifier with Low Noise Compensation

SINGLE-SUPPLY BRIDGE AMPLIFIER

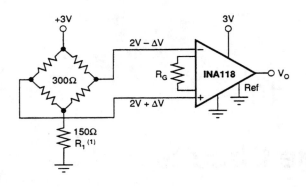

DESIRED GAIN	R_G (Ω)	NEAREST 1% R_G (Ω)
1	NC	NC
2	50.00k	49.9k
5	12.50k	12.4k
10	5.556k	5.62k
20	2.632k	2.61k
50	1.02k	1.02k
100	505.1	511
200	251.3	249
500	100.2	100
1000	50.05	49.9
2000	25.01	24.9
5000	10.00	10
10000	5.001	4.99

NOTE: (1) R_1 required to create proper common-mode voltage, only for low voltage operation — see text.

BURR-BROWN

Fig. 15-1

The INA118 can be used on single-power supplies of +2.7 to +36 V. The figure shown is a basic single-supply circuit. The output Ref terminal is connected to ground. Zero differential input voltage will demand an output voltage of 0 V (ground). Actual output voltage swing is limited to approximately 35 mV above ground, when the load is referred to ground as shown. The typical performance curve "Output Voltage vs. Output Current" shows how the output voltage swing varies with output current.

With single-supply operation, $+V_{IN}$ and $-V_{IN}$ must both be 1.1 V above ground for linear operation. You cannot, for instance, connect the inverting input to ground and measure a voltage connected to the noninverting input.

To illustrate the issues affecting low-voltage operation, consider the circuit in the figure. It shows the INA118, operating from a single 3-V supply. A resistor in series with the low side of the bridge ensures that the bridge output voltage is within the common-mode range of the amplifier's inputs.

WHEATSTONE BRIDGE

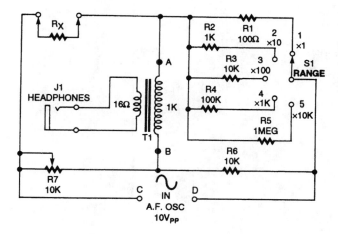

Fig. 15-2

This circuit can be used to measure resistances. R7 is calibrated and fitted with an indicator dial, then:

$$\frac{R_x}{(R_1 \text{ through } R_5)} = \frac{R_7}{R_6} \text{ or } R_x = \frac{R_7}{R_6} \times (R_1 \text{ through } R_5)$$

A frequency of 1 kHz for the audio oscillator is usually used.

BRIDGE AMPLIFIER WITH LOW NOISE COMPENSATION

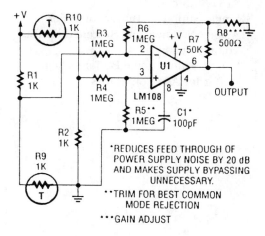

*REDUCES FEED THROUGH OF
POWER SUPPLY NOISE BY 20 dB
AND MAKES SUPPLY BYPASSING
UNNECESSARY.

**TRIM FOR BEST COMMON
MODE REJECTION

***GAIN ADJUST

Fig. 15-3

16

Buffer Circuits

The sources of the following circuits are contained in the Sources section, which begins on page 707. The figure number in the box of each circuit correlates to the entry in the Sources section.

Unity-Gain ADC Buffer
HI-*Z* Microphone Buffer Amplifier
Wideband General-Purpose Buffer
ADC Buffer
Single-Supply ac Buffer Amplifier
Analog Noninverting Switched Buffer
Voltage Follower
Simple Bidirectional Buffer Design
Buffer for A/D Converters

UNITY-GAIN ADC BUFFER

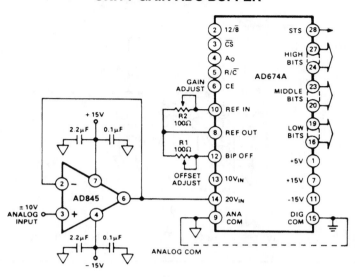

ANALOG DEVICES

Fig. 16-1

This buffer is suitable for ADCs of 12 bits with conversion times of 5 µs or greater. The wide bandwidth of the AD845 ensures a low output impedance at higher frequencies in the voltage follower (buffer) configuration.

HI-*Z* MICROPHONE BUFFER AMPLIFIER

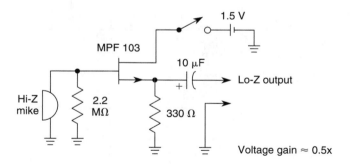

WILLIAM SHEETS

Fig. 16-2

A low impedance output from a high-*Z* microphone can be obtained with this circuit. No voltage gain is obtained, but a power gain is obtained because the output impedance is much lower (300 Ω), with −6-dB voltage gain.

WIDEBAND GENERAL-PURPOSE BUFFER

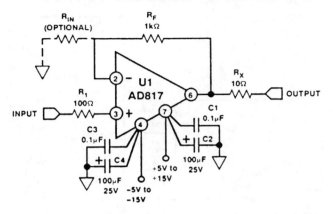

ANALOG DEVICES

Fig. 16-3

This circuit has unity gain and response up to 70 MHz. U1 is an Analog Devices AD817.

ADC BUFFER

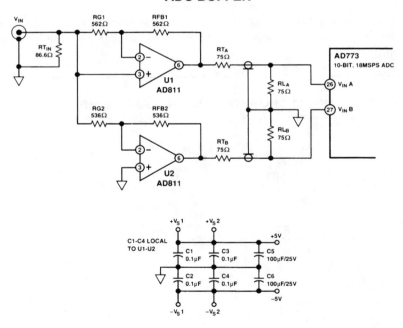

ANALOG DEVICES

Fig. 16-4

Useful for driving high-speed, 10-bit ADCs, this circuit was developed to drive an 18-MSPS 10-bit ADS. It works from ±5-V supplies.

SINGLE-SUPPLY ac BUFFER AMPLIFIER

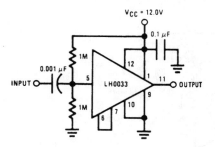

This buffer might be used with a single supply without special considerations. The input is dc biased to mid-operating point and is ac coupled. Its input impedance is approximately 500 kΩ at low frequencies. Note that for dc loads referenced to ground, this quiescent current is increased by the load current set at the input dc bias voltage.

NATIONAL SEMICONDUCTOR   *Fig. 16-5*

ANALOG NONINVERTING SWITCHED BUFFER

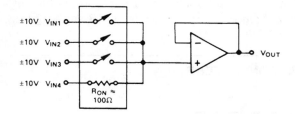

Here is noninverting solution.

ANALOG DEVICES   *Fig. 16-6*

VOLTAGE FOLLOWER

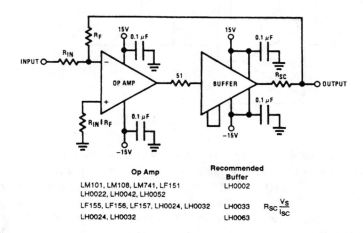

Op Amp	Recommended Buffer	
LM101, LM108, LM741, LF151 LH0022, LH0042, LH0052	LH0002	
LF155, LF156, LF157, LH0024, LH0032	LH0033	$R_{SC} \frac{V_S}{I_{SC}}$
LH0024, LH0032	LH0063	

NATIONAL SEMICONDUCTOR   *Fig. 16-7*

SIMPLE BIDIRECTIONAL BUFFER DESIGN

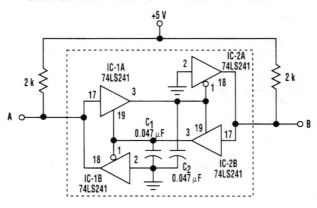

ELECTRONIC DESIGN

Fig. 16-8

This circuit shown in the figure uses two 74LS241s. When both input/output lines are high, IC-1A and IC-2B turn on, and C1 and C2 are charged to high voltage. Meanwhile, IC-1B and IC-2A are off to prevent a logic "1" latch.

BUFFER FOR A/D CONVERTERS

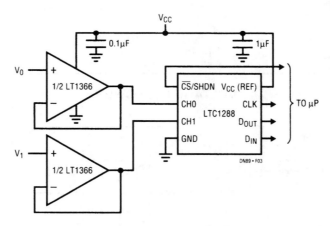

LINEAR TECHNOLOGY

Fig. 16-9

This circuit uses an LT1366 driving an LTC1288 two-channel micropower A/D. The LTC1288 can accommodate voltage references and input signals equal to the supply rails. The sampling nature of this A/D eliminates the need for an external sample-and-hold, but might call for a drive amplifier because of the A/D's 12-μs settling requirement. The LT1366's rail-to-rail operation and low-input offset voltage make it well suited for low-power, low-frequency A/D applications. In addition, the op-amp's output settles to 1% in response to a 3-mA load step through 100 pF in less than 1.5 μs.

17

Clock Circuits

The sources of the following circuits are contained in the Sources section, which begins on page 707. The figure number in the box of each circuit correlates to the entry in the Sources section.

Set Time Windows within a Clock
Low-Frequency Clock

SET TIME WINDOWS WITHIN A CLOCK

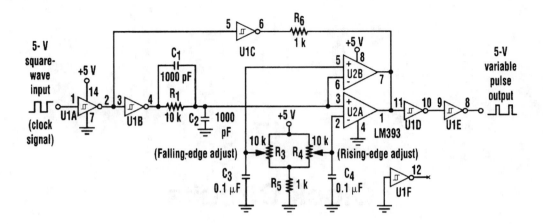

ELECTRONIC DESIGN

Fig. 17-1

At times, it is necessary to produce pulses of adjustable width and whose start times might vary with reference to a master clock. The input signal is inverted and buffered by U1A, while U1B and U1C reinvert the signal to produce square, buffered renditions of the input signal. Potentiometers R3 and R4 set references for the comparators.

The input polarity of U2A keeps its output transistor turned on until the voltage at the noninverting input exceeds the reference set by R4 (the rising edge adjustment). When this reference voltage is exceeded, the output transistor is turned off and the output signal is pulled up via R6. Meanwhile, the input polarity of U2B keeps its output transistor turned off until the voltage at the inverting input exceeds the reference set by R3 (the falling edge adjustment). When this reference voltage is surpassed, the output transistor of U2B is turned on, pulling the output signal low through the wired-OR configuration of U2. The output of U2 is then double-inverted and buffered by U1D and U1E. What results is a pulse whose start time (rising edge) can be adjusted by R4, and whose stop time (falling edge) can be adjusted by R3.

The output of the comparators is pulled up to the input waveform through resistor R6 to U1C. This prevents the comparators from switching during the low cycle of the input waveform, regardless of the positions of R3 and R4. This has the effect of "locking out" changes during the low period of the input signal, and would probably require additional logic if it were done strictly in the digital domain.

The circuit, with the component values shown, works well between about 50 and 150 kHz.

LOW-FREQUENCY CLOCK

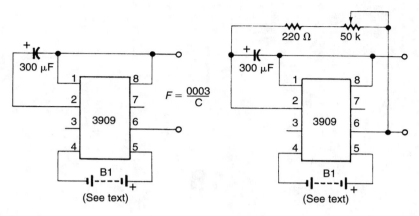

$$F = \frac{0003}{C}$$

ELECTRONICS NOW

Fig. 17-2

The LM3090 is an LED flasher IC that is designed to oscillate at low frequencies. The clock output of the first circuit can be changed by changing the value of the capacitor, and the second circuit lets you adjust the frequency with the trimmer. B1 can be one or two alkaline 1.5-V cells. The LM3909 can supply up to 45-mA pulses at greater than 2 V.

18

Computer-Related Circuits

The sources of the following circuits are contained in the Sources section, which begins on page 707. The figure number in the box of each circuit correlates to the entry in the Sources section.

RS-422 to RS-232 Converter
Printer Port
PC Password Protector
Key Wireless RTS with Data
Microprocessor Supervisory Circuit
Computer-Powered RS-232
+12-V Flash Memory Programming Supply
EEPROM Programming Doubler Circuit
Monitor Power Saver for Computers

RS-422 TO RS-232 CONVERTER

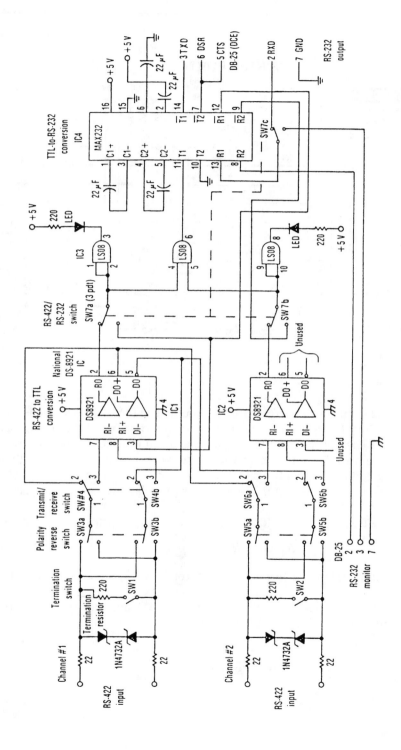

Fig. 18-1

ELECTRONIC DESIGN

The circuit supplies two LEDs for visual indication of line activity and terminating resistors when needed. The 220-Ω resistors and 5-V zeners at the RS-422 line inputs supply circuit protection.

Switch SW7 allows the circuit monitor both transmitted and received signals when tee-connected into an RS-232 line. One function of this optional feature is the ability to test a software-locking device that connects to the COM1 port on an IBM PC.

133

PRINTER PORT

REGISTOR SELECTION CODES	
Data	Selected resistor
0X00	R_1
0X01	R_2
0X02	R_3
0X03	R_4
0X04	R_5
0X05	R_6
0X06	R_7
0X07	R_8
0X08	R_9
0X09	R_{10}
0X0A	R_{11}
0X0B	R_{12}
0X0C	R_{13}
0X0D	R_{14}
0X0E	R_{15}
0X0F	R_{16}

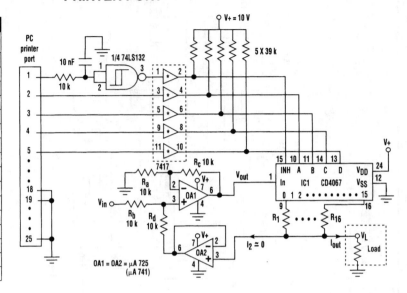

ELECTRONIC DESIGN

Fig. 18-2

A 16-step programmable current generator can be modified so that it's controllable by a printer port. This is done by switching the resistor connected between the output of the generator's OA1 op amp and the input of OA2. The CMOS single 16-channel analog multiplexer (IC1) chooses one resistor at a time, in accordance with the code sent by the printer port through four of its eight data-output lines (pins 2 to 9). In addition, one control line (pin 1) is used to enable the operation. As a result, 16 outputs can be selected by a 4-bit word (the table shows the relationship between data word and selected resistor).

The following must be fulfilled in order for the circuit to work as a true current generator:

$$R_2 \times R_d - R_b \times R_c = 0$$

The smaller the resistors' tolerance (especially R1 through R16), the greater the output resistance of the generator.

Because the OA2 is connected as a repeater, the current $I_2 = 0$, and only the load current flows through one of the R1 through R16 resistors. Therefore:

$$I_{out} = \frac{V_{out}}{(R_x + R_{on})}$$

PRINTER PORT (*Cont.*)

where $V_{out} = V_{in}$; $X = 1...16$; and $R_{on} \leq 150\ \Omega$ (for $V_{DD} = 10$ V) is the resistance of one analog switch (CD4067) in conduction.

Therefore, the values of resistors R_1 through R_{16} can be inferred from the needed currents:

$$R_x = \left(\frac{V_{in}}{I_{out}}\right) - R_{on}$$

The Turbo C++ program also controls the current through the load.

TURBO C++ CONTROL PROGRAM

```
#include <stdio.h>
#include <dos.h>

#define   OUT_PORT    0x378    /*printer output port address   */
#define   CTRL_PORT   0x37A    /*printer control port address  */

int main(void)
{
  int data;
  outport(CTRL_PORT , 0x01);    // enable operation
  delay(1);
  outportb(OUT_PORT) , data); // one of R1 -R16 selected (table)
  printf("\n\aR%d selected.",data+1);
  return 0;
}
```

PC PASSWORD PROTECTOR

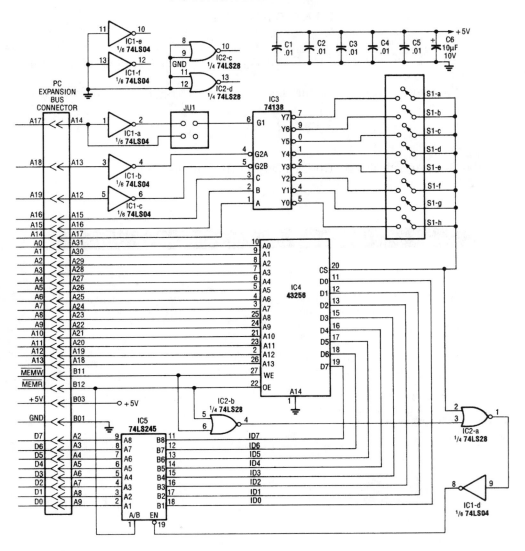

Fig. 18-3

IC4, a static RAM, is mounted in a "smart" built-in switch over circuitry. This retains SRAM contents when power is off. The rest of the circuitry consists of address decoding logic and jumper JU1, used to decode a 16K address space for the 32K static RAM. Software is necessary and this is contained in the original article (see reference).

KEY WIRELESS RTS WITH DATA

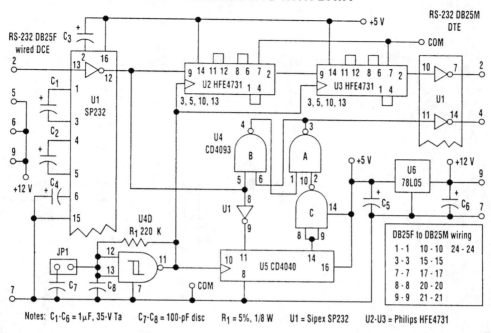

Notes: C_1-C_6 = 1μF, 35-V Ta C_7-C_8 = 100-pF disc R_1 = 5%, 1/8 W   U1 = Sipex SP232   U2-U3 = Philips HFE4731

ELECTRONIC DESIGN

Fig. 18-4

This simple keyer supplies both the RTS control and data delay needed to interface a digital radio with an RS-232, data-only system. It supports speeds to 19.2 kbits/s sync or async.

MICROPROCESSOR SUPERVISORY CIRCUIT

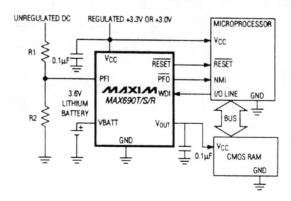

MAXIM

Fig. 18-5

This circuit provides a reset pulse during power up, down, or low-voltage battery back up, switching, a watchdog timer and a 1.25-V threshold detector for power failure or power fail warning, or to monitor another supply.

137

COMPUTER-POWERED RS-232

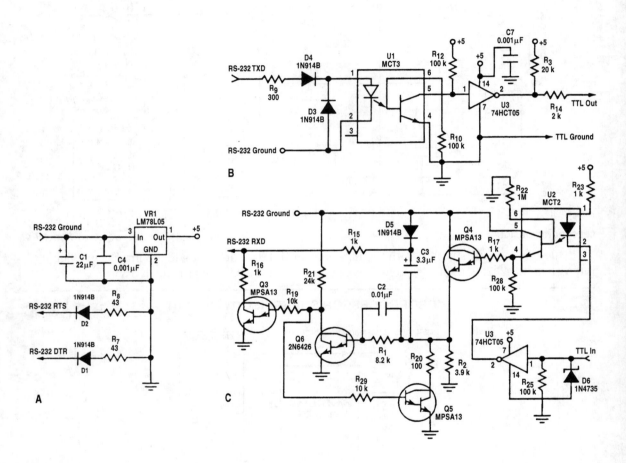

COMPUTER-POWERED RS-232 (*Cont.*)

Commercializing battery-operated equipment that must interface to a computer via the RS-232 port runs into the problem of power consumption. To load the system batteries strictly to power the interface is unacceptable. An alternative is to let the computer that the device is connected to provide the interface's power. One snag is that the RS-232 specification doesn't have a power tap on the connector, but it does provide RTS and DTR (request to send and data terminal ready) signals that assert a negative voltage in their quiescent state.

Figure 18-6A shows a simple scheme of deriving a 5-V potential from the RTS and DTR signals. R7 and R8 and diodes D1 and D2 mix the return current to the RS-232 port so that the RTS and DTR drivers split the current drawn by the interface. This scheme, even from a laptop computer, can supply 12 mA to the interface. The only drawback is that the TTL device must be isolated from the computer's ground (earth ground) because the interface treats the RS-232 ground as a positive voltage.

A modified optocoupler system shifts the RS-232 level to TTL voltages (Fig. 18-6B). It will support up to 9600 bits/s. C3 is charged up to about 1 V less than the RTS voltage while the TTL line asserts a marking state. As the capacitor is charging, Q3 is biased into saturation, thus providing a negative voltage (with respect to RS-232 ground) to the RS-232 RXD line. When a spacing bit is driven from the TTL line, Q3 switches off and Q4 switches on. This biases Q4's emitter up to the RS-232 ground. That ground potential is summed with C3's charge to create an RS-232-compatible spacing signal (approximately 1 V less than a $-V_{RTS}$).

The discharge rate on C3 is limited by R15 to prevent the signal sag from becoming a problem down to 110 bits/s. The C2/R1 time constant must be fairly close (within 4 times) to the C3/R15 time constant to ensure that Q3 turns off correctly.

+12-V FLASH MEMORY PROGRAMMING SUPPLY

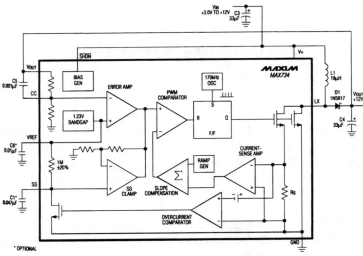

MAXIM

Fig. 18-7

The MAX734 can deliver up to 120 mA @12 V from a +5-V supply, using few external components. This supply can also be used for other applications than memory programming. A logic level is used for shutdown. Efficiency is about 85%.

EEPROM PROGRAMMING DOUBLER CIRCUIT

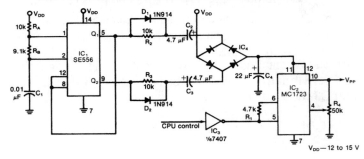

ELECTRONIC DESIGN

Fig. 18-8

Even though electrically erasable PROMs offer the convenience of single-byte write and erase operations, the parts require a somewhat nonstandard programming voltage—21 V. A simple circuit that develops the appropriate voltage from a computer system's standard 12-to-15-Vdc supply, remedies the problem nicely. Moreover, it permits the programming voltage to be pulsed under the control of an external CPU.

As shown in the figure, the chip uses its complementary outputs, Q1 and Q2 to trigger a bridge rectifier through capacitors C2 and C3. Resistors R2 and R3 and diodes D1 and D2 limit the current and protect IC1 from spikes from C2 and C3. If required, the regulator will deliver up to 150 mA.

Circuit IC3 is an open-collector TTL gate whose output, when low, disables IC2 and causes it to put out 5 Vdc. The regulator delivers the 21-V programming pulse.

MONITOR POWER SAVER FOR COMPUTERS

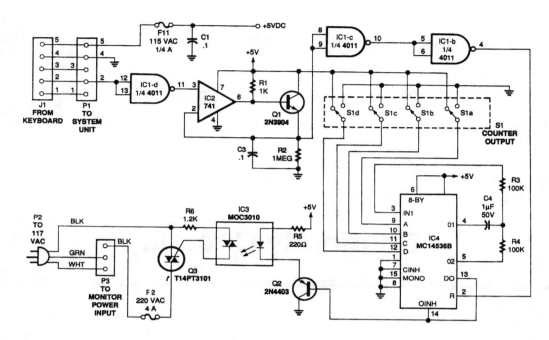

ELECTRONICS NOW

Fig. 18-9

The circuit monitors PC keyboard activity through five-pin DIN connector J1. When the user presses a key, the keyboard sends a series of negative-going pulses on pin 2. In conjunction with Q1 and C3, the op amp essentially functions as an integrator, which stretches the continually varying periods of the input pulses to a relatively constant period with a higher average dc value.

Inverters IC1-c and IC1-b buffer the peak detector's output to trigger IC4, an MC14536B programmable timer.

19

Continuity Circuits

The sources of the following circuits are contained in the Sources section, which begins on page 707. The figure number in the box of each circuit correlates to the entry in the Sources section.

Short-Circuit Beeper
Adjustable Continuity Tester
Simple Audio/Video Cable Tester
Audible Continuity Tester

SHORT-CIRCUIT BEEPER

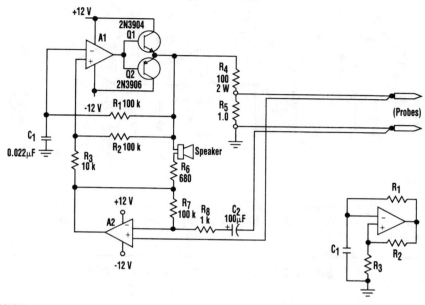

ELECTRONIC DESIGN

Fig. 19-1

This design offers a way to trace resistance in the milliohm range, right to a short between bridged traces beneath a solder mask. It simply translates resistance into an audible tone, which increases in pitch as the measured value approaches zero.

In the classic op-amp multivibrator (shown in the inset), oscillation frequency is determined not only by the R1/C1 time constant, but also by the hysteresis set by the R_2/R_3 resistance ratio. A1 in the main figure, with current boosters Q1 and Q2, is this same configuration.

Assuming a virtual ground at the output of A2, free-run frequency is about 1 kHz—quite audible through a tiny 8-Ω speaker. Q1 and Q2 deliver a ±10-V squarewave to R4, dumping a ±100 mA through a short circuit placed across the probe tips. R5 ensures that open circuit voltage never exceeds ±0.1 V.

A2 monitors the voltage between the probes. The differential input must have its own separate path to the probe tips to eliminate test lead resistance from the measurement. Miniature "zip-cord" sold as loudspeaker wire makes a tidy two-conductor test lead.

When the probes are open, A2's gain equals the R4/R5 divider loss, and the output of both amplifiers is identical. This has two effects: first, hysteresis is greatly increased and the frequency falls to a low growl; second, the loudspeaker that bridges the two in-phase outputs is effectively silenced.

The dead short across the probe tips will return nothing to A2 and the circuit will squeal at its nominal 1-kHz rate. Anything less than a perfect short produces some output from A2, increasing multivibrator hysteresis and lowering the pitch. The circuit has so much "leverage," and the ear is so sensitive to pitch changes in this range, that it's easy to resolve minute resistance differences.

Any general-purpose op amp will suffice in this circuit—a couple of 741s or an equivalent dual.

ADJUSTABLE CONTINUITY TESTER

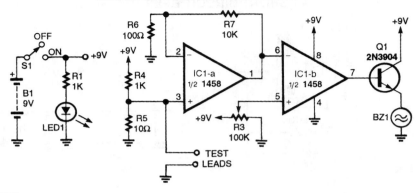

Fig. 19-2

A problem with most continuity testers is that the exact decision point (circuit resistance) between continuity and open is indefinite. This circuit allows setting of this point to a known resistance between 1 and 50 Ω.

SIMPLE AUDIO/VIDEO CABLE TESTER

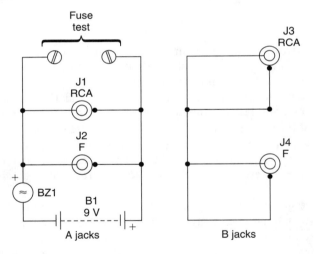

Fig. 19-3

As simple as it looks, the audio/video cable tester tests cables while they are plugged in. You can flex them vigorously while listening to the built-in buzzer.

Simply plug the ends of the cable in the appropriate jack. Only one end of the cable need be plugged in for a complete short test; the other end is left free. If the buzzer sounds, there is a short circuit somewhere in the cable. If nothing is heard, test for an intermittent short by flexing the cable several times, particularly in the plug area of both the free and plugged-in ends.

AUDIBLE CONTINUITY TESTER

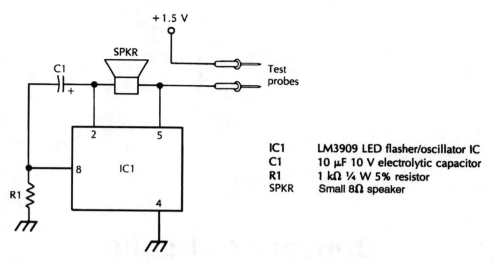

IC1	LM3909 LED flasher/oscillator IC
C1	10 µF 10 V electrolytic capacitor
R1	1 kΩ ¼ W 5% resistor
SPKR	Small 8Ω speaker

Fig. 19-4

20

Converter Circuits

The sources of the following circuits are contained in the Sources section, which begins on page 707. The figure number in the box of each circuit correlates to the entry in the Sources section.

WWV Converter
Simple HF Receive Converter
225-W, 15-V Output Converter
12-Bit DAC
Driven Flyback Converter
Sine-Wave Converter
SCR Converter
5-V, 5-A Step-Down Converter
Sync-to-Async Converter
Differential Voltage-to-Current Converter
Direct-Conversion 7-MHz Receiver
Low-Frequency Converter
Programmable Current-to-Voltage Converter
Current-to-Voltage Converter with Boost Transistor
Current-to-Voltage Converter for Grounded Loads
Output-to-Current Converter

WWV CONVERTER

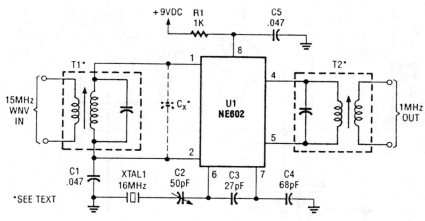

RADIO CRAFT

Fig. 20-1

This converter heterodynes the 15-MHz WWV signal with a 16-MHz oscillator so that it can be heard at 1 MHz on an AM broadcast receiver. T1 and T2 are a modified 10.7-MHz IF transformer and AM BC oscillator coil, respectively.

SIMPLE HF RECEIVE CONVERTER

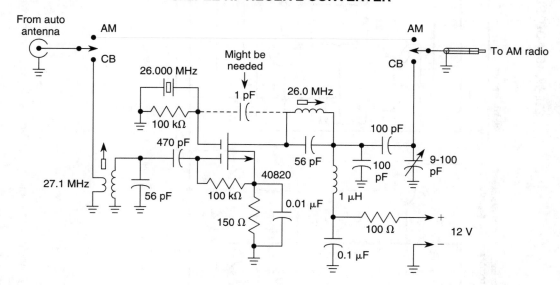

WILLIAM SHEETS

Fig. 20-2

Designed for CB reception, this crystal-controlled converter uses one 40820 dual-gate MOSFET. The circuit will work with any crystal either 3rd overtone or fundamental, over 1 to 50 MHz.

225-W, 15-V OUTPUT CONVERTER

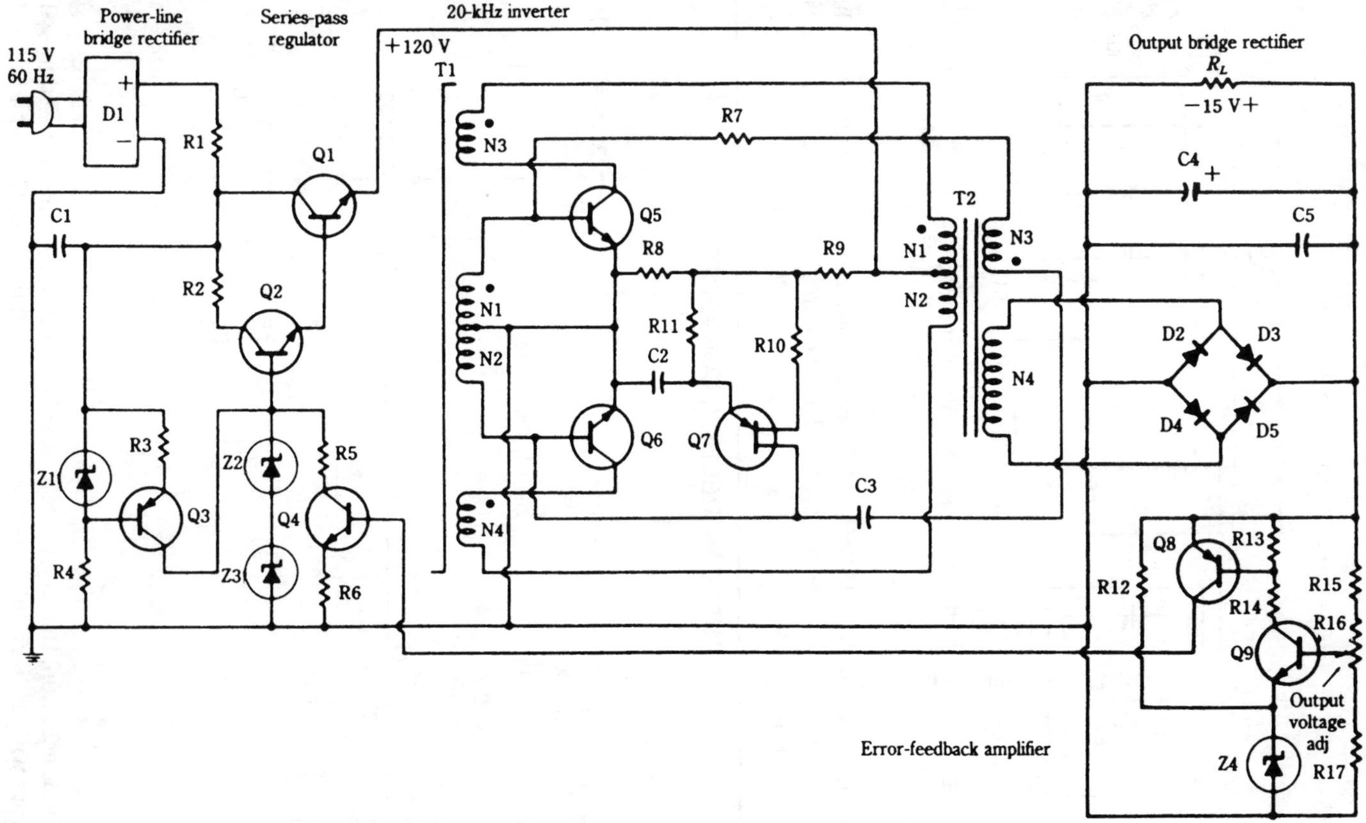

McGRAW-HILL

Fig. 20-3

A converter designed to supply ±15 Vdc is shown. This converter is several times lighter in weight than an equivalent.

225-W, 15-V OUTPUT CONVERTER (*Cont.*)

C1 — 2500 µF, 350 V Electrolyic
C2 — 0.1 µF Disc Ceramic
C3 — 0.1 µF Paper
C4 — 10 µF Electrolyic
C5 — 0.25 µF Paper
D1 — MDA-980-4 Bridge Rectifier Assembly
D2, D3, D4, D5 — 1N5826, 20 V 15 A

Q1, Q5, Q6, — 2N6307
Q2, Q4 — 2N5052
Q3 — 2N5345
Q7 — 2N4870
Q8 — 2N3905
Q9 — 2N3903

All Resistors in Ohms and 1/2 W Unless Otherwise Noted

R1 — 1, 10 W
R2 — 100
R3 — 82
R4 — 22K
R5 — 1.5K, 15 W
R6 — 200
R7 — 15
R8 — 4.7K
R9 — 51

R10 — 1K
R11 — 10K
R12 — 270
R13 — 1K
R14 — 7.5K
R15 — 2.5K
R16 — 5K
R17 — 3.5K

T1 — Core — Magnetics Inc. 80623 — 1/2 D — 080
   N1, N2 — 20 Turns Each, No. 30 AWG (Bifilar)
   N3, N4 — 3 Turns Each, No. 20 AWG

T2 — Core — Arnold 6T 5800 D1
   N1, N2 — 100 Turns Each, No. 20 AWG (Bifilar)
   N3 — 7 Turns No. 26 AWG
   N4 — 12 Turns Each, No. 12 AWG (No. 16 AWG, 3 in Parallel)

Z1 — 1N4733, 5.1 V
Z2, Z3 — 1N4760, 68 V
Z4 — 1N4736

12-BIT DAC

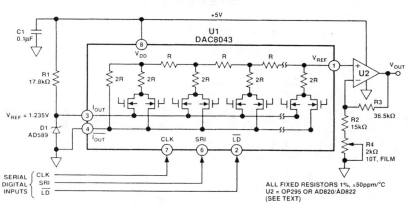

ANALOG DEVICES

Fig. 20-4

This circuit uses an Analog Devices DAC-8043 12-bit multiplying DAC. The output voltage will be $D/4096 \times V_{ref}$ where D is the numerical value of the digital input word (0 to 4095). V_{ref} is 1.235 volts in this circuit.

DRIVEN FLYBACK CONVERTER

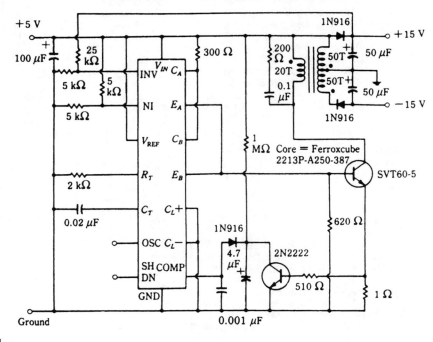

McGRAW-HILL

Fig. 20-5

This circuit uses an SG1524 Silicon General regulating pulse width modulator and provides ±15 V from a 5-V supply rail.

SINE-WAVE CONVERTER

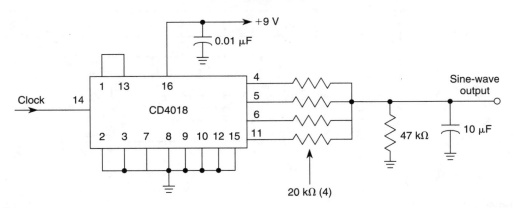

WILLIAM SHEETS

Fig. 20-6

This circuit produces a sine wave with a low-frequency clock input. The clock rate should be 100 Hz or less.

SCR CONVERTER

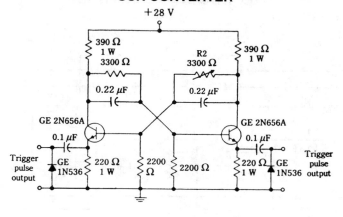

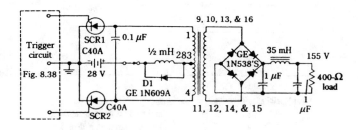

McGRAW-HILL

Fig. 20-7

Two SCR devices are used in a push-pull driver to convert 28 Vdc to 155 Vdc, using the transformer and bridge shown. A center-tapped transformer with 24 V:120 V could be used for 60-Hz applications. The trigger circuit supplies a push-pull drive signal.

5-V, 5-A STEP-DOWN CONVERTER

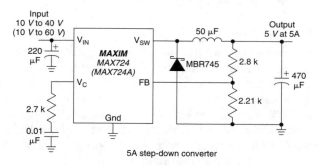

5A step-down converter

MAXIM

Fig. 20-8

This circuit is useful where power must be distributed by a higher (10 to 60 V) bus. The circuit reduces power dissipation and eliminates inefficient passive linear regulators. The switching frequency is in the 100-kHz region.

SYNC-TO-ASYNC CONVERTER

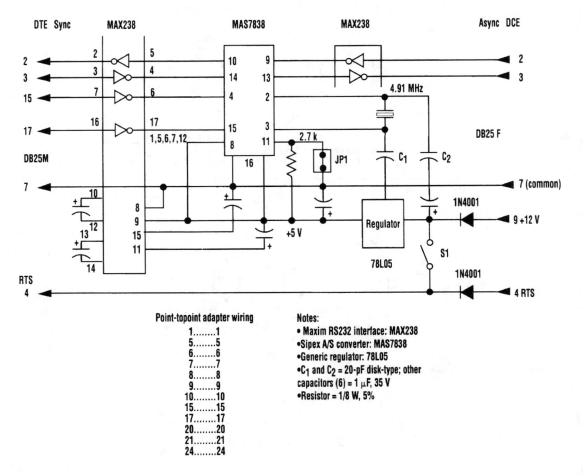

Point-topoint adapter wiring

1	1
5	5
6	6
7	7
8	8
9	9
10	10
15	15
17	17
20	20
21	21
24	24

Notes:
- Maxim RS232 interface: MAX238
- Sipex A/S converter: MAS7838
- Generic regulator: 78L05
- C_1 and C_2 = 20-pF disk-type; other capacitors (6) = 1 μF, 35 V
- Resistor = 1/8 W, 5%

ELECTRONIC DESIGN

Fig. 20-9

This simple converter consists of two ICs and a voltage regulator. The Sipex MAS7838, which acts as the converter, selects the conversion speed to that of the synchronized data clock. It has internal switches and registers to perform the async-to-sync, or sync-to-async, conversion. The Maxim MAX238 provides the RS-232 drivers and receivers for interfacing with the data bus. These chips require a 5-Vdc power supply; a generic 78L05 reduces the +12 V at the DB25 pin 9 to the +5 V needed. A crystal frequency of 4.91 MHz is suitable for converting to 19.2 kbits or a sub-multiple (9.6, 4.8, 2.4, etc.). Two 1N4001 diodes protect the external RTS (ready to send) control circuitry if the RTS is enabled by S1. When JP1 is removed, the converter is transparent in the sync mode and no conversion will occur.

The completed unit is mounted atop a universal breakout adapter, and the control lines are jumpered according to the chart in the figure. The physical size is approximately $1 \times 2.25 \times 2.5$ inches and will easily plug into the DB25 socket on a synchronized data communications equipment (DCE) communication device.

DIFFERENTIAL VOLTAGE-TO-CURRENT CONVERTER

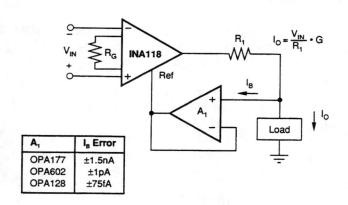

$$I_O = \frac{V_{IN}}{R_1} \cdot G$$

A₁	I_B Error
OPA177	±1.5nA
OPA602	±1pA
OPA128	±75fA

DESIRED GAIN	R_G (Ω)	NEAREST 1% R_G (Ω)
1	NC	NC
2	50.00k	49.9k
5	12.50k	12.4k
10	5.556k	5.62k
20	2.632k	2.61k
50	1.02k	1.02k
100	505.1	511
200	251.3	249
500	100.2	100
1000	50.05	49.9
2000	25.01	24.9
5000	10.00	10
10000	5.001	4.99

BURR-BROWN

Fig. 20-10

DIRECT-CONVERSION 7-MHz RECEIVER

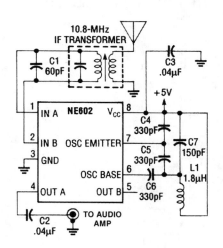

RADIO-ELECTRONICS

Fig. 20-11

An NE602 is used to mix signals in the 7-MHz range with an LO and to produce audio output.

LOW-FREQUENCY CONVERTER

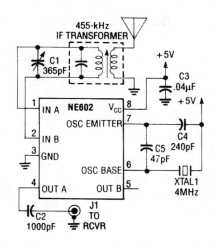

RADIO-ELECTRONICS

Fig. 20-12

This converter circuit translates the 350- to 500-kHz range to 4.35 to 4.50 MHz, enabling the frequency range to be received on a conventional shortwave receiver.

PROGRAMMABLE CURRENT-TO-VOLTAGE CONVERTER

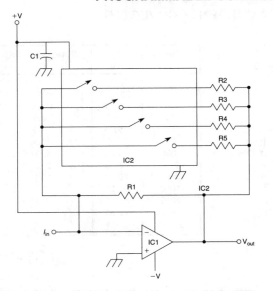

Programmable current-to-voltage converter permits you to electrically select from 16 resistor values using bilateral switches.

IC1	741 op amp (or similar)
IC2	CD4066 quad bilateral switch
C1	0.1-µF capacitor
R1	10-kΩ, ¼-W 5% resistor
R2	4.7-kΩ, ¼-W 5% resistor
R3	2.2-kΩ, ¼-W 5% resistor
R4	1.2-kΩ, ¼-W 5% resistor
R5	100-Ω, ¼-W 5% resistor

McGRAW-HILL

Fig. 20-13

CURRENT-TO-VOLTAGE CONVERTER WITH BOOST TRANSISTOR

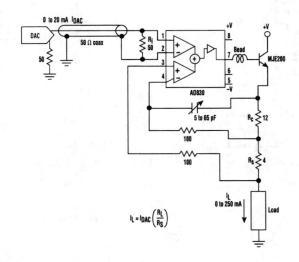

$$I_L = I_{DAC}\left(\frac{R_L}{R_S}\right)$$

ELECTRONIC DESIGN

Fig. 20-14

A transistor such as the MJE200 can be added to an Analog Devices AD830 to produce this current to voltage converter. Loads to 250 mA can be driven. The 5- to 65-pF trimmer is for compensation.

CURRENT-TO-VOLTAGE CONVERTER FOR GROUNDED LOADS

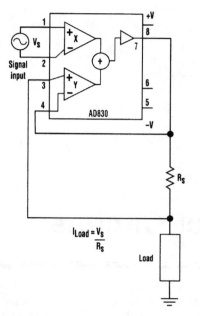

This circuit uses an Analog Devices AD830 video difference amplifier. The circuit consists of two differential inputs. Unlike a conventional op amp, the AD830's output is nulled when the sum of the differences of the two inputs is zero.

The AD830's stated unity-gain bandwidth is 60 MHz, and the device is capable of driving up to ±30 mA directly. The differential input voltage is limited to ±2 V, while the maximum power supply is ±15 V.

If more output current is desired, the AD830 can drive a bipolar transistor (such as an MJE200) directly. This will produce a one-sided output.

A ferrite bead can be placed on the base to prevent oscillation under some conditions. Compensation can be added by splitting R_s and adding a variable capacitor. A resistor can be positioned at the input to match the amplifier's input to a transmission line.

ELECTRONIC DESIGN

Fig. 20-15

OUTPUT-TO-CURRENT CONVERTER

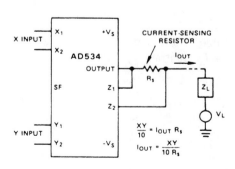

Occasionally, it is preferable to generate a current, rather than a voltage, output into the load. The availability of differential inputs allows this to be accomplished in any of the four basic modes.

If the output is to integrated, Z_L can be simple high-quality capacitor, unloaded by an op amp connected as a high-impedance follower. Note that, if desired, one side of a rest switch can be grounded.

The compliance constraint for this configuration, where V_L is an arbitrary common-mode potential, is:

$$|V_L + I_{OUT} (Z_L + R_S)| \leq 12 \text{ V}$$

ANALOG DEVICES

Fig. 20-16

21

Crystal Oscillator Circuits

The sources of the following circuits are contained in the Sources section, which begins on page 707. The figure number in the box of each circuit correlates to the entry in the Sources section.

FET QUARTZ CRYSTAL OSCILLATOR

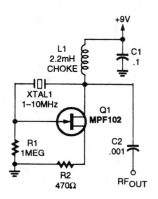

POPULAR ELECTRONICS

Fig. 21-1

This oscillator uses an MPF102 JFET as an active element.

CRYSTAL OSCILLATOR I

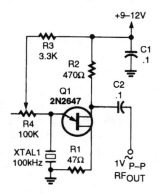

POPULAR ELECTRONICS

Fig. 21-2

In this circuit, series-resonant crystal XTAL1 is used as a frequency-determining element. XTAL1 is between 0.1 to 10 MHz.

FET VXO CIRCUIT

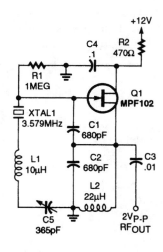

POPULAR ELECTRONICS

Fig. 21-3

An MPF 102 is used in a Colpitts-type oscillator in order to pull the crystal frequency slightly.

UJT 100-kHz CALIBRATION OSCILLATOR

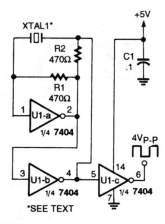

POPULAR ELECTRONICS

Fig. 21-4

This unusual 100-kHz oscillator (whose frequency is determined by XTAL1) can be used as a marker generator to calibrate the analog dial of a communication receiver, or its output can be fed to a divider counter to produce a stable lower-frequency output for use as a clock-signal generator.

157

CRYSTAL OSCILLATOR WITH FM CAPABILITY

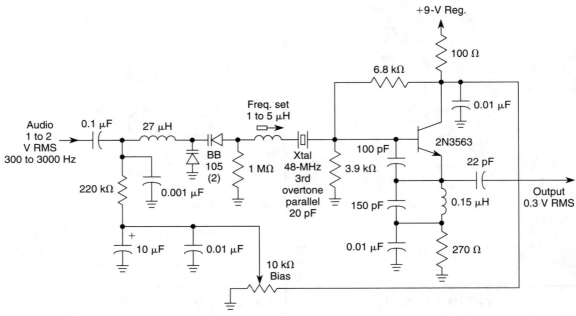

WILLIAM SHEETS

Fig. 21-5

This crystal oscillator produces a good FM signal that can be tripled to 146 MHz and produces a clean 5-kHz deviation signal for FM voice. The bias control is adjusted for cleanest audio while the 1- to 5-µH coil is adjusted to set the oscillator frequency to the exact setting required.

CRYSTAL OSCILLATOR II

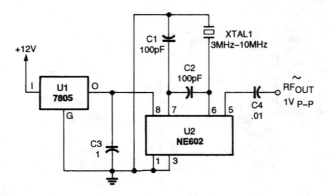

POPULAR ELECTRONICS

Fig. 21-6

An NE602 can be used as a crystal oscillator.

dc-SWITCHED CRYSTAL OSCILLATOR

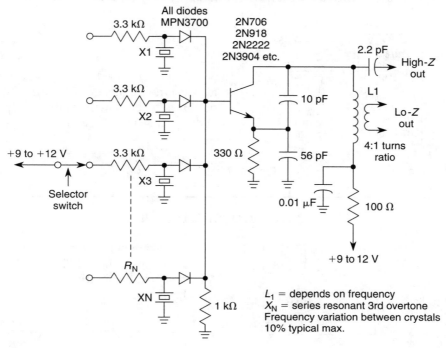

WILLIAM SHEETS

Fig. 21-7

This circuit is useful where several different crystal frequencies must be switched using a dc source. The values shown are typical for 40- to 60-MHz third-overtone crystals. Limitation on number of crystals depends on PIN diode capacitance and layout factors, but up to 5 or 10 crystals is possible.

CRYSTAL OSCILLATOR WITH ADJUSTABLE FREQUENCY

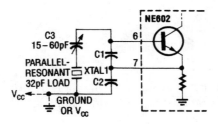

RADIO-ELECTRONICS

Fig. 21-8

In this crystal oscillator circuit, C3 adjusts the frequency of the oscillator for exact netting. The crystal is a fundamental type. $C_1 = 100$ pF and $C_2 = 1000$ pF are typical.

FREQUENCY DOUBLER AND CRYSTAL OSCILLATOR

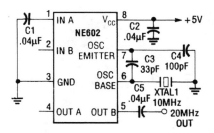

This frequency doubler produces a sine wave at twice the frequency of XTAL1. Notice that the output is taken only from OUT B (pin 5), while OUT A (pin 4) is left open.

RADIO-ELECTRONICS   *Fig. 21-9*

CRYSTAL OSCILLATOR III

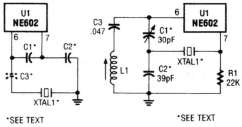

*SEE TEXT   *SEE TEXT

These circuits are for use with a crystal-controlled LO using the NE602. C1, C2, and C3 are for crystals in the 5-MHz region and are approximately chosen from

$$C_1 = \frac{100}{\sqrt{f_{MHz}}} \text{ pF}, \; C_2 = \frac{1000}{\sqrt{f_{MHz}}} \text{ pF}$$

C3 is for fine tuning the crystal frequency and will be 20 to 50 pF typically.

RADIO CRAFT   *Fig. 21-10*

COLPITTS OSCILLATOR

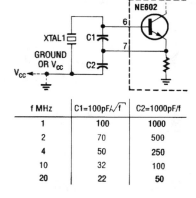

f MHz	C1=100pF√/f	C2=1000pF/f
1	100	1000
2	70	500
4	50	250
10	32	100
20	22	50

Here: $L_1 \approx 7 \; \mu H/f$, $C_1 \approx C_2 \approx C_3 \approx 2400$ pF/f, where f is in MHz. In this circuit, the oscillator is free-running.

RADIO-ELECTRONICS   *Fig. 21-11*

22

Current Source and Sink Circuits

The sources of the following circuits are contained in the Sources section, which begins on page 707. The figure number in the box of each circuit correlates to the entry in the Sources section.

Current Generator
Voltage-Controlled Current Source
Voltage-Controlled Current Sink
Multiple Fixed Current Source

CURRENT GENERATOR

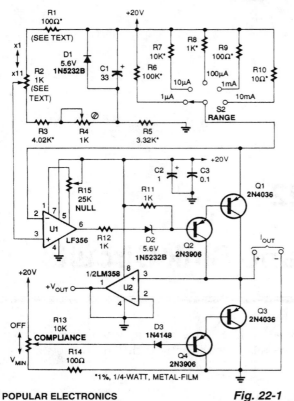

This circuit is useful for supplying constant current to test semiconductors. V_{OUT} from U2 reads the voltage across the load connected to I_{OUT}. R13 adjusts the supply compliance from 1 to about 18 V.

POPULAR ELECTRONICS

Fig. 22-1

VOLTAGE-CONTROLLED CURRENT SOURCE

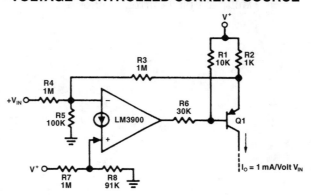

NATIONAL SEMICONDUCTOR

Fig. 22-2

A voltage-variable current source is shown in the figure. The transconductance is $-(1/R_2)$ as the voltage gain from the input terminal to the emitter of Q1 is -1. For $V_{in} = 0$ Vdc, the output current is essentially 0 mA dc. Resistors R1 and R6 guarantee that the amplifier can turn OFF transistor Q1.

VOLTAGE-CONTROLLED CURRENT SINK

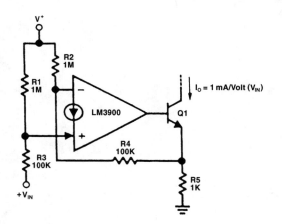

$I_O = 1$ mA/Volt (V_{IN})

A voltage-variable current sink is shown in the figure. The output current is 1 mA per volt of V_{in} (as $R_5 = 1$ kΩ and the gain is +1). This circuit provides approximately 0 mA output current for $V_{in} = 0$ V DCL.

Fig. 22-3

MULTIPLE FIXED CURRENT SOURCE

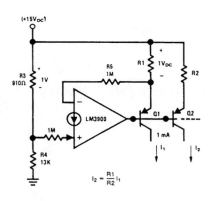

$$I_2 = \frac{R1}{R2} I_1$$

A multiple fixed current source is provided by the circuit. A reference voltage (1 Vdc) is established across resistor R3 by the resistive divider (R3 and R4). Negative feedback is used to cause the voltage drop across R1 to also be 1 Vdc. This controls the emitter current of transistor Q1 and if we neglect the small current diverted into the (−) input via the 1-MΩ input resistor (13.5 μA) and the base current of Q1 and Q2 (an additional 2% loss if the β of these transistors is 100), essentially this same current is available out of the collector of Q1.

Larger input resistors can be used to reduce current loss and a Darlington connection can be used to reduce errors caused by the β of Q1.

The resistor, R2, can be used to scale the collector current of Q2 either above or below the 1-mA reference value.

Fig. 22-4

23

dc-to-dc Converter Circuits

The sources of the following circuits are contained in the Sources section, which begins on page 707. The figure number in the box of each circuit correlates to the entry in the Sources section.

ISOLATED dc-to-dc CONVERTER

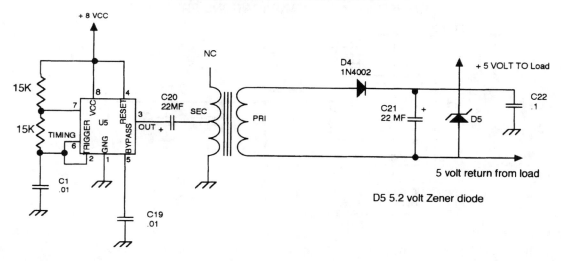

73 AMATEUR RADIO TODAY

Fig. 23-1

A NE555 timer is used to drive a small transformer to change the 5- to 7-Vp-p output of the NE555 to a suitable value to drive a rectifier/Zener combination. This method is useful where a small isolated power source is needed.

dc-to-dc CONVERTER

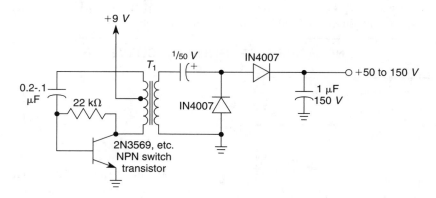

WILLIAM SHEETS

Fig. 23-2

This converter should be useful where a few milliamperes of dc at a higher voltage than available supplies can deliver is needed. T1 is typically a 1 kΩ CT:10-kΩ transistor audio transformer. Depending on T1, about 50 to 150 Vdc can be obtained at a few milliamperes.

ULTRA LOW-POWER dc-to-dc CONVERTER
FOR PERSONAL COMMUNICATIONS PRODUCTS

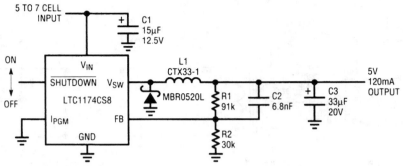

C1: PANASONIC SP SERIES (201) 348-4630
C3: AVX TAJ SERIES (803) 956-0690
L1: COILTRONICS OCTAPAK (407) 241-7876

LINEAR TECHNOLOGY   *Fig. 23-3*

The LTC1174 step-down converter is designed specifically to eliminate noise at audio frequencies while maintaining high efficiency at low output currents. This circuit shows a 5-V, 120-mA output derived from 5 to 7 NiCad or NiMH cells. Small input and output capacitors that are capable of handling the necessary ripple currents help conserve space. In applications where shutdown is desired, this feature is available (otherwise short this pin to V_{in}).

The LTC1174's internal switch, connected between V_{in} and V_{sw}, is current controlled at a peak of approximately 340 mA. Low peak switch current is one of the key features that allows the LTC1174 to minimize system noise compared to other chips that carry significantly higher peak currents, easing shielding and filtering requirements, and decreasing component stresses. Output current of up to 450 mA is possible with this device by connecting the 1 pgm pin to V_{in}. This increases the peak current to 600 mA, allowing for a high average output current.

NEGATIVE STEP-UP dc-to-dc CONVERTER

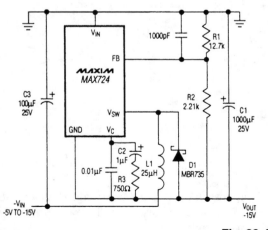

A Maxim MAX724 is used in a step-up switching converter to provide –15-V output from a –5- to –15-V input.

MAXIM   *Fig. 23-4*

166

dc-to-dc CONVERTER II

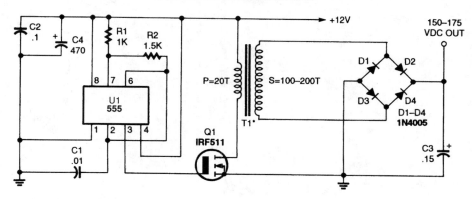

POPULAR ELECTRONICS

Fig. 23-5

In this dc-to-dc converter, the 555 is used to produce a rising and collapsing field in T1's primary, generating a high voltage in T1's secondary winding. That voltage is then full-wave, bridge rectified by D1 through D4, and filtered by C3. T1 is an Amidon Associates EA-775-375E core and nylon bobbin, with #26 wire for the primary and #28 or #30 for the secondary. About 5 W of power is available.

167

24

Decoder Circuits

The sources of the following circuits are contained in the Sources section, which begins on page 707. The figure number in the box of each circuit correlates to the entry in the Sources section.

DTMF Decoder
FM Stereo Decoder
Typical NE567 Tone Decoder Circuit
Video Line Decoder I
Video Line Decoder II

DTMF DECODER

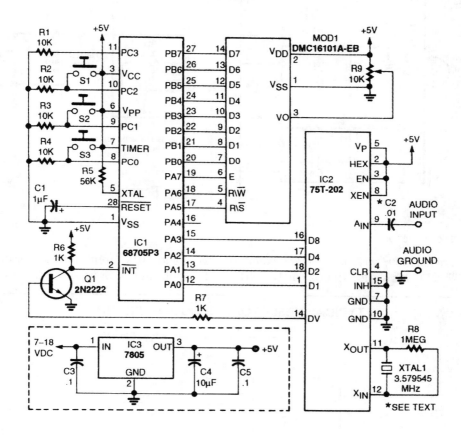

Fig. 24-1

This DTMF decoder uses a Motorola 68705P3 microcontroller and a 75T202 DTMF receiver (Silicon Systems, Inc.). An LCD module is used for the display (MOD1). Switch S1 is used to scroll the display, S2 clears the display, and S3 clears the memory.

FM STEREO DECODER

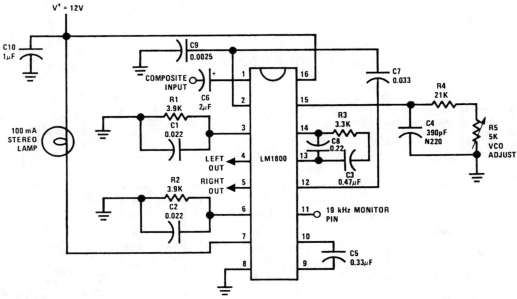

NATIONAL SEMICONDUCTOR

Fig. 24-2

Using an LM1800, this circuit takes composite baseband MPX input and recovers L7R audio channels. The VCO is set for 19 kHz (or 15.7 kHz for TV applications) or as needed.

TYPICAL NE567 TONE DECODER CIRCUIT

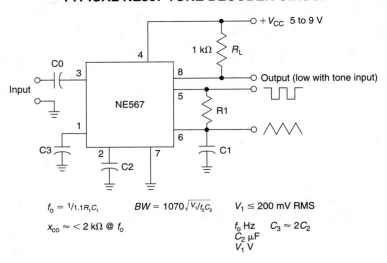

$$f_o = 1/1.1R_1C_1 \qquad BW = 1070\sqrt{V_1/f_o C_2} \qquad V_1 \le 200 \text{ mV RMS}$$

$$x_{co} \approx < 2 \text{ k}\Omega @ f_o \qquad \begin{array}{l} f_o \text{ Hz} \\ C_2 \text{ }\mu\text{F} \\ V_1 \text{ V} \end{array} \quad C_3 \approx 2C_2$$

WILLIAM SHEETS

Fig. 24-3

This circuit illustrates use of NE567 as a tone decoder.

VIDEO LINE DECODER I

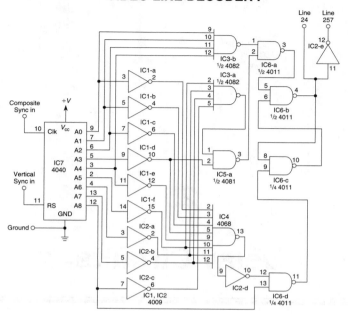

Fig. 24-4

This line decoder uses only one pin for the line indicator.

VIDEO LINE DECODER II

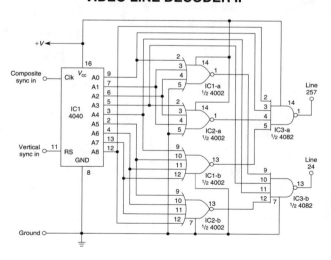

Fig. 24-5

This circuit will produce pulses useful for gating lines 24 and 257 of a video signal, but by changing the decoding logic, other lines can be decoded.

25

Delay Circuits

The sources of the following circuits are contained in the Sources section, which begins on page 707. The figure number in the box of each circuit correlates to the entry in the Sources section.

Time-Delay Generator
Simple Time Delay Circuit

TIME-DELAY GENERATOR

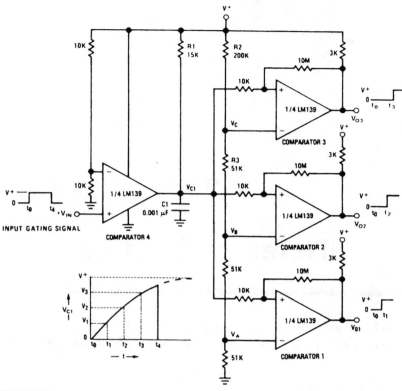

NATIONAL SEMICONDUCTOR

Fig. 25-1

This circuit uses a charging capacitor and three comparators to read the voltage across it.

SIMPLE TIME DELAY CIRCUIT

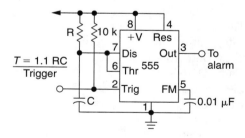

ELECTRONICS NOW

Fig. 25-2

Rotating the potentiometer wiper will change the time delay from the 555 IC.

26

Detector Circuits

The sources of the following circuits are contained in the Sources section, which begins on page 707. The figure number in the box of each circuit correlates to the entry in the Sources section.

High-Speed Peak Detector with Hold and Reset Controls
Lock Detector
Linearized RF Detector
Glitch Detector
VCR Video Detector Controller
Grid-Leak Detector
Negative Peak Detector
Double-Ended Limit Detector
Positive Peak Detector
LM556 Timer Frequency Detector
Single-Comparator Window Detector
15-kHz Tone Detector
Crystal Radio Detector
Switch Closure Circuit
Air Flow Detector
Low Drift Peak Detector
Negative Peak Detector
Positive Peak Detector
455-kHz AM Detector
ac Noise Detector

HIGH-SPEED PEAK DETECTOR WITH HOLD AND RESET CONTROLS

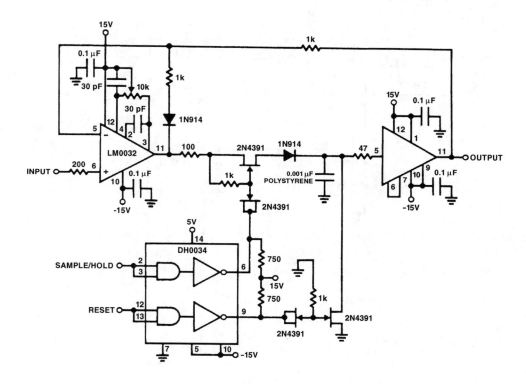

NATIONAL SEMICONDUCTOR

Fig. 26-1

The LH0033 and LH0063 are useful in high-speed sample-and-hold or peak detector circuits because of their very high speed and low-bias-current FET input stages. The high-speed peak detector circuit shown could be changed to a sample-and-hold circuit simply by removing the detector diode and reset circuitry. For best accuracy, the circuit can be trimmed with the 10-kΩ offset adjustment pot shown. The circuit has a typical acquisition time of 900 ns, to 0.1% of the final value for the 10-V input step signal, and a droop rate of 100 μV/ms. Even faster acquisition time can be achieved by reducing the hold capacitor value.

LOCK DETECTOR

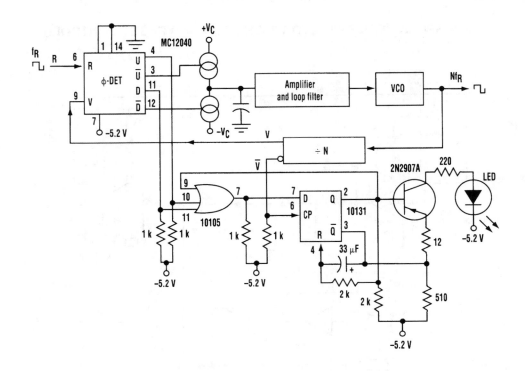

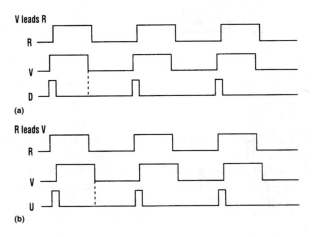

Fig. 26-2

LOCK DETECTOR (*Cont.*)

This PLL lock indicator not only can detect a "locked" or "off-of-lock" condition, but also even if a single pulse or transition has been missed.

When being sampled by the flip-flop, if the V signal leads the input reference signal R and the rising edge of R is lost, the D signal will remain high throughout the interval, allowing the flip-flop to be clocked high (Fig. 26-2A). If the R signal leads the V signal when the transition is missed, the rising edge of the V signal will trigger the D signal of the phase detector, causing the LED to blink (Fig. 26-2B).

A "lock" detector often is used with a phase-locked loop (PLL) or synthesizer to indicate when the loop is phase-locked with an input signal. This circuit can be helpful, but single cycle skips usually will go undetected because of the presence of the low-pass filter.

LINEARIZED RF DETECTOR

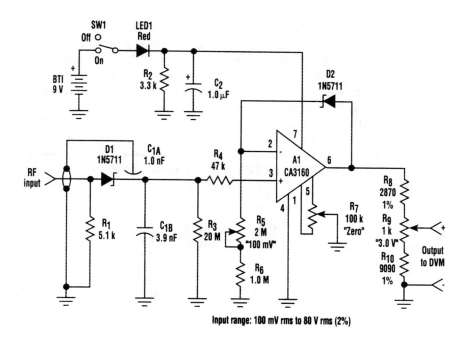

Input range: 100 mV rms to 80 V rms (2%)

ELECTRONIC DESIGN

Fig. 26-3

The circuit produces an extremely linear dc output for RF inputs between 80 mV rms and 4.0 mV rms. For inputs below 50 V rms, the dc output quickly drops to 0 V.

GLITCH DETECTOR

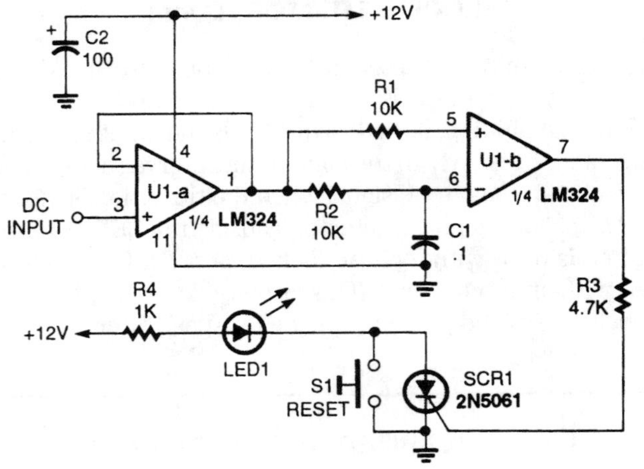

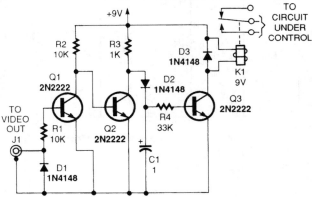

POPULAR ELECTRONICS

Fig. 26-4

In the circuit, two op amps (half of an LM324 quad op amp) and an SCR are direct coupled in a dc-voltage monitoring circuit. Op-amp U1-a is configured as a voltage follower, which feeds the bridged inputs of the second op amp, U1-b. A resistor/capacitor combination (R2/C1) connected to the negative input of UJ1-b forms an RC time-delay circuit. As long as there is no change in the dc-voltage level at either of U1-b's inputs, its output is near zero. If a voltage glitch occurs, the RC timing circuit will delay the voltage change at the op amp's inverting input, causing its output to go high, triggering SCR1 and causing LED1 to light. The circuit's sensitivity allows it to detect voltage changes in the millivolt range. Pressing S1 diverts the SCR's holding current to ground, causing it to turn off and reset the circuit.

VCR VIDEO DETECTOR CONTROLLER

POPULAR ELECTRONICS

Fig. 26-5

This circuit uses the video output from a VCR or camera to control a relay. Video turns on Q1, cutting off Q2, allowing Q3 to be forward biased, activating relay K1. You can use the timer in your VCR and this unit to generate long time delays as well.

GRID-LEAK DETECTOR

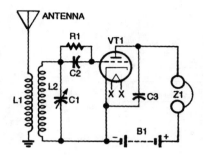

ELECTRONICS NOW

Fig. 26-6

Tuned-circuit receiver with grid-leak detection.

NEGATIVE PEAK DETECTOR

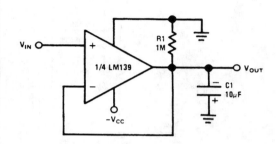

NATIONAL SEMICONDUCTOR

Fig. 26-7

DOUBLE-ENDED LIMIT DETECTOR

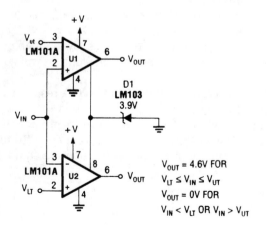

V_{OUT} = 4.6V FOR

$V_{LT} \leq V_{IN} \leq V_{UT}$

V_{OUT} = 0V FOR

$V_{IN} < V_{LT}$ OR $V_{IN} > V_{UT}$

POPULAR ELECTRONICS

Fig. 26-8

POSITIVE PEAK DETECTOR

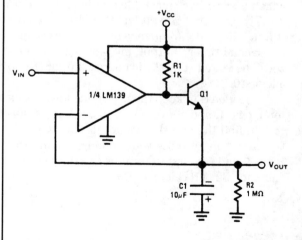

NATIONAL SEMICONDUCTOR

Fig. 26-9

LM556 FREQUENCY DETECTOR

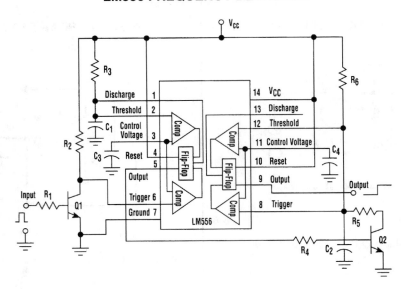

ELECTRONIC DESIGN

Fig. 26-10

The circuit (see the figure), is built around an LM556 dual-timer IC. The 556's first timer is wired as a one-shot and is used to stretch the incoming pulses into fixed-length pulses. The second timer, which is similar to an astable multivibrator (pin 13 remains disconnected), functions as follows:

The one-shot's fixed-length pulses, which are output on pin 5, turn on Q1 and discharge C2 through R5. If the frequency of the pulses is high enough, the voltage on C2 will fall below ⅓ VCC and the second timer's output, pin 9, will go to a logic 1. Conversely, if the frequency is low enough or is zero, the voltage on C2 will charge through R6 to a level above ⅔ VCC, and the pin 9 output will go to a logic "0."

The idea is to keep the upper and lower peak voltage on C2 below ⅔ VCC and ⅓ VCC, respectively for a logic 1, and above ⅔ VCC and ⅓ VCC, respectively, for a logic 0.

To find the one-shot values, R3 and C1, select a pulse width $(1.1 \times R_3 \times C_1)$ that's greater than the largest input pulse width and less than twice the inverse of the highest input frequency. To find R5, R6, and C2, first determine the duty cycle (t_{on}/t_{off}) of the input signal. Next, choose a standard value for C2 and calculate R6:

$$R_6 = \frac{[(t_{off} \times 0.61)^2 + t_{off}]}{C_2}$$

Also, $R_5 = R_6$ (t_{on}/t_{off}). A tweak of resistors R5 and R6 might be needed to get the preferred response. Input signals with low duty cycles work the best. Finally, notice that capacitors C3 and C4 can be any value between 0.01 and 0.1 μF.

SINGLE COMPARATOR WINDOW DETECTOR

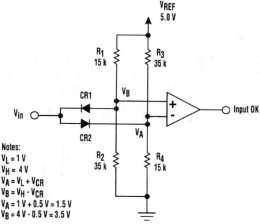

Notes:
$V_L = 1\,V$
$V_H = 4\,V$
$V_A = V_L + V_{CR}$
$V_B = V_H - V_{CR}$
$V_A = 1\,V + 0.5\,V = 1.5\,V$
$V_B = 4\,V - 0.5\,V = 3.5\,V$

ELECTRONIC DESIGN

Fig. 26-11

Simply by adding two steering diodes, a window detector can be built using only a single comparator. The detector performs well for windows of about 1 V or greater, but it isn't suitable where extreme precision is required because the forward drops of the diodes vary.

In the basic circuit, two resistive dividers set threshold voltage levels at both the inverting and noninverting inputs of the comparator by dividing the reference voltage. The input voltage is steered to the appropriate comparator input by diodes CR1 and CR2.

When the input voltage is within the window, neither diode conducts, and the comparator is biased for a high output. When the input goes above the window, CR2 conducts and pulls the inverting input high, causing the comparator output to go low. When the input voltage goes below the window, CR1 conducts, pulling the noninverting input low, again causing the comparator output to go low. The source resistance of V_{in} must be low compared to the equivalent parallel resistance of each divider.

15-kHz TONE DETECTOR

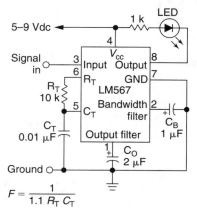

$$F = \frac{1}{1.1\,R_T\,C_T}$$

This circuit detects the presence of a 15-kHz audio signal and light the LED when it does so.

ELECTRONICS NOW!

Fig. 26-12

CRYSTAL RADIO DETECTOR

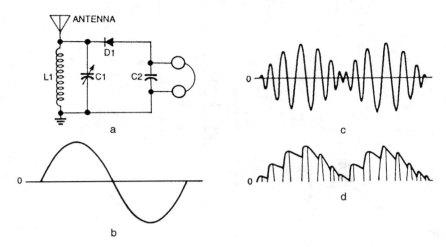

a

b

c

d

ELECTRONICS NOW!

Fig. 26-13

This is a crystal detector receiver with headphones (Fig. 26-13A), audio-frequency signal (Fig. 26-13B), modulated signal (Fig. 26-13C), and a demodulated wave (Fig. 26-13D).

SWITCH CLOSURE CIRCUIT

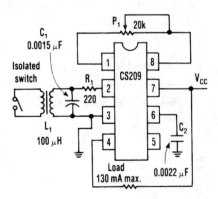

ELECTRONIC DESIGN

Fig. 26-14

A standard proximity detector circuit for the Cherry Semiconductor CS209 IC can detect an isolated switch closure by adding a few turns of wire around the circuit's inductor (Radio Shack 273-102). Moreover, the technique doesn't require any isolated power (see the figure). With the switch open, the potentiometer P1 is adjusted until the output switches off. When the switch is closed, the Q of the circuit changes and the output turns on. Capacitor C1 should be silvered mica, and potentiometer P1 should be a multiturn type such as the Bourns 3006P-1-203. A 9-V supply can be used for Vcc.

AIR FLOW DETECTOR

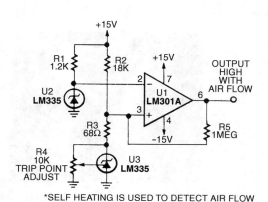

*SELF HEATING IS USED TO DETECT AIR FLOW

Fig. 26-15

LOW DRIFT PEAK DETECTOR

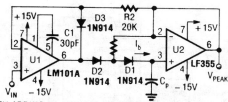

BY ADDING D1 AND R_f, V_{D1}-0 DURING HOLD MODE. LEAKAGE OF D2 PROVIDED BY FEEDBACK PATH THROUGH R_f.

LEAKAGE OF CIRCUIT IS ESSENTIALLY I_b(LF155,LF156) PLUS CAPACITOR LEAKAGE OF Cp.

DIODE D3 CLAMPS V_{OUT} (A1) TO V_{IN}-V_{D3} TO IMPROVE SPEED AND TO LIMIT REVERSE BIAS OF D2.

MAXIMUM INPUT FREQUENCY SHOULD BE $<< 1/2\pi R_f C_{D2}$ WHERE C_{D2} IS SHUNT CAPACITANCE OF D2.

Fig. 26-16

NEGATIVE PEAK DETECTOR

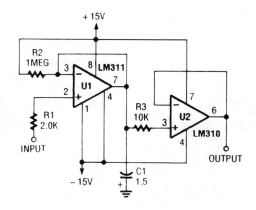

Fig. 26-17

POSITIVE PEAK DETECTOR

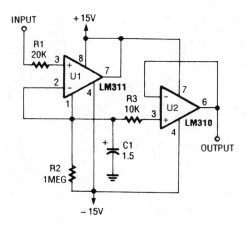

Fig. 26-18

455-kHz AM DETECTOR

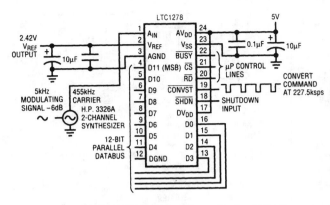

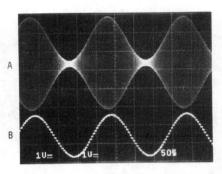

Figure A. The LTC1278 undersamples the 455-kHz
carrier to recover the 5-kHz modulating signal

Figure B. Demodulating an IF by undersampling

LINEAR TECHNOLOGY

Fig. 26-19

The LTC1278 undersamples the 455-kHz carrier to recover the 5-kHz modulating signal. The application shown uses the LTC1278 to undersample (at 227.5 ksps) a 455-kHz IF amplitude-modulated by a 5-kHz sine wave. Figures 26-19A and 26-19B show, respectively, the 455-kHz IF carrier and the recovered 5-kHz sine wave that results from a 12-bit DAC reconstruction.

ac NOISE DETECTOR

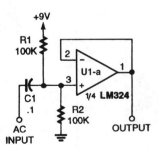

POPULAR ELECTRONICS

Fig. 26-20

This circuit can be added to the glitch detector to trigger on ac noise.

27

Differential Amplifier Circuits

The sources of the following circuits are contained in the Sources section, which begins on page 707. The figure number in the box of each circuit correlates to the entry in the Sources section.

Basic Op-Amp Differential Amplifier
Precision High Gain Differential Amp

BASIC OP-AMP DIFFERENTIAL AMPLIFIER

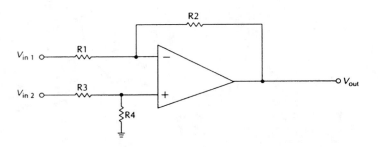

Fig. 27-1

In most cases, R_1 is equal to R_2, and R_3 has the same value as R_4. These equalities don't always have to be true, but they do significantly simplify the circuit design in most practical applications. In any case, for a true differential amplifier, the $R_3{:}R_1$ and $R_4{:}R_2$ ratios must be equal. That is:

$$\frac{R_3}{R_1} = \frac{R_4}{R_2}$$

The circuit still functions even if these ratios are not maintained, but the signals at the inverting and noninverting inputs are subjected to differing amounts of gain, which would be undesirable in most practical applications.

$$R_1 = R_2$$
$$R_3 = R_4$$

These resistance ratios determine the gain of the amplifier:

$$G = \frac{R_3}{R_1} = \frac{R_4}{R_2}$$

Assuming that the resistance ratios are maintained, the output voltage is equal to the differences between the two input voltages, multiplied by the gain. That is,

$$V_{\text{OUT}} = G \times (V_1 - V_2)$$

PRECISION HIGH-GAIN DIFFERENTIAL AMP

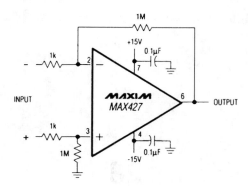

MAXIM

Fig. 27-2

This circuit has a gain of 60 dB and a gain bandwidth of 8 MHz.

28

Display Circuits

The sources of the following circuits are contained in the Sources section, which begins on page 707. The figure number in the box of each circuit correlates to the entry in the Sources section.

Multiplexed BCD Decoder-Driver Circuit
Color-Shifting LED Display
Stereo Level Display
High-Efficiency Display Contrast and Backlight Control
Bar-Graph Level Gauge
Simple Color Organ
Voice Level Meter
LCD Contrast Temperature Compensator
LED Bargraph Driver Circuit

MULTIPLEXED BCD DECODER-DRIVER CIRCUIT

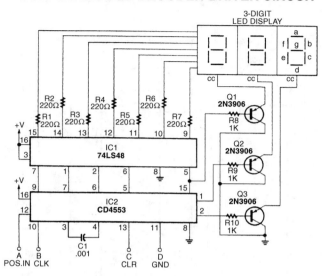

ELECTRONICS NOW

Fig. 28-1

The BCD decoder-driver circuit will interface with any standard BCD output to produce a digital display.

COLOR-SHIFTING LED DISPLAY

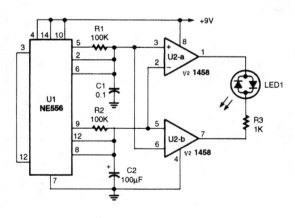

POPULAR ELECTRONICS

Fig. 28-2

This circuit is used to make a tricolor LED gradually change color from yellow to red to yellow to green, and then back to yellow, where the cycle repeats. It is very simple to make, and the theory of operation is also simple. Both of the timers in the 556 dual oscillator/timer are configured for astable operation with a 50% duty cycle. One timer is set to oscillate much faster than the other. The timing capacitor voltage of each is sent to two comparators, which apply a voltage across the tri-color LED whose polarity depends on which capacitor voltage is higher. The rapidly changing capacitor's voltage causes the red and green elements of the LED to be alternately lit, thus giving the illusion of yellow light. As the slowly rising and falling voltage from the slower trimming capacitor changes in average value, it shifts the duty cycle to favor one color or the other. That gives the transition between colors a smooth appearance.

189

STEREO LEVEL DISPLAY

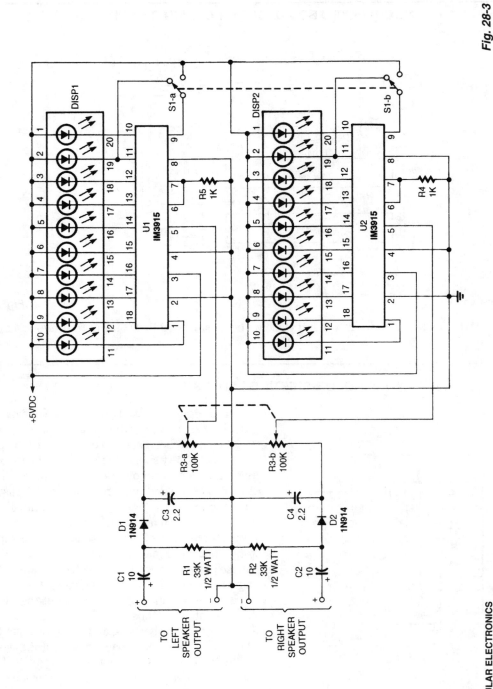

Fig. 28-3

POPULAR ELECTRONICS

Two bar graph drivers and LEDs are used in this volume (level) indicator. R3a and R3b set the sensitivity of the circuit. The LEDs can be either bar graph units or individual LEDs can be used.

HIGH-EFFICIENCY DISPLAY CONTRAST AND BACKLIGHT CONTROL

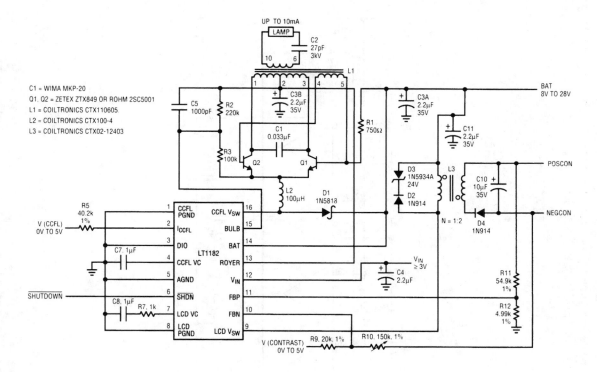

C1 = WIMA MKP-20
Q1, Q2 = ZETEX ZTX849 OR ROHM 2SC5001
L1 = COILTRONICS CTX110605
L2 = COILTRONICS CTX100-4
L3 = COILTRONICS CTX02-12403

LINEAR TECHNOLOGY

Fig. 28-4

The LT1182 and LT1183 are compact high-performance solutions for powering LCD screens used in portable computers and instruments. Backlight control using a Cold Cathode Fluorescent lamp (CCFL) is accomplished with a switching regulator at efficiencies up to 90%. A second switching regulator converts the positive input to either positive or negative bias voltages used for LCD contrast control. Both regulators allow full range of adjustment using a D/A converter, PWM or potentiometer control. Grounded bulb configurations are also easily controlled with minimal parts count. The 200-kHz switching frequency minimizes the size of transformers and external components. A shutdown mode powers down both regulators and reduces supply current to just 35 µA.

BAR-GRAPH LEVEL GAUGE

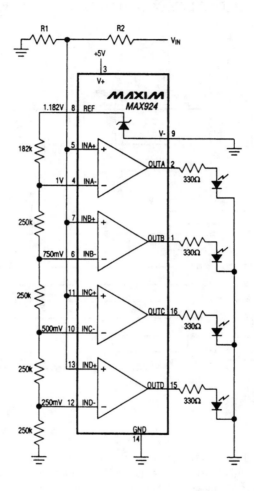

Fig. 28-5

A quad comparator and divider network form a 4-level comparator that drives 4 LEDs. R1 and R2 can be used to scale the basic sensitivity of 250 mV/LED to higher voltages than the basic 1 volt (for all 4 LEDs lit).

SIMPLE COLOR ORGAN

IC1	op amp
IC2, IC3, IC4	LM3909 LED flasher/oscillator
D1, D2, D3	diode (1N914, 1N4148, or similar)
D4, D5, D6	LED
T1	impedance-matching transformer (8 Ω:1 kΩ— see text)
C1, C2, C3, C4	0.1 μF capacitor
C5	0.047 μF capacitor
C6	0.01 μF capacitor
C7, C8, C9	47 μF 6 V electrolytic capacitor
R1	100 kΩ potentiometer
R2	47 kΩ ¼ W 5% resistor
R3	2.2 kΩ ¼ W 5% resistor
R4	680 Ω ¼ W 5% resistor
R5	220 kΩ ¼ W 5% resistor
R6	390 Ω ¼ W 5% resistor
R7	1.2 kΩ ¼ W 5% resistor
R8, R10, R12, R14, R16	10 kΩ ¼ W 5% resistor
R9	3.3 kΩ ¼ W 5% resistor
R11, R13, R15	33 kΩ ¼ W 5% resistor

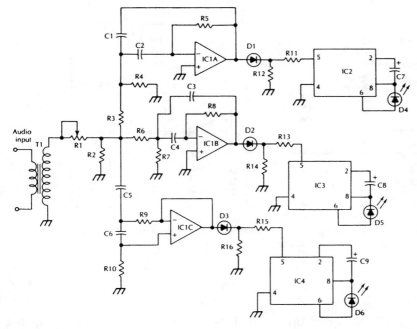

McGRAW-HILL

Fig. 28-6

Three active filters that divide the audio spectrum into three bands drive rectifiers and then drive IC2, 3, and 4, flashing the LEDs at 6 Hz. D4, D5, and D6 should be three different colors for best effect.

193

VOICE LEVEL METER

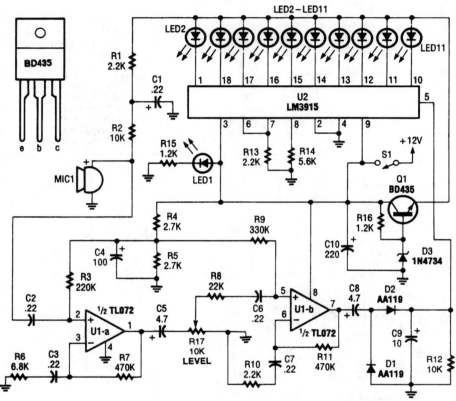

This volume meter can be handy anywhere you need to measure the relative sound level in a room. It is readily adjustable to increase its usefulness.

POPULAR ELECTRONICS

Fig. 28-7

Using an LM3915 VU meter LED bar graph driver, 10 LEDs are driven. A simple audio amplifier drives a detector circuit, which provides dc drive for the LM3915.

LCD CONTRAST TEMPERATURE COMPENSATOR

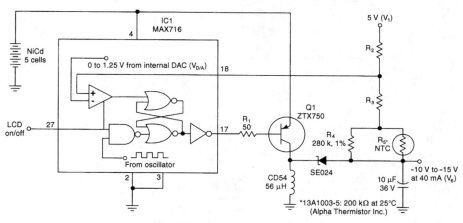

ELECTRONIC DESIGN

Fig. 28-8

Negative temperature-coefficient resistor R5 modifies feedback in this switching regulator, which results in a negative output voltage that varies with temperature. With properly chosen resistor values, the circuit produces a temperature-compensated bias voltage that ensures constant contrast in an LCD.

LED BARGRAPH DRIVER CIRCUIT

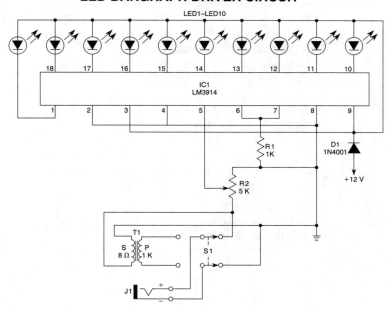

ELECTRONICS NOW

Fig. 28-9

This circuit is used as an audio indicator. S1 selects direct input or a 1-kΩ (high impedance) audio input. R2 is a sensitivity control.

29

Driver Circuits

The sources of the following circuits are contained in the Sources section, which begins on page 707. The figure number in the box of each circuit correlates to the entry in the Sources section.

Two-Input Video MUX Cable Driver
Impedance-Matched Line Driver with 75-Ω Load
Tests Driver for Hobby Servos
Stereo Line Driver
High-Speed Shield/Line Driver
Simple Neon Light Driver
High-Side MOSFET Driver
TTL-Based Speaker Driver
Low-Distortion Composite ±100-mA Line Driver
Video Cable Driver
Coax Cable Driver
Ultra Low Distortion ±50-mA Driver
Very Efficient Solenoid Driver

TWO-INPUT VIDEO MUX CABLE DRIVER

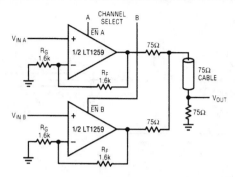

LINEAR TECHNOLOGY

Fig. 29-1

IMPEDANCE-MATCHED LINE DRIVER WITH 75-Ω LOAD

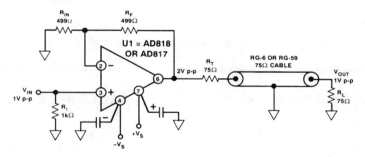

ANALOG DIALOG

Fig. 29-2

This circuit is a wideband 75-Ω line driver, for video applications (1 V p-p into 75 Ω).

TESTS DRIVER FOR HOBBY SERVOS

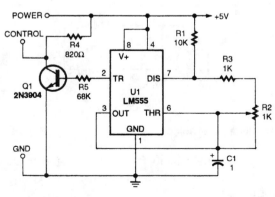

This circuit will generate the pulse used to control hobby servos. With the components shown the servo should produce a 90° total rotation.

POPULAR ELECTRONICS

Fig. 29-3

STEREO LINE DRIVER

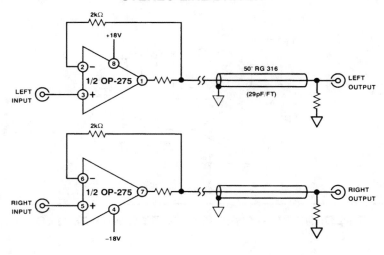

ANALOG DEVICES

Fig. 29-4

One Analog Devices OP-275 can be used for stereo line driver applications.

HIGH-SPEED SHIELD/LINE DRIVER

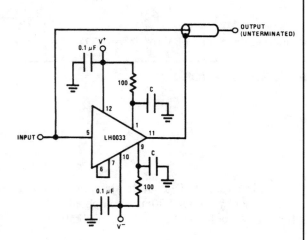

NATIONAL SEMICONDUCTOR

Fig. 29-5

SIMPLE NEON LIGHT DRIVER

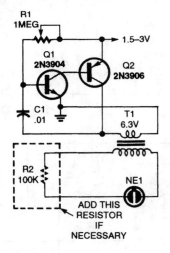

POPULAR ELECTRONICS

Fig. 29-6

NE1, a neon lamp, is lit by this simple inverter circuit. T1 is a 20:1 turn ratio transformer (transistor radio output, etc.).

HIGH-SIDE MOSFET DRIVER

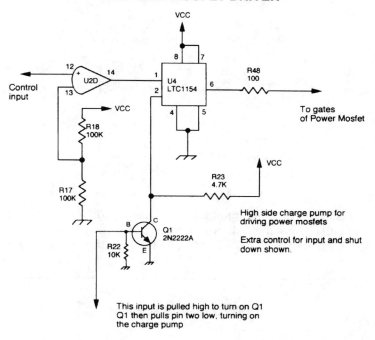

73 AMATEUR RADIO TODAY

Fig. 29-7

A Linear Technology LTC1154 is used as a charge pump to drive the gate of a high-side power MOSFET.

TTL-BASED SPEAKER DRIVER

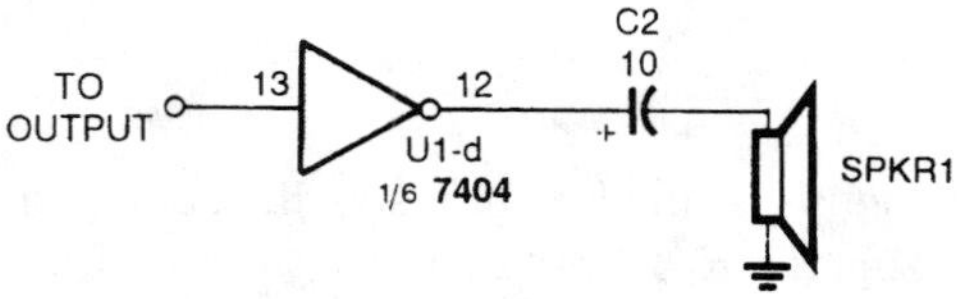

POPULAR ELECTRONICS

Fig. 29-8

A TTL IC, such as a 7404, can drive a small speaker with enough audio to be used as an alarm or annunciator. The speaker can be a 32- or 100-Ω unit.

LOW-DISTORTION COMPOSITE ± 100-mA LINE DRIVER

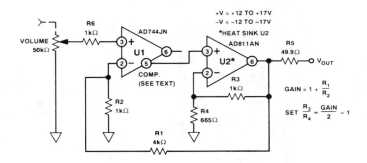

ANALOG DEVICES

Fig. 29-9

This line driver combines the high input impedance of an FET-input IC and a 100-mA op amp. U1's output is left open. The compensation terminal (pins) drive U2's high-Z input for increased overall phase margin. Gain is 14 dB, THD +N at 5 V, and RMS output is around 0.001% below 20 kHz.

VIDEO CABLE DRIVER

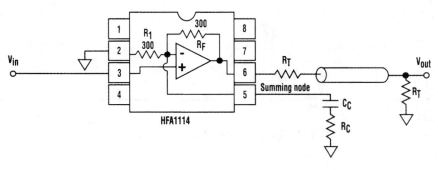

ELECTRONIC DESIGN

Fig. 29-10

The driver's frequency response is tunable for a specific cable length via components connected to the summing node. By shunting R1, R_c acts to increase the amplifier's gain, and C_c controls the cut-in frequency of the compensation.

These three components peak the amplifier's frequency response to counteract the cable's roll-off characteristic. By squeezing more bandwidth out of a given cable, higher-performance cables aren't needed.

COAX CABLE DRIVER

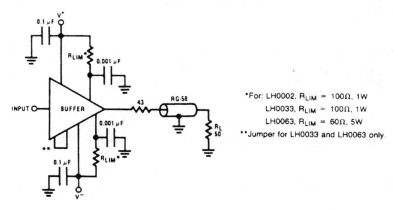

*For: LH0002, R$_{LIM}$ = 100Ω, 1W
      LH0033, R$_{LIM}$ = 100Ω, 1W
      LH0063, R$_{LIM}$ = 60Ω, 5W
**Jumper for LH0033 and LH0063 only.

NATIONAL SEMICONDUCTOR   *Fig. 29-11*

Because of their high-current drive capability, the LH0002, LH0033, and LH0063 buffer amplifiers are suitable for driving terminated or unterminated coaxial cables, and high-current or reactive loads. Current-limiting resistors should be used to protect the device from excessive peak load currents or accidental short circuit. No current limiting is built into the devices other than that imposed by the limited beta of the output transistors. This figure shows a coaxial-cable drive circuit. The 43-Ω resistor is included, the output voltage to the load is about half what it would be without the near-end termination.

ULTRA LOW DISTORTION ±50-mA DRIVER

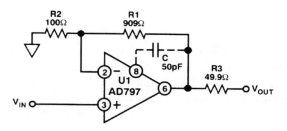

ANALOG DEVICES   *Fig. 29-12*

For a 600-Ω load, THD is typically –115 dB at 20 kHz, 3-V RMS output, with ±15-V supplies. The –3-dB BW is 6 MHz.

VERY EFFICIENT SOLENOID DRIVER

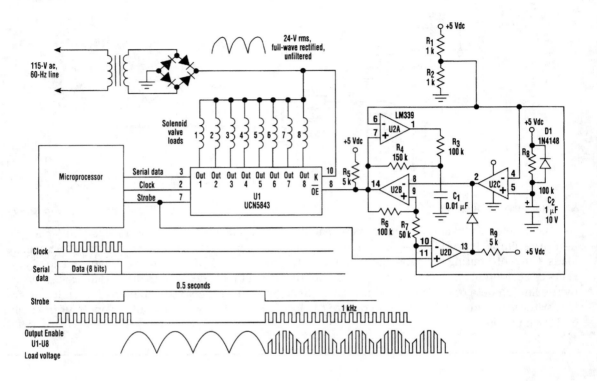

Fig. 29-13

In the circuit shown, the IC in the center, U1, contains a microprocessor-compatible serial-in, parallel-out shift register, with strobed latch and high-level solenoid drives (see the figure).

The strobe pulse turns on the selected loads at full power for its pulse width, which is 0.5 second. Following the strobe pulse, the driver outputs are pulse-width modulated by the multivibrator, which is set for a duty cycle of 25%. The solenoids are therefore held in the engaged position with a voltage that's 25% of the nominal supply voltage.

U2B acts as the basic multivibrator in the circuit, and U2A sets the duty cycle by setting the discharge current of capacitor C1. U2D overrides the oscillation during the strobe pulse. And U2C provides the Power-On Reset (POR), inhibiting the solenoid loads from turning on during the initialization period of the microprocessor.

30

Electronic Lock Circuits

The sources of the following circuits are contained in the Sources section, which begins on page 707. The figure number in the box of each circuit correlates to the entry in the Sources section.

ELECTRONIC LOCK

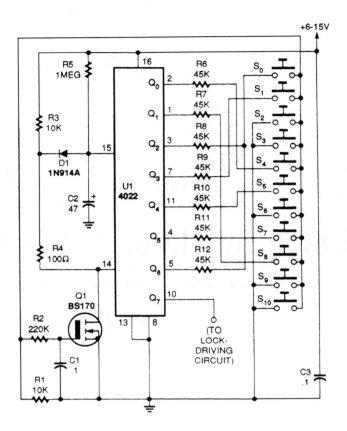

Fig. 30-1

The heart of the circuit is a 4022 octal counter. When first powered up, C2 is charged via R5, so the reset input of the counter is kept high. That causes output Q_o to go high while all other outputs are low. With the switches wired as shown, when S4 is pressed, the BS170 is switched on via debouncing network R2/C1, and U1 receives a clock pulse. Also, C2 is discharged via R4 and D1, removing the reset signal of the counter, allowing it to advance. The time required for C2 to charge via R5 (i.e., to reset the counter), is the maximum time that can lapse before the next key is pressed. The above cycle is therefore repeated only if S8 (connected to the Q1 output) is pressed in time. When all keys have been pressed in time and in the correct order, Q7 goes high for about four seconds to drive the "unlock" circuitry (e.g., a relay driver for an automatic door opener. A builder can change the code by reviewing the switches. The code for the lock shown in the circuit diagram is 4-8-0-1-5-7-0. However, the 4022 octal counter can be replaced by a 4017 divide-by-10 counter. That will make it possible to add two more digits to the combination.

FREQUENCY-BASED LOCK

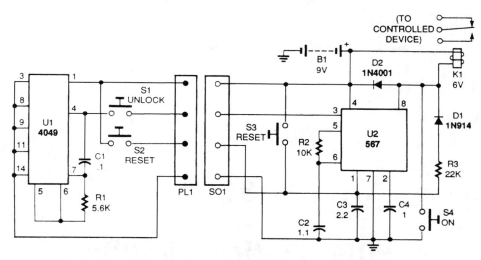

POPULAR ELECTRONICS

Fig. 30-2

The system is formed by two separate circuits—a key and a keyhole. The key engages the keyhole with a mating pair of connectors. The key is a tone-generator circuit consisting of a 4049 hex inverter CMOS IC (U1), switches (S1 and S2), a resistor (R1), and a capacitor (C1). The value of the tone generated by that circuit in Hz is determined by;

$$\frac{1}{(1.4 \times R_1 C_1)}$$

The keyhole is a 567 tone-decoder circuit that can be configured to detect any frequency from 0.01 Hz to 500 kHz. The frequency it detects (f_o), via the 567 IC, turns on the relay (K1). Components R3 and D1 are used to latch the circuit, so the output stays on even after the input tone is removed. When S2 is pressed, the system is reset. Switch S3 resets the circuit from inside.

SIMPLE LOCK

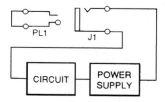

POPULAR ELECTRONICS

Fig. 30-3

Only an appropriately wired plug of the right size will activate circuits with a nonshorting jack in their power supply circuit.

31

Fiber-Optics Circuits

The sources of the following circuits are contained in the Sources section, which begins on page 707. The figure number in the box of each circuit correlates to the entry in the Sources section.

EXPERIMENTAL DATA TRANSMITTER FOR FIBER OPTICS

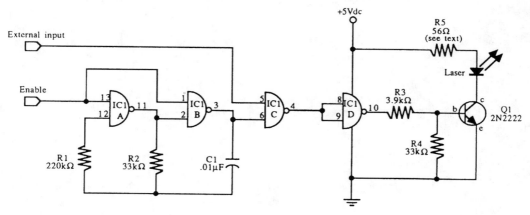

McGRAW-HILL

Fig. 31-1

This schematic for an experimental data transmitter uses optical fibers and a laser diode. Transmission frequency of the free-running oscillator is approximately 3 kHz. R5 might have to be varied to suit your laser diode. IC1 is a CD4093.

EXPERIMENTAL FIBER-OPTIC DATA RECEIVER

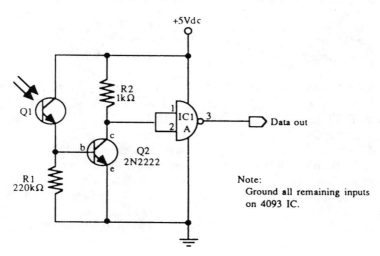

McGRAW-HILL

Fig. 31-2

An infrared phototransistor acts as the sensor for this receiver. IC1a is a section of a CD4093 CMOS NAND gate.

32

Filter Circuits

The sources of the following circuits are contained in the Sources section, which begins on page 707. The figure number in the box of each circuit correlates to the entry in the Sources section.

STATE VARIABLE FILTER

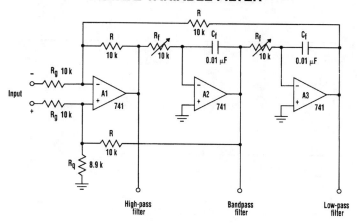

ELECTRONIC DESIGN

Fig. 32-1

The state variable filter shown consists of only three op amps and a few passive components. It provides several key features. These include the ability to simultaneously provide low-pass, high-pass, and bandpass filter functions, and adjust bandwidth in a wide range by changing the values of C_f and R_f. The device also is easy to tune and simple to construct, while the quality factor (Q) of each filter is independent of each other.

SALLEN-KEY HIGH-PASS FILTER

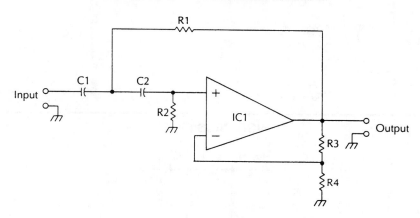

McGRAW-HILL

Fig. 32-2

R3 and R4 set the circuit gain

$$f_c = \frac{1}{2\pi\sqrt{R_1 R_2 C_1 C_2}}$$

usually $C_1 = C_2$, $R_1 = R_2$

$$R_3 = 0.586\, R_4$$

ACTIVE BANDPASS FILTER CIRCUIT

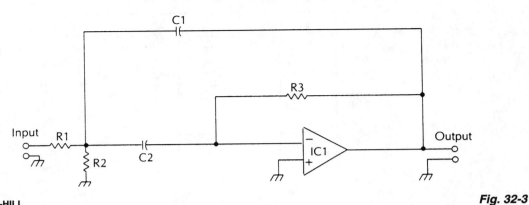

Fig. 32-3

In this circuit,

$$C_1 = C_2 = C$$

$$Q = \frac{f_o}{BW} \quad K = circuit\ gain,\ f_o = center\ frequency$$

$$R_1 = \frac{Q}{2\pi f_o C K}$$

$$R_2 = \frac{2Q}{2\pi f_o C}$$

$$R_3 = \frac{Q}{2\pi f_o C(2Q - K)}$$

HIGH-PASS FILTER

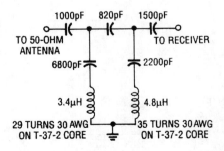

Fig. 32-4

This high-pass filter will attenuate AM stations by 40 dB. Its low-frequency cutoff is about 2.2 MHz. This filter is useful for SW listening in areas of high AM radio signal strength.

SECOND-ORDER VOLTAGE-CONTROLLED FILTER

(a)

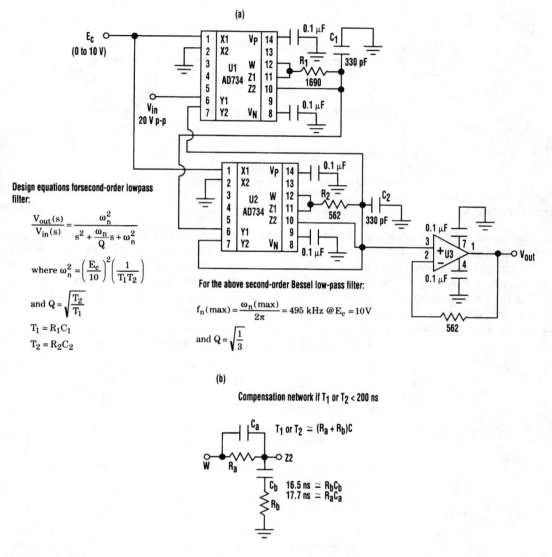

Design equations for second-order lowpass filter:

$$\frac{V_{out}(s)}{V_{in}(s)} = \frac{\omega_n^2}{s^2 + \frac{\omega_n}{Q}s + \omega_n^2}$$

where $\omega_n^2 = \left(\frac{E_c}{10}\right)^2\left(\frac{1}{T_1 T_2}\right)$

and $Q = \sqrt{\frac{T_2}{T_1}}$

$T_1 = R_1 C_1$

$T_2 = R_2 C_2$

For the above second-order Bessel low-pass filter:

$$f_n(max) = \frac{\omega_n(max)}{2\pi} = 495 \text{ kHz } @E_c = 10V$$

and $Q = \sqrt{\frac{1}{3}}$

(b)

Compensation network if T_1 or T_2 < 200 ns

T_1 or $T_2 \simeq (R_a + R_b)C$

16.5 ns $\simeq R_b C_b$
17.7 ns $\simeq R_a C_a$

ELECTRONIC DESIGN

Fig. 32-5

Desirable second-order voltage-controlled low-pass filter response can be achieved with this voltage-controlled filter (A). By using low-distortion, wide-bandwidth multipliers, it achieves higher cutoff frequencies than switched-capacitor filters. If the circuit's RC network has a time constant less than 200 ns, it should be replaced by a lag compensator network (B).

COMBINATION FILTER

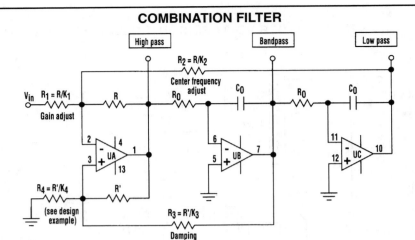

ELECTRONIC DESIGN

Fig. 32-6

The classic "state variable" two-integrator filter is known for its insensitivity to component variations, and its ability to provide three separate simultaneous outputs—low pass, high pass, and bandpass.

Typically, a quad op amp is used to implement the state-variable filter. The classic configuration uses two integrating amplifiers, a filter input amplifier, and a filter feedback amplifier.

The design described here combines both input and feedback amplifiers into one adder/subtractor amplifier, achieving a three op-amp filter design (see the figure).

SHORTWAVE RECEIVER IF FILTER

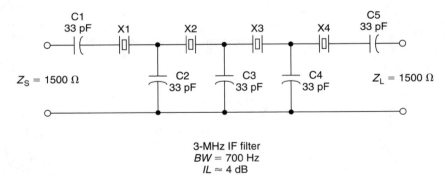

3-MHz IF filter
$BW = 700$ Hz
$IL \approx 4$ dB

X1 through X4   3.000 MHz $\pm$50 Hz
C_X = C1 through C5   33 pF $\pm$5% NPO

WILLIAM SHEETS

Fig. 32-7

An inexpensive filter can be made from microprocessor crystals. This filter has 700 Hz BW (3 dB) and has a flat response (<1 dB) for about 400 to 500 Hz. Although a 3-MHz crystal was used, any frequency from 2 to 15 MHz (using fundamental crystal) should work, with appropriate scaling of components. Crystal resonant frequencies should match within 20% and preferably 10% of expected bandwidth (which is narrower as C_x increases. Impedance is reduced with wider bandwidths.

PIN DIODE FILTER SELECTION CIRCUIT

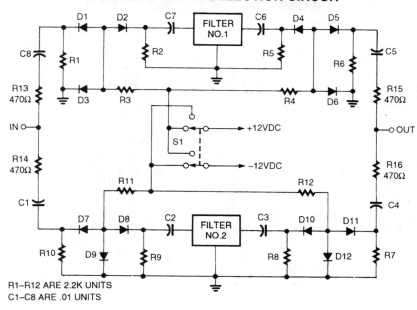

R1–R12 ARE 2.2K UNITS
C1–C8 ARE .01 UNITS

POPULAR ELECTRONICS

Fig. 32-8

Selecting IF bandpass filters via series/shunt PIN-diode switching can be accomplished with this circuit.

HIGH-PASS ACTIVE FILTER

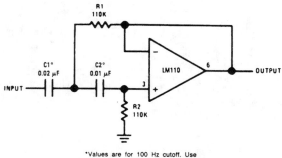

*Values are for 100 Hz cutoff. Use metalized polycarbonate capacitors for good temperature stability

NATIONAL SEMICONDUCTOR

Fig. 32-9

AM BROADCAST TRAP FOR SIMPLE SW RECEIVERS

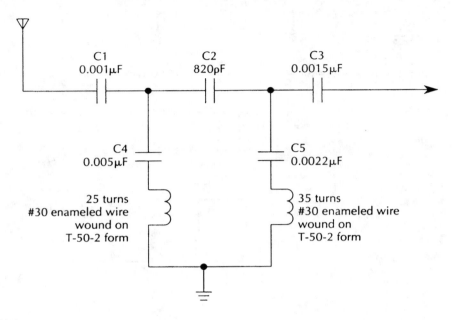

McGRAW-HILL

Fig. 32-10

SHORTWAVE INTERFERENCE TRAP

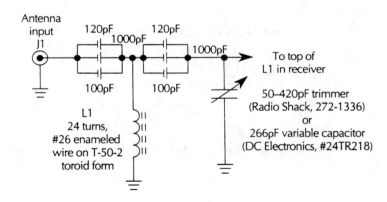

McGRAW-HILL

Fig. 32-11

Build this interference trap to help block strong shortwave, broadcast, and FM stations from coming in on the shortwave bands.

PROGRAMMABLE ANALOG FILTER

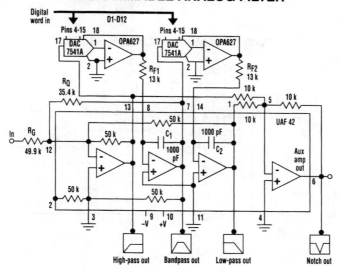

ELECTRONIC DESIGN

Fig. 32-12

The circuit in the figure shows how an analog, digitally programmable filter can be built using a UAF42. This monolithic, state-variable active filter chip provides a two-pole filter building block with low sensitivity to external component variations. It eliminates aliasing errors and clock feed though noise common to switched-capacitor filters. Low-pass, high-pass, bandpass, and notch (band-reject) outputs are available.

ACTIVE LOW-PASS FILTER

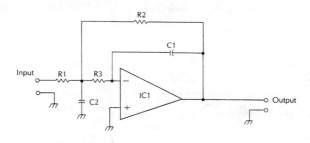

McGRAW-HILL

Fig. 32-13

In this circuit, $R_1 = 2$
$$R_2 = R_4$$
$$R_3 = 2R_1$$
$$C_1 = C_2$$
$$f = \frac{1}{2\pi RC}$$

This circuit has a rolloff of 6 dB/Octave.

TWO OP-AMP BANDPASS FILTER

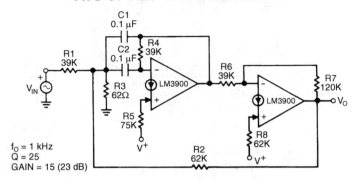

NATIONAL SEMICONDUCTOR

Fig. 32-14

This circuit uses only two capacitors. The amplifier on the right supplies a controlled amount of positive feedback for improved response characteristics. Resistors R5 and R8 are used to bias the outvoltage of the amplifiers at $V+/2$.

SINGLE 3.3-V SUPPLY 4-POLE STATE VARIABLE FILTER

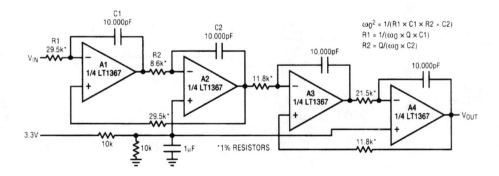

LINEAR TECHNOLOGY

Fig. 32-15

HIGH-Q NOTCH FILTER

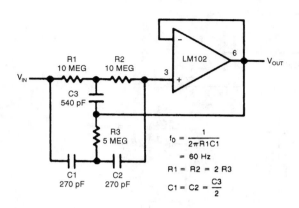

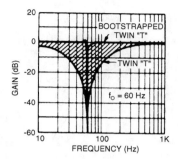

$$f_0 = \frac{1}{2\pi R1 C1}$$

$$= 60 \text{ Hz}$$

$$R1 = R2 = 2\ R3$$

$$C1 = C2 = \frac{C3}{2}$$

Response of High and Low Q Notch Filter

NATIONAL SEMICONDUCTOR

Fig. 32-16

This shows a twin "T" network connected to an LM102 to form a high Q, 60-Hz notch filter. The junction of R3 and C3, which is normally connected to ground, is bootstrapped to the output of the follower. Because the output of the follower is a very low impedance, neither the depth nor the frequency of the notch change; however, the Q is raised in proportion to the amount of signal fed back to R3 and C3.

ADJUSTABLE-Q NOTCH FILTER

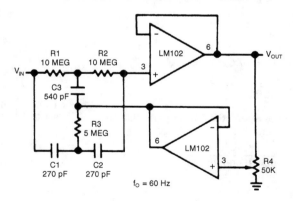

$$f_0 = 60 \text{ Hz}$$

NATIONAL SEMICONDUCTOR

Fig. 32-17

This figure shows a circuit where the Q can be varied from 0.3 to 50. A fraction of the output is fed back to R3 and C3 by a second voltage follower, and the notch Q is dependent on the amount of signal fed back. A second follower is necessary to drive the twin "T" from a low-resistance source so that the notch frequency and depth will not change with the potentiometer setting.

DIGITAL COMB FILTER

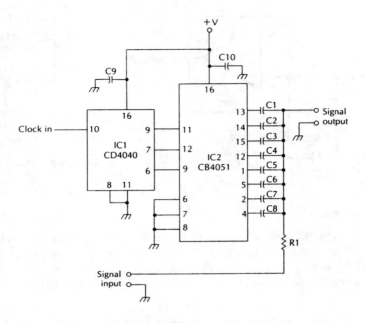

This circuit uses an eight-step switching sequence, so n=8. This makes the center frequency equal to:

$$F_c = \frac{1}{(2nRC)}$$

$$= \frac{1}{(2 \times 8)RC}$$

$$= \frac{1}{16RC}$$

Using the component values suggested in the parts list, the circuit has a main center frequency of:

$$F_c = \frac{1}{(16 \times 1000 \times 0.00000001)}$$

$$= \frac{1}{0.00016}$$

$$= 6250 \text{ Hz}$$

Suggested parts list for the digital comb filter	
IC1	CD4040 BCD-ripple counter
IC2	CD4051 BCD-to-decimal decoder (SP8T rotary bilateral switch)
C1–C8	0.01 µF close tolerance capacitor
C9, C10	0.1 µF capacitor
R1	1K ¼ W 5% resistor

Fig. 32-18

VOLTAGE-CONTROLLED LOW-PASS FILTER

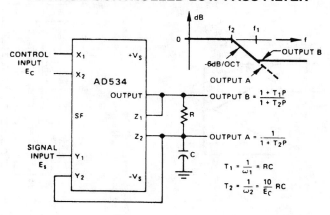

ANALOG DEVICES   *Fig. 32-19*

The voltage at Output A, which should be unloaded by a follower, responds as though E_s were directed to the RC filter, but the filter's break frequency were proportional to E_c [i.e,. $= E_c/(20\pi RC)$]. The frequency response has a break at f_2 and the 6-dB/octave rolloff. The voltage at Output B has the same response, up to $[f_1 \; (f_1 = 1/(2\pi RC))]$, then levels off at a constant attenuation of $f_2/f_1 = E_c/10$. For example, if $R = 8$ kΩ, C = 0.002 μF, Output A has a pole at 100 Hz to 10 kHz and can be loaded. The circuit can be converted to high-pass by interchanging C and R.

VSB FILTER FOR LM2889

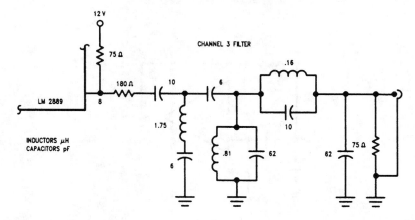

NATIONAL SEMICONDUCTOR   *Fig. 32-20*

This filter is for CH3, in order to get a vestigial sideband TV signal. It is designed for 75-Ω impedance levels.

20-kHz BUTTERWORTH ACTIVE FILTER

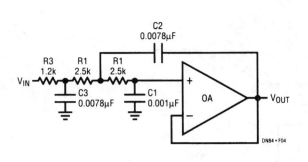

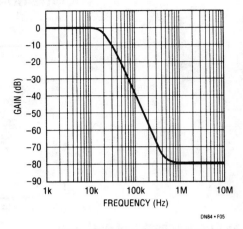

Filter Frequency Response

LINEAR TECHNOLOGY

Fig. 32-21

This filter will be useful for anti-aliasing or band limiting in an audio system. The op amp is a Linear Technology, LT1124, LT1355, or LT1169.

BANDPASS FILTER

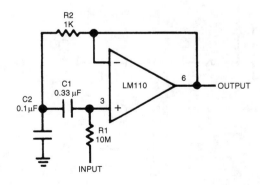

NATIONAL SEMICONDUCTOR

Fig. 32-22

SALLEN-KEY LOW-PASS FILTER

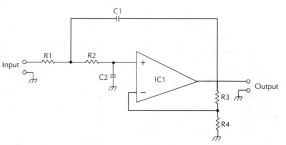

R3 and R4 set the circuit gain

$$f_c = \frac{1}{2\pi\sqrt{R_1 R_2 C_1 C_2}}$$

usually $C_1 = C_2$, $R_1 = R_2$, $R_3 = 0.586\,R_4$

McGRAW-HILL   *Fig. 32-23*

ACTIVE HIGH-PASS FILTER

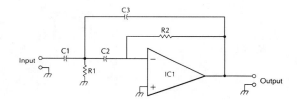

In this circuit,

$$f_{CO} = \frac{1}{2\pi RC}$$

McGRAW-HILL   *Fig. 32-24*

RC NOTCH FILTER

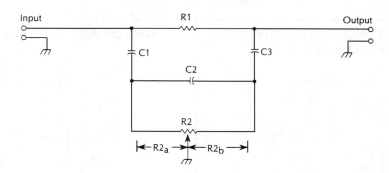

$C_1 = C_2 = C_3$
$R_1 = 6R_2$

$$R_2 = R_{2a} + R_{2b}$$

Reject frequency (notch), $F_c = \dfrac{1}{2\pi C\sqrt{3R_{2a}R_{2b}}}$

McGRAW-HILL   *Fig. 32-25*

1-kHz 4TH-ORDER BUTTERWORTH FILTER

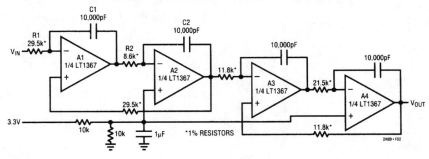

LINEAR TECHNOLOGY

Fig. 32-26

The filter is a simplified state variable architecture consisting of two cascaded 2nd-order sections. Each section uses the 360° phase shift around the two op-amp loop to create a negative summing junction at A1's positive input. The circuit has low sensitivities for center frequency and Q, which are set with the following equations:

$$\omega 0^2 = \frac{1}{(R_1 \times C_1 \times R_2 \times C_2)}$$

where,

$$R_1 = \frac{1}{(\omega 0 \times Q \times C_1)} \text{ and } R_2 = \frac{Q}{(\omega 0 \times C_2)}$$

The dc bias applied to A2 and A4, half supply, is not needed when split supplies are available. The circuit swings rail-to-rail in the passband making it an excellent anti-aliasing filter for A/Ds. The amplitude response is flat to 1 kHz then rolls off at 80 dB/decade.

SAW-FILTER IMPEDANCE-MATCHING PREAMPLIFIER

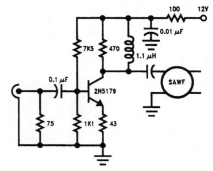

This circuit matches a saw filter to an IF amplifier.

NATIONAL SEMICONDUCTOR   *Fig. 32-27*

ONE OP-AMP BANDPASS FILTER

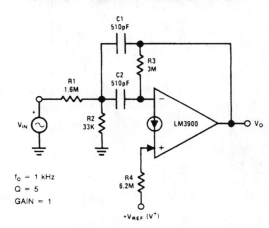

f_o = 1 kHz
Q = 5
GAIN = 1

NATIONAL SEMICONDUCTOR   *Fig. 32-28*

33

Flasher Circuits and Blinker

The sources of the following circuits are contained in the Sources section, which begins on page 707. The figure number in the box of each circuit correlates to the entry in the Sources section.

RANDOM LED STROBE

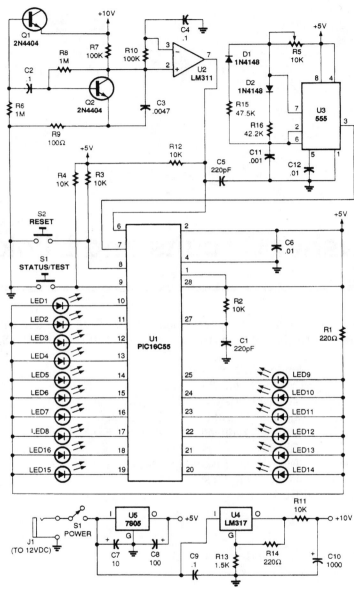

POPULAR ELECTRONICS

Fig. 33-1

This circuit generates a random output that is translated into LED "movement" by a prepro-grammed PIC16C55 microcontroller, U1. That PIC also senses and records the bias of the LED's movement. This device was originally used for an application involving psychokinesis testing where the person was asked to "think" the lights in either a clockwise or counterclockwise direction.

FLASHING NEON CHRISTMAS LIGHTS

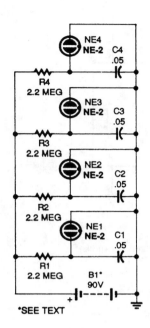

POPULAR ELECTRONICS   *Fig. 33-2*

This flashing set of neon Christmas lights will make an attractive decoration for any time of year. B1 is made up of ten 9-V transistor radio batteries in series. The battery life can be measured in months.

FLASHING CHRISTMAS LED DISPLAY

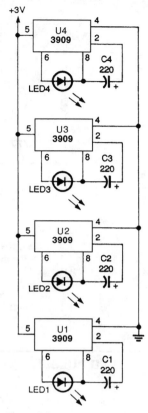

POPULAR ELECTRONICS   *Fig. 33-3*

Using LEDs and 3909 ICs, you can make a flashing-light circuit that will run for months on two AA batteries.

DUAL FLASHER ADD-ON FOR 555 CIRCUITS

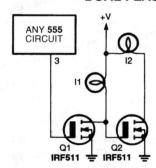

A pair of hex FETs drive two incandescent lamps in an alternating flasher circuit. The lamps can be 12-V automotive types, etc.

POPULAR ELECTRONICS   *Fig. 33-4*

VARIABLE-FREQUENCY HIGH-POWER LED FLASHER

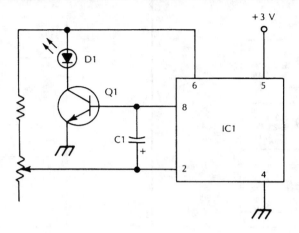

IC1	LM3909 LED flasher/oscillator IC
Q1	NPN transistor (2N3904, Radio Shack RS2009 or similar)
D1	LED
C1	100 µF 5 V electrolytic capacitor
R1	470 Ω ¼ W 5% resistor
R2	50 kΩ potentiometer

McGRAW-HILL

Fig. 33-5

LED PULSER

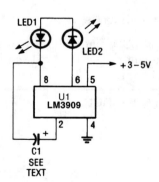

POPULAR ELECTRONICS

Fig. 33-6

In this circuit, the LM3909 is used to drive a pair of series-connected LEDs.

LED PULSER WITH AUDIBLE OUTPUT

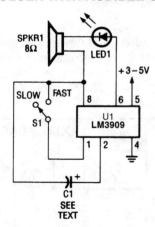

POPULAR ELECTRONICS

Fig. 33-7

The LM3090 can also be used to drive both an LED and a speaker. In this circuit, each time that LED1 blinks, SPKR1 (an 8-Ω speaker) emits a sharp click sound.

SIMPLE LAMP PULSER

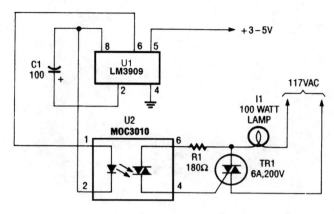

POPULAR ELECTRONICS

Fig. 33-8

Here, the LM3090 (configured as a timing oscillator) is used to control a 117-Vac lamp through an MOC3010 optoisolator/coupler.

LED FLASHER

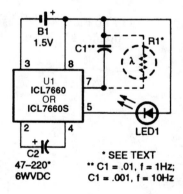

POPULAR ELECTRONICS

Fig. 33-9

This circuit provides a low-cost way to flash an LED from a single 1.5-V source. Based on the ICL7660 dc-to-dc voltage converter, the circuit makes use of an external capacitance (C1) on the oscillator rate-control pin to decrease the charge/dump time to the desired flash rate. A dc resistance (R1) on the same pin can also be used to disable the oscillator and extend the power-cell's life. That optional dc resistance (in the form of a photoconductive cell) will shut off the oscillator in daylight.

34

Flip-Flop Circuits

The sources of the following circuits are contained in the Sources section, which begins on page 707. The figure number in the box of each circuit correlates to the entry in the Sources section.

Trigger Flip Flop
Two-Amplifier Flip Flop

TRIGGER FLIP FLOP

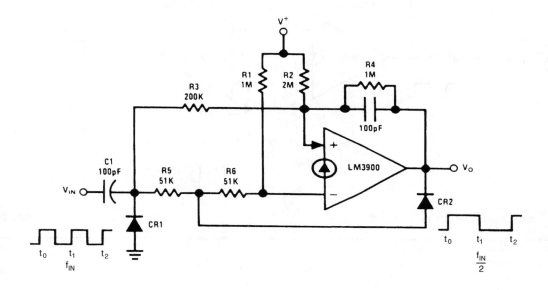

NATIONAL SEMICONDUCTOR

Fig. 34-1

Trigger flip flops are useful to divide an input frequency as each input pulse will cause the output of a trigger flip flop to change state. Due to the absence of a clocking signal input, this is for an asynchronous logic application. A circuit that uses only one amplifier is shown. Steering of the differentiated positive input trigger is provided by diode CR2. For a low-output voltage state, CR2 shunts the trigger away from the (−) input and resistor R3 couples this positive input trigger to the (+) input terminal. This causes the output to switch high. The high-voltage output state now keeps CR2 off and the smaller value of $(R_5 + R_6)$ compared with R3 causes a larger positive input trigger to be coupled to the (−) input, which causes the output to switch to the low-voltage state.

TWO-AMPLIFIER FLIP FLOP

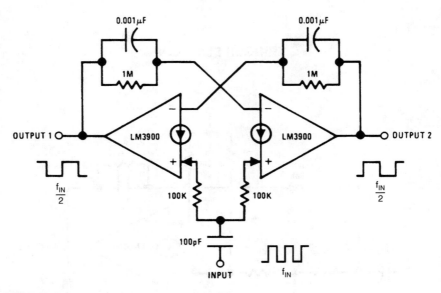

Fig. 34-2

35

Frequency-to-Voltage Converter Circuits

The sources of the following circuits are contained in the Sources section, which begins on page 707. The figure number in the box of each circuit correlates to the entry in the Sources section.

INDUSTRIAL FREQUENCY-TO-VOLTAGE CONVERTER

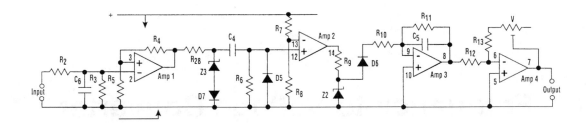

ELECTRONIC DESIGN

Fig. 35-1

Control and process equipment often require the indication of frequency (speed or rate) of linear or rotary mechanical movement. Motion can be detected using various pulse-generating pickups and proximity detectors that output ac or dc pulsed signals.

This industrial converter can serve in a wide variety of applications. The circuit operates around a quad-FET input op amp and is designed to be self-contained or run from a bipolar supply. The input signal of dc pulses or ac waveforms is applied to R2.

Amplifier 1, which acts as a Schmitt trigger, supplies a squarewave output of fixed amplitude to C4. Resistor R3 and capacitor C6 filter out input noise, and R4 and R5 determine the switching levels, and Zener diode Z3 sets the amplitude.

Amplifier 2 gives a fixed-duration pulse on the positive transition of C4, with a time constant set by C4 and R6 and the switching level set by R7 and R8. Resistor R9 and Zener diode Z2 fix the amplitude of the pulses and amplifier 3 integrates them via R10 and C5. Diode D6 blocks negative integration and R11 discharges C5 with a long-time constant.

Hence, the dc output of amplifier 3 is proportional to the frequency applied to the input. Amplifier 4 inverts and buffers the negative output of amplifier 3 and provides amplitude adjustment voltage.

The complete circuit is linear and sufficiently accurate providing that C4 is chosen to give a pulse duration less than the maximum input frequency and that R11 > R10.

FREQUENCY-TO-VOLTAGE CONVERTER

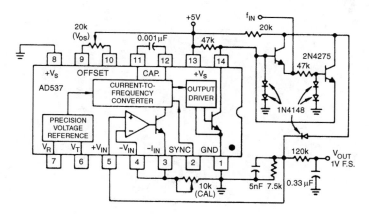

ANALOG DEVICES

Fig. 35-2

The AD537 can also be used to perform frequency-to-voltage conversion. The transistor pair shown here operates as an exclusive-or gate to perform the phase comparison. It locks onto the input frequency within two cycles. The configuration requires only 3 mA for frequencies up to 10 kHz. In most situations, an output buffer will be required to unload the filter. Use 0- to 5-V pulses or square waves with 40-μs minimum pulsewidth.

36

Function Generator Circuits

The sources of the following circuits are contained in the Sources section, which begins on page 707. The figure number in the box of each circuit correlates to the entry in the Sources section.

Function Generator
Sweep/Function Generator
Simple Function Generator
Accurate, Stable Function Generator
Wide-Range Function Generator

FUNCTION GENERATOR

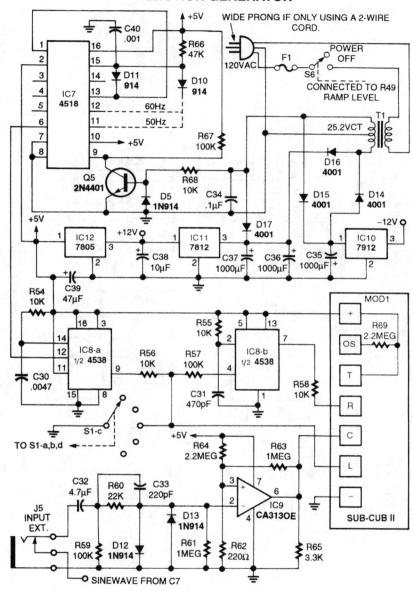

THE COUNTER MODULE (MOD1) has a 0.35-inch high, six-digit liquid crystal display. Pulses at 1-second intervals are derived from the AC power line which has a typical accuracy of 99.99 %.

Fig. 36-1

These three circuits make up an audio frequency function generator and can be individually used for custom applications.

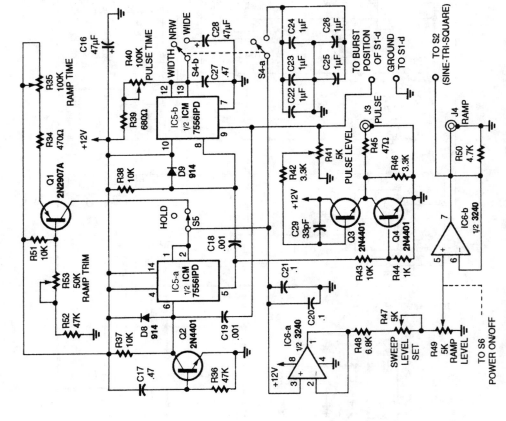

FUNCTION GENERATOR (Cont.)

A DUAL CMOS TIMER'S (IC5) output triggers another timer's input. Once Q2 starts IC5-a on initial power-up, the circuit continues to oscillate.

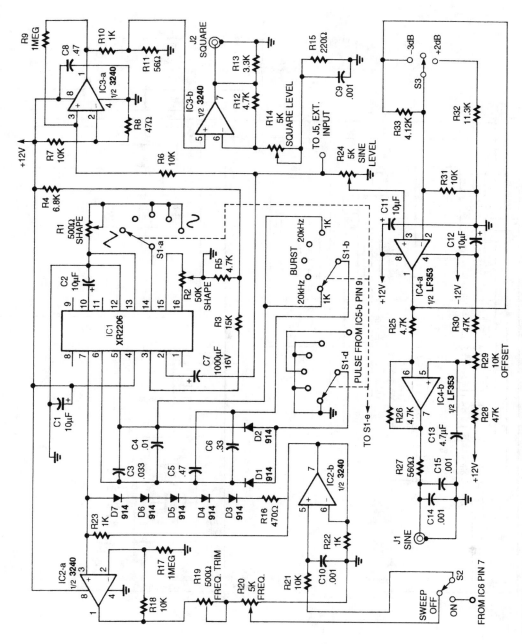

AN XR 2206 FUNCTION-GENERATOR CHIP provides a triangle output at pin 2 when S1-a is open.

SWEEP/FUNCTION GENERATOR

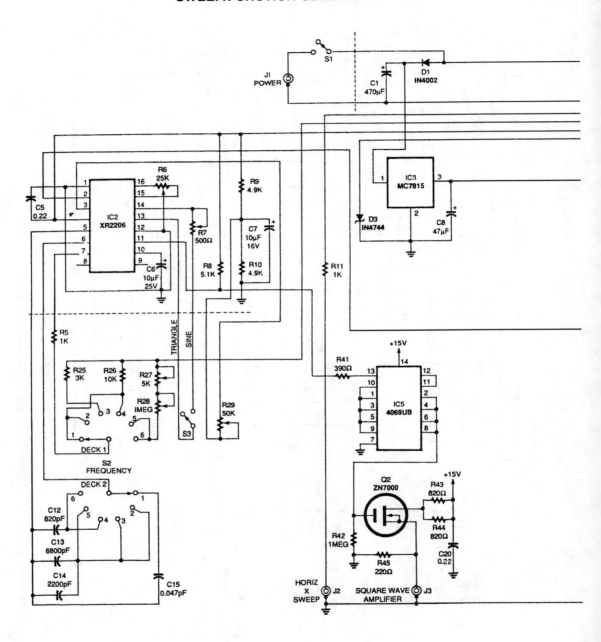

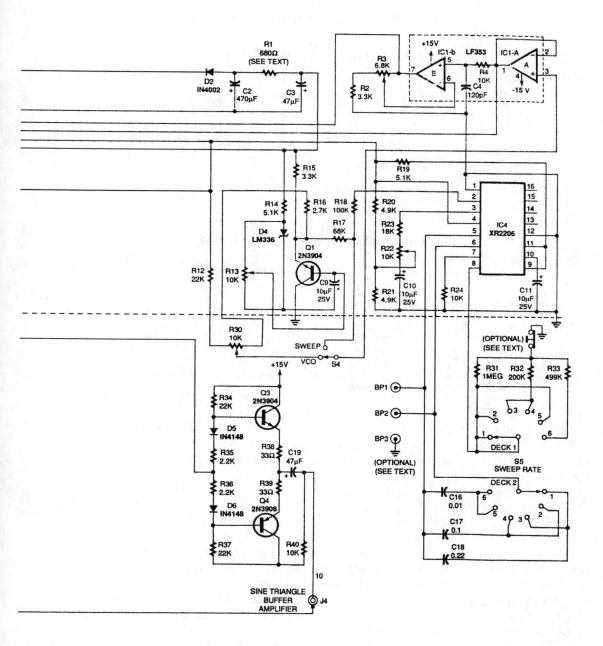

Fig. 36-2

239

SWEEP/FUNCTION GENERATOR (*Cont.*)

TABLE 1
FUNCTION GENERATOR CHARACTERISTICS

Waveform output	Maximum P-to-P	Frequency	Conditions
Sine (1)	5V	10 Hz-100 kHz	1 V@800 kHz
Triangle (1)	8 V	10 Hz- 50 kHz	1 V>500 kHz
Square (2)	5 V		Positive output DC-coupled, ground ref: rise/fall >50 ns
Ramp (3)			Descending, 6 rates

(1) Output level variable frim min. to max.
(2) Output level not adjustable.
(3) X and Y amplitude internally adjustable.

TABLE 2
SWEEP RANGES OF THE FUNCTION GENERATOR

Switch	Condition	Frequency range
1	Preset	20Hz to >2kHz
2	Preset	<400Hz to >10kHz
3	Preset	<1kHz to >25kHz
4	Preset	5kHz to >100kHz
5*	Resistance tuned	2kHz to 100kHz
	Resistance & VCO tuned	<10Hz to >100kHz
6*	Resistance tuned	<40kHz to >800kHz
	Resistance & VCO tuned	<100Hz to >800kHz

* Ranges show for positions 5 and 6 represent the total tuning range of the function generator and do not imply one continuous sweep.

Both IC2 and IC4 are Exar XR2206 monolithic function generators; IC4 functions as a ramp generator, and IC2 functions as a generator of sine, triangular, and square waveforms. Dual operational amplifier IC1 produces a scaled, level-shifted ramp output that is capable of deflecting an oscilloscope's horizontal sweep.

Any frequency of interest along the horizontal axis of an oscilloscope that is coupled to this function generator can be measured with an external frequency counter by manually tuning the function generator's VCO instead of sweeping it. The performance characteristics of the sweep/function generator are summarized in the Table.

The generator's sweep rate and frequency can be set by front-panel rotary six-position switches, Sweep Rate Switch S5 and Frequency Switch S2. The VCO control R30 manually tunes the VCO. Table 2 lists the sweep ranges of the function generator. Sweep ranges not covered in ranges 1 to 4 can be set up as required on positions 5 and 6. Selecting the VCO setting on the front panel toggle switch S4 permits tuning any fixed frequency within the total frequency range of the instrument with both frequency switch S2 and VCO control R30.

The sweep rate or duration of the sweep ramp is selected by the rotary six-position Sweep Rate Switch S5. Table 3 lists the sweep rate durations for each of the six positions. Longer periods should be used for lower frequency sweeps.

SWEEP/FUNCTION GENERATOR (*Cont.*)

TABLE 3
SWEEP RATE OR DURATION

Sweep position	Period (milliseconds)
1	~130
2	~ 60
3	~ 30
4	~ 15
5	~ 6
6	~ 3

SIMPLE FUNCTION GENERATOR

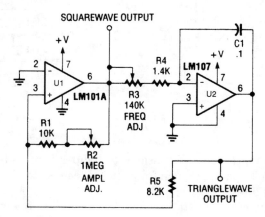

Fig. 36-3

ACCURATE, STABLE FUNCTION GENERATOR

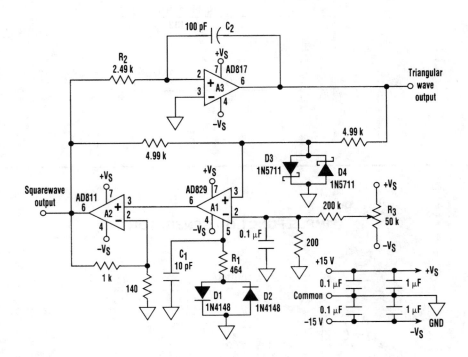

ELECTRONIC DESIGN

Fig. 36-4

Supply-limited oscillators usually are sensitive to temperature and power-supply changes, are never symmetrical, and don't operate at high frequencies because the amplifier's output is saturated when it reaches the supply lines.

The circuit shown, a function generator, can alleviate these problems. Its square-wave output boasts a rapid rise time, quick settling time, and an amplitude that's temperature insensitive. Also, its triangular output waveform features a perfectly constant rate of change throughout its range.

Amplifier A1 together with A2 generates a stable +10 V. This signal, which is integrated using A3, C2, and R2, makes a negative-going ramp. When the peak output of A3 equals −10 V, the output of A1 and A2 change state and the A3's output ramps up. When A3's output equals +10 V, the outputs of A1 and A2 change state again and new cycle starts.

WIDE-RANGE FUNCTION GENERATOR

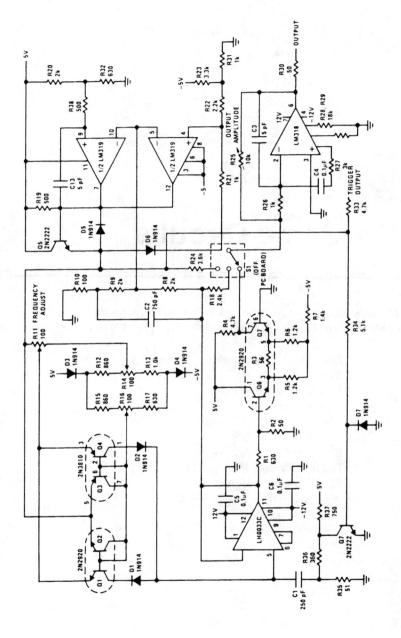

Fig. 36-5

NATIONAL SEMICONDUCTOR

The sine, square, triangle function generator is exceptionally useful. Various IC circuits have been published for generating square and triangle waveforms in an attempt to duplicate the general-purpose function generator. However, these simple circuits are usually limited to about 10 kHz and have no sine-wave output. The function generator shown here provides all three waveforms and operates from below 10 Hz to 1 MHz with usable output to about 2 MHz.

37

Game Circuits

The sources of the following circuits are contained in the Sources section, which begins on page 707. The figure number in the box of each circuit correlates to the entry in the Sources section.

Electronic Craps Game
21 Game
First-Response Monitor I
Z-Dice Game
Three-Input First-Response Monitor
Electronic Coin Toss
Electronic One-Arm Bandit
Digital "First-to-Respond" Box
First-Response Monitor II
Analog First-Response Monitor
Wheel of Fortune

ELECTRONIC CRAPS GAME

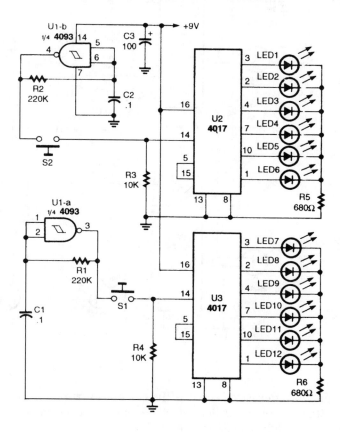

Fig. 37-1

Two gates of a 4093 quad, 2-input NAND, Schmitt-trigger CMOS IC are connected in astable-oscillator circuits as clocks. The two 4017 ICs have six LEDs connected to its first six outputs. As the clock pulses enter pin 14 of the 4017s, the ICs count from one to six over and over as long as the clock pulses are present. When S1 and/or S2 are released, one of the LEDs in each circuit will remain on, indicating a number from one to six.

The circuit is set up so that you can roll the dice together by pressing S1 and S2 at the same time, or roll each die one at a time.

21 GAME

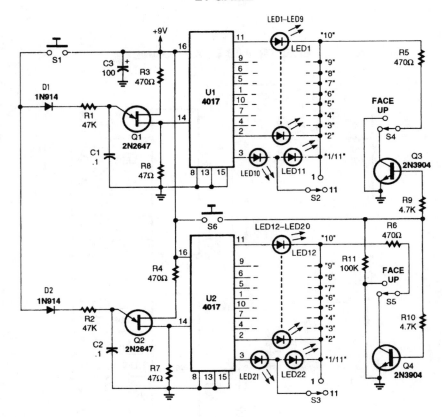

POPULAR ELECTRONICS

Fig. 37-2

Two 2N2647 unijunction transistors serve as the clock generators for the two 4017 ICs. A single "deal" push-button switch, S1, operates both clock generators at the same time. Diodes D1 and D2 isolate the two clock circuits, allowing S1 to operate both.

The 4017 counter/readout circuits are identical in circuitry and operation. As long as clock pulses enter pin 14 of each 4017, the ICs count from 1 to 10 over and over until the clock pulses stop. When S1 is released, the clock pulses stop and one LED from each IC remains on to indicate a card with a number value of 1 (1 or 11) to 10.

The position of switches S2 and S3 determines whether the number 1 ("Ace") output of the 4017s count as an 11 or a 1. Both S2 and S3 can be switched in either position before or after the cards are played.

The cards can be played either face up or face down. When switches S4 and S5 are in the position shown in the figure, the cards are dealt face down. Transistors Q3 and Q4 are turned off in this position and no current can flow through the LEDs. Pressing S6 turns both transistors on, lighting the LEDs.

FIRST-RESPONSE MONITOR I

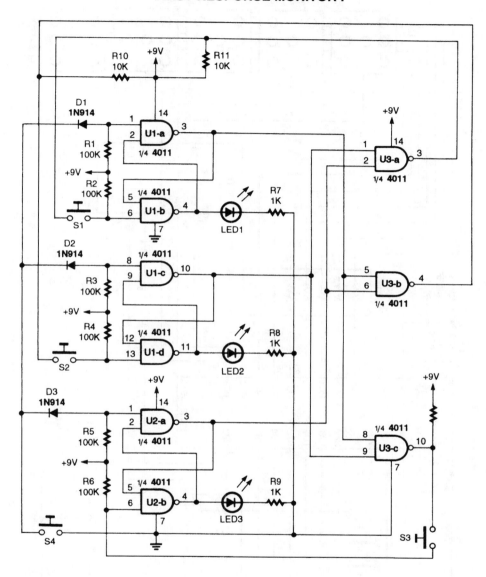

Fig. 37-3

Three interlocked flip-flops enable the detection of the first input. S1, S2, and S3 are inputs. Analog switches controlled by logic gates, or other logic circuitry could be sub-statement for S1, S2, and S3.

Z-DICE GAME

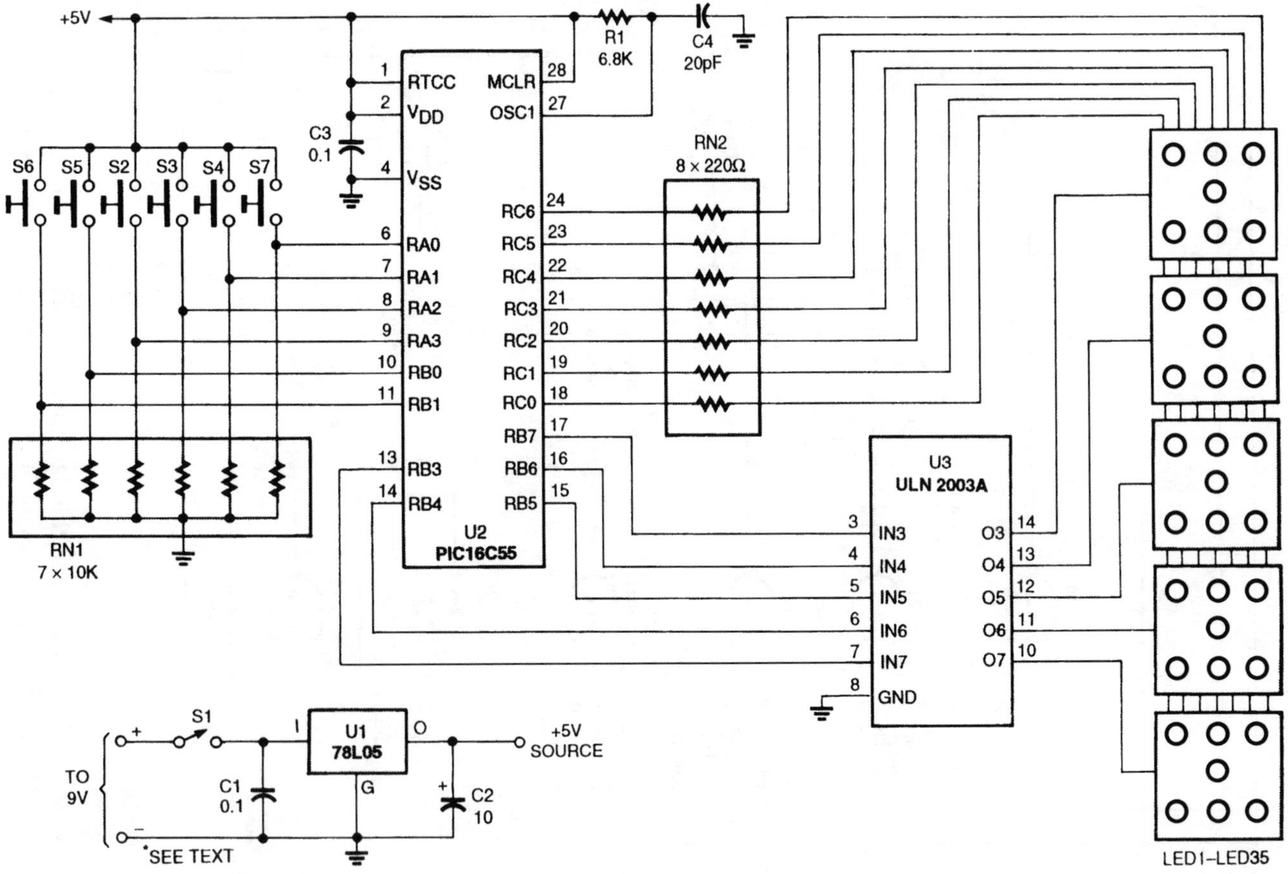

LED1–LED35

U3
ULN 2003A

U2
PIC16C55

RN2
8 × 220Ω

RN1
7 × 10K

R1
6.8K

C4
20pF

C3
0.1

U1
78L05

C2
10

C1
0.1

SEE TEXT

S1

TO
9V

+5V
SOURCE

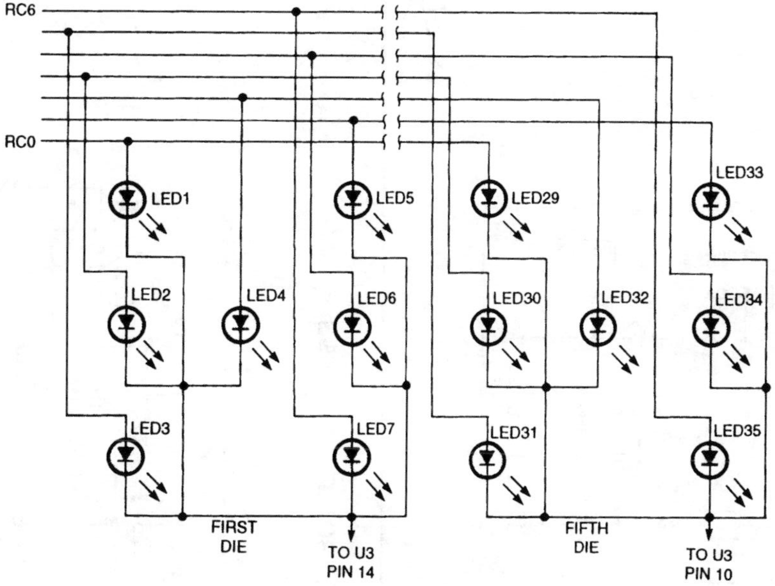

Fig. 37-4

POPULAR ELECTRONICS

Using a microcontroller (U2) keeps the parts count and the cost of this 5-dice LED display relatively low. Z-dice uses five clusters of seven LEDs to represent the marks or "pips" on five dice. Buttons below each of the LED dice let the player mark a die to be rolled on the next throw. Marked dice show up as dimmed LEDs. Pressing the button to the right of the display rolls the marked dice. If the player changes his or her mind about rolling a particular die before pressing the roll button, he or she can un-mark it by pressing its button a second time. If no dice are marked at the time the player presses the roll button, then all of the dice are marked to be rolled. A second press starts them rolling, animating the LEDs of the marked dice for a second or so before displaying the results of the roll. Z-Dice doesn't count rolls or keep score, so it's still up to the players to make sure that nobody cheats!

This diagram shows the wiring details of the dice display. For space and simplicity, only the first and last dice are shown. A programmed microcontroller is needed for this circuit. Refer to the original article for software.

249

THREE-INPUT FIRST-RESPONSE MONITOR

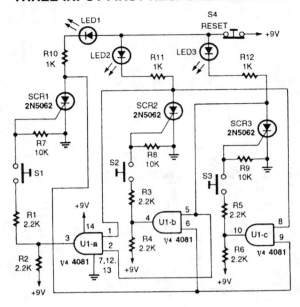

Fig. 37-5

ELECTRONIC COIN TOSS

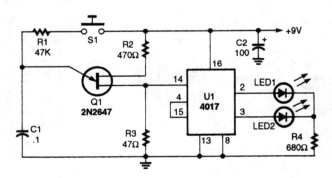

Fig. 37-6

Integrated circuit U1 is connected in a two-stage counter circuit that counts "one-two" over and over as long as clock pulses enter pin 14 of the 4017. When the clock pulses stop, one of the LEDs will remain on, indicating the last even or odd count. Designate one LED as "heads" and the other as "tails" and you have an electronic coin flipper.

ELECTRONIC ONE-ARM BANDIT

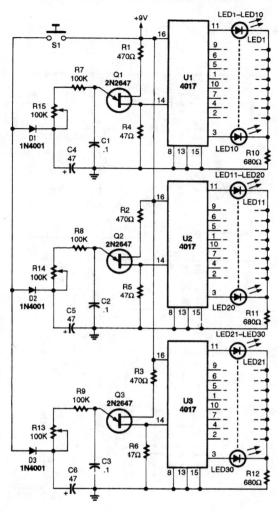

Fig. 37-7

The one-arm bandit circuit is made up of three clock circuits and three counter/readout circuits. A single roll switch, S1, turns on all three clocks at the same time. When S1 is closed, capacitors C4, C5, and C6 are charged through D31, D32, and D33 to about 8 V. After S1 is released, the three clocks run, taking energy from the three charged capacitors. As the capacitors discharge, the three clocks begin to slow down, producing the effect of the drums in a mechanical bandit slowing to a stop.

The 4017's 10-output LEDs can be numbered or designated as apples, cherries, bells, wild cards, or anything you like to make the game more interesting. Additional logic circuitry can be added to the 4017 outputs to sound an alert or turn on a light when any three numbers or output items match.

Three potentiometers, R12, R13, and R14, can be varied for each roll to change the clock's frequency and the roll rate.

DIGITAL "FIRST-TO-RESPOND" BOX

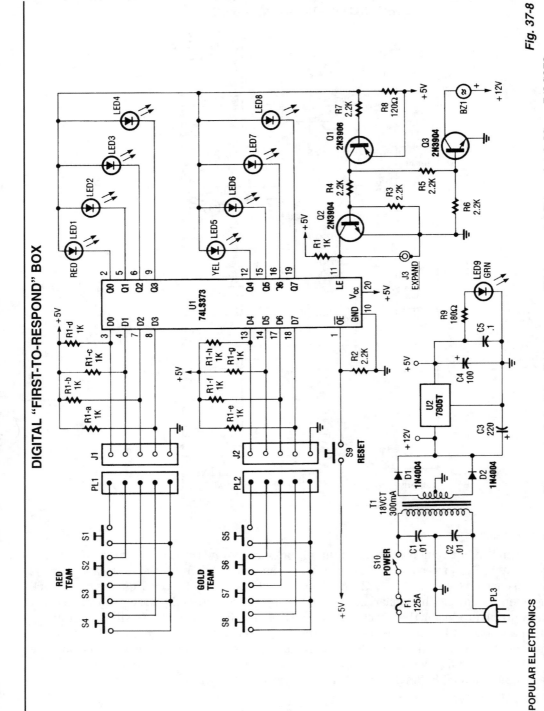

Fig. 37-8

POPULAR ELECTRONICS

This device is useful for quizzes and games to determine first response. U1 is an octal D type latch IC, an 74LS373.
When a button is pushed, this circuit lights the corresponding LED. Q1 conducts, sounding an alarm (BZ1) connected to driver Q3, and Q1 supplies bias to Q2, disabling the rest of the latches in U1.

FIRST-RESPONSE MONITOR II

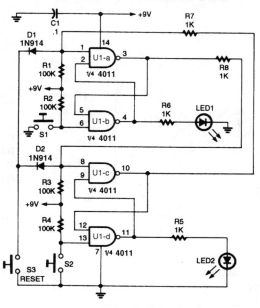

POPULAR ELECTRONICS

Fig. 37-9

Two interlocking flip flops are used to detect the first of two inputs. S1 and S2 are input devices, but a logical-level signal can be substituted.

ANALOG FIRST-RESPONSE MONITOR

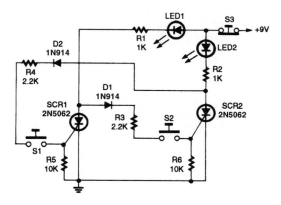

POPULAR ELECTRONICS

Fig. 37-10

The analog first-response monitor is built around a pair of cross-coupled SCRs, each of which receives its gate trigger current from the anode of the other SCR.

WHEEL OF FORTUNE

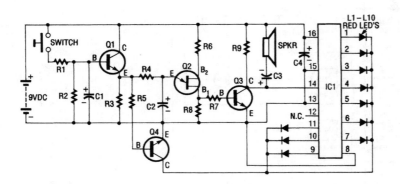

C1,C3 .... 10 µF Capacitor
C2 .... 1 µF Capacitor
C4 .... 47 µF Capacitor
IC1 .... MC14017BCP
L1-L10 .... Jumbo Red LEDs
Q1,Q3,Q4 .. 2N3904 Transistor
Q2 .... MU10UJT Transistor
R1,R5 .... 33K Resistor
R2 .... 2.2 Meg Resistor
R3 .... 82K Resistor
R4 .... 47K Resistor
R6 .... 2.2K Resistor
R7 .... 390 ohm Resistor
R8 .... 100 ohm Resistor
R9 .... 680 ohm Resistor
S1 .... Pushbutton Switch

POPULAR ELECTRONICS

Fig. 37-11

The oscillation of Q2 is amplified by Q3 and fed to Johnson counter IC1. The output of IC1 drives the LEDs in sequence to give the impression of a spinning red ball.

38

Humidity Sensor Circuits

The sources of the following circuits are contained in the Sources section, which begins on page 707. The figure number in the box of each circuit correlates to the entry in the Sources section.

Humidity Monitor
Digital Relative Humidity Gauge

HUMIDITY MONITOR

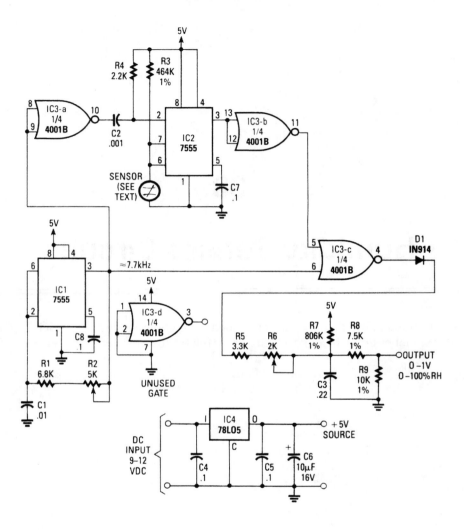

Fig. 38-1

This circuit uses a Phillips capacitive humidity sensor that has a ΔC variation of 45 pF over 0 to 100 pF RH. IC2 is an oscillator whose frequency is determined by the RH sensor. It is compared to fixed oscillator, and the difference frequency is taken by IC3C and rectified, outputting a 0- to 1-V signal for RH between 0 and 100%.

DIGITAL RELATIVE HUMIDITY GAUGE

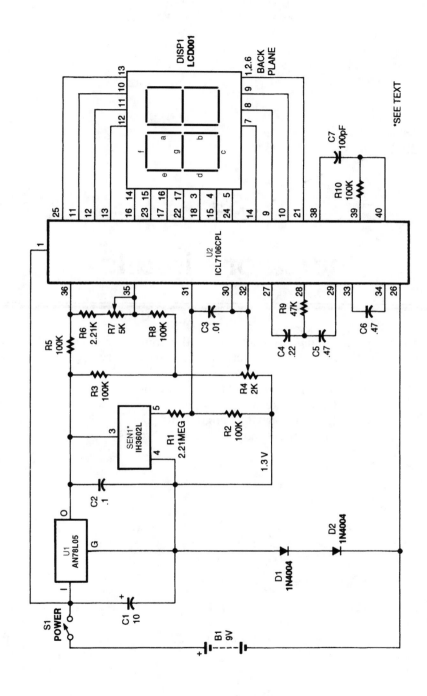

POPULAR ELECTRONICS

Fig. 38-2

Sensor SEN1 outputs a dc voltage that varies linearly with relative humidity. This dc voltage is fed through R1 and R2 to A/D converter chip U2. The LCD display is calibrated with R4. Zero set is performed with R4. The LCD display is calibrated with R7 to read 0 to 100 percent.

257

39

Indicator Circuits

The sources of the following circuits are contained in the Sources section, which begins on page 707. The figure number in the box of each circuit correlates to the entry in the Sources section.

Model Car Derby Winner Indicator
Current Indicator
Receiver Signal-Strength Indicator
LED Output Indicator for 555 Circuits

MODEL CAR DERBY WINNER INDICATOR

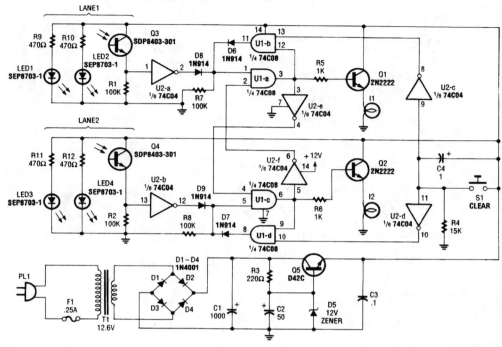

POPULAR ELECTRONICS

Fig. 39-1

This derby-winner indicator uses infrared emitters and sensors to detect a car crossing the finish line. The first car to finish locks out the data from the second car, and the system can be reset by pressing S1.

CURRENT INDICATOR

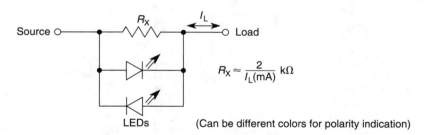

$$R_X \approx \frac{2}{I_L(mA)} \text{ k}\Omega$$

(Can be different colors for polarity indication)

WILLIAM SHEETS

Fig. 39-2

An LED requires 1.5 to 3 V across its terminals to light. This circuit uses a resistor shunt in series with source and load to produce this drop and cause the LED to light. At higher currents (>100 mA) use limiting resistors in series with LEDs to limit current to a safe value.

RECEIVER SIGNAL-STRENGTH INDICATOR

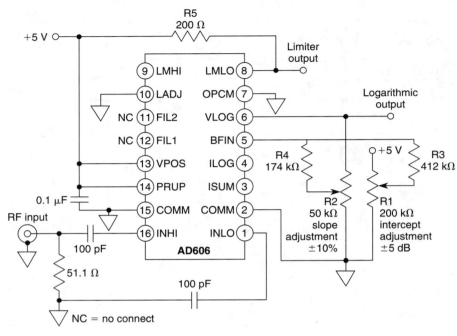

ANALOG DEVICES

Fig. 39-3

Using an AD606 log amplifier, this indicator gives a logarithmic output of +0.3 V at –80-dBm input to +3.5 V at 10-dBm input. Frequency range is to 50 MHz for this IC device.

LED OUTPUT INDICATOR FOR 555 CIRCUITS

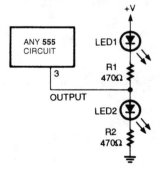

POPULAR ELECTRONICS, JANUARY 1994, P. 73

Fig. 39-4

A pair of LEDs connected as shown here can be used with just about any low-frequency 555 oscillator to give high-/low-output indications. When the output goes high LED2 turns on, and when the output goes low LED1 turns on.

40

Infrared Circuits

The sources of the following circuits are contained in the Sources section, which begins on page 707. The figure number in the box of each circuit correlates to the entry in the Sources section.

Audio-Modulated IR Transmitter
Audible IR Detector
Wireless IR Headphone Transmitter
TV Remote-Control Relay
Single-Tone Infrared Control Transmitter
IR Illuminator for Night-Vision TV Cameras and Scopes
Low-Power Infrared Data-Link Receiver
Infrared Body Heat Detector
IR Detector Circuit
Steady-Tone Infrared Transmitter
FM Infrared Receiver for Audio Reception
General-Purpose IR Receiver
Wireless IR Headphone Receiver
Pulse Frequency-Modulated IR Transmitter
Single-Tone Infrared Receiver
Audible-Output Infrared Receiver

AUDIO-MODULATED IR TRANSMITTER

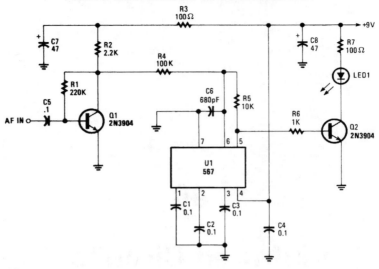

POPULAR ELECTRONICS

Fig. 40-1

This circuit produces an effect similar to frequency modulation (FM) by varying the voltage at pin 6 of the PLL using an audio signal. The FM IR signal can be picked up by a receiver with an FM detector suitably tuned.

AUDIBLE IR DETECTOR

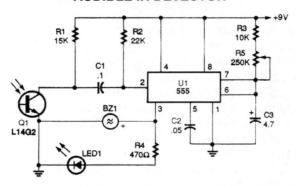

POPULAR ELECTRONICS

Fig. 40-2

An IR-detector circuit offers an audible (as well as a visual) output, and also stretches the on time of the detected pulse to make the output easier to see, as shown.

Photoresistor Q1 detects a remote's IR output pulse and sends a negative-going pulse to the trigger input (pin) of the 555 IC, U1. The 555 is connected in a one-shot timer circuit; the output (pin 3) on time is set by the values of C_3, R_3, and R_5. When an input pulse is detected, pin 3 goes high, lighting LED1 and activating the piezo buzzer, BZ1.

For longer output pulses, set R_5 to its maximum resistance value. To lengthen the circuit's on-time range, increase the value of C_3, and to shorten the on-time range lower the value of C_3.

WIRELESS IR HEADPHONE TRANSMITTER

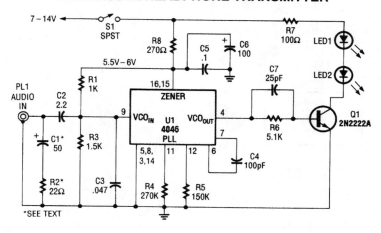

POPULAR ELECTRONICS

Fig. 40-3

Audio input from PL1 frequency modulates the VCO section of a 4046 PLL chip. The VCO output drives Q1, a switching transistor. Q1 drives two IR LEDs. The signal produced is around 100 kHz, FM carrier VCO sensitivity is around 7.5 kHz/V.

TV REMOTE-CONTROL RELAY

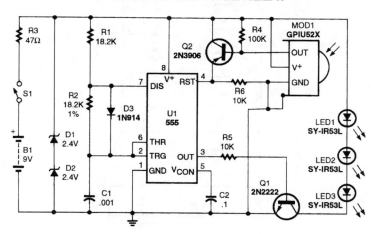

POPULAR ELECTRONICS

Fig. 40-4

This circuit functions as an IR "repeater" to extend the range of your TV remote control. MOD1 is a P/N GP1U52X IR detector and the receiver is available as Radio Shack P/N 276-137.

263

SINGLE-TONE INFRARED CONTROL TRANSMITTER

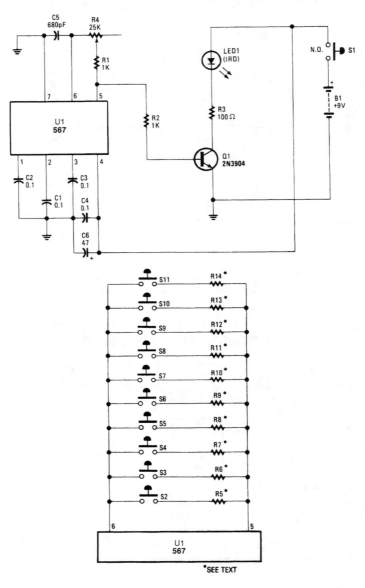

Fig. 40-5

A modulated beam of IR light is produced by this transmitter. This circuit can be used for on/off controls or tone (CW) communications. The pot can be replaced by several pushbuttons and resistors, as shown for multitone applications.

IR ILLUMINATOR FOR NIGHT-VISION TV CAMERAS AND SCOPES

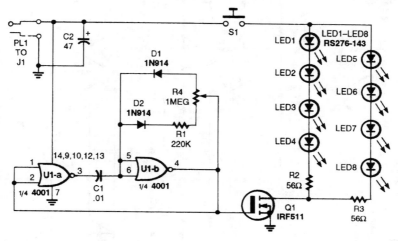

POPULAR ELECTRONICS

Fig. 40-6

This source uses LEDs and an astable oscillator to control the switch, duty-cycle, and effective IR illumination output.

LOW-POWER INFRARED DATA-LINK RECEIVER

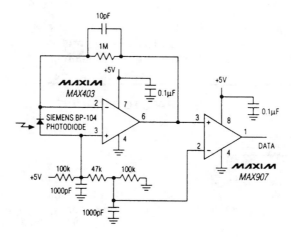

MAXIM

Fig. 40-7

The Maxim MAX403 in this circuit consumes only 1 mA and is capable of speeds over 1 MBPS.

INFRARED BODY-HEAT DETECTOR

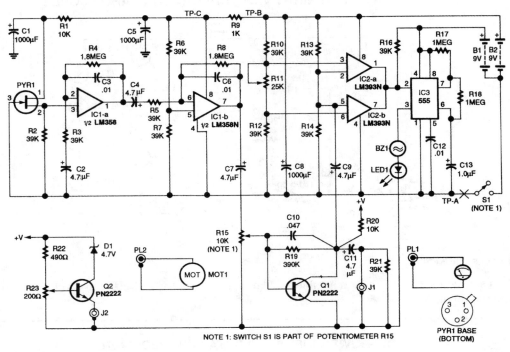

ELECTRONICS NOW

Fig. 40-8

This circuit uses a pyroelectric detector to detect IR emissions in the 6- to 14-micron range. It is useful for security or infrared experiments. PYR1 is a pyroelectric IR detector. The unit should be mounted in a case with an IR lens to focus energy on the detector.

IR DETECTOR CIRCUIT

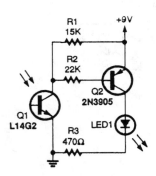

The circuit uses an IR phototransistor, Q1, to detect a remote control's IR output signal. A PNP transistor, Q2, then amplifies Q1's output and lights LED1. That indicates that an infrared signal has been detected by the phototransistor, or in other words, that your remote control works.

POPULAR ELECTRONICS

Fig. 40-9

STEADY-TONE INFRARED TRANSMITTER

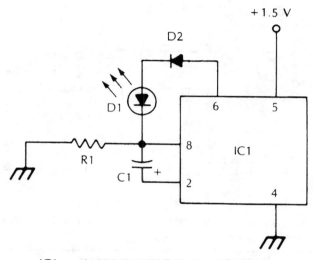

IC1	LM3909 LED flasher/oscillator IC
D1	infrared LED
D2	diode (1N4148, 1N914, or similar)
C1	1 μF 5 V electrolytic capacitor
R1	1.5 kΩ ¼ W 5% resistor

McGRAW-HILL

Fig. 40-10

This oscillator pulses an IR LED at about 1000 Hz. It should be useful as a test for lining up IR communications links or setting up fiber-optic cables, etc.

FM INFRARED RECEIVER FOR AUDIO RECEPTION

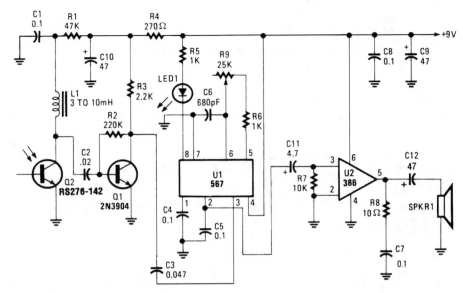

POPULAR ELECTRONICS

Fig. 40-11

Modulated IR energy strikes Q2, a phototransistor. Q1 is a tuned amplifier, and feeds PLL detector U1. U2 is an audio amplifier that drives a speaker.

GENERAL-PURPOSE IR RECEIVER

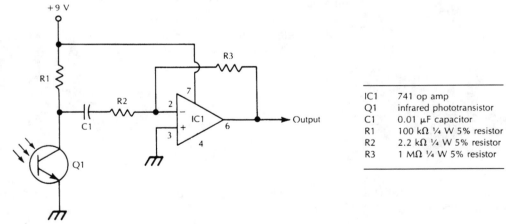

IC1	741 op amp
Q1	infrared phototransistor
C1	0.01 μF capacitor
R1	100 kΩ ¼ W 5% resistor
R2	2.2 kΩ ¼ W 5% resistor
R3	1 MΩ ¼ W 5% resistor

McGRAW-HILL

Fig. 40-12

Suitable for amplitude-modulated IR beams, this receiver provides an audio signal that corresponds to the modulation envelope. Phototransistor Q1 should be properly mounted and shielded from stray light. This receiver should drive a small earphone directly.

WIRELESS IR HEADPHONE RECEIVER

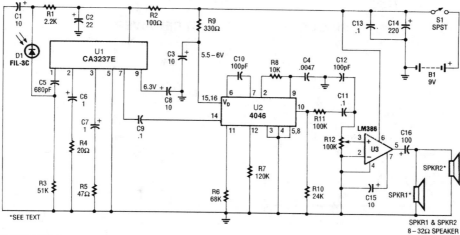

POPULAR ELECTRONICS

Fig. 40-13

A photodiode D1 feeds high gain IR remote control preamp IC, a CA3237E. U2 is a PLL FM detector tuned to around 100 kHz. The detector output is amplified by U3 and it can drive a speaker or a set of headphones.

PULSE FREQUENCY-MODULATED IR TRANSMITTER

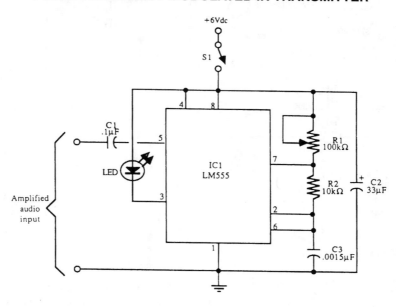

McGRAW-HILL

Fig. 40-14

Schematic diagram for the pulse frequency-modulated LED transmitter. Adjust the frequency by rotating R1. With components shown, the frequency range is between 8 and 48 kHz.

SINGLE-TONE INFRARED RECEIVER

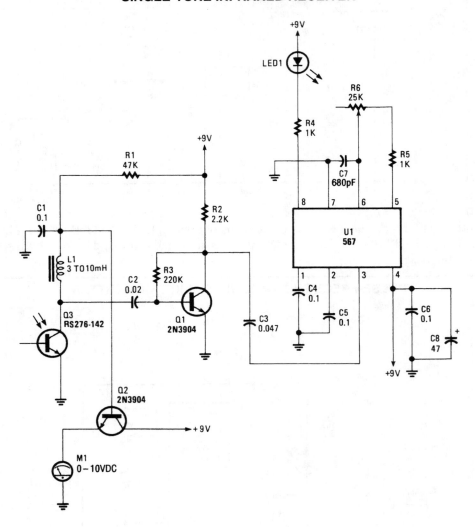

Fig. 40-15

Phototransistor Q3 acts as a sensor that detects modulated IR energy. Q1 is an amplifier and U1 is a tone decoder. LED1 lights on reception of an IR signal with proper tone modulation.

AUDIBLE-OUTPUT INFRARED RECEIVER

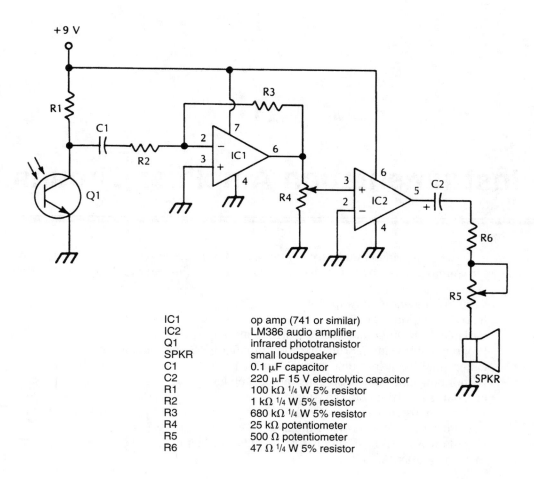

IC1	op amp (741 or similar)
IC2	LM386 audio amplifier
Q1	infrared phototransistor
SPKR	small loudspeaker
C1	0.1 μF capacitor
C2	220 μF 15 V electrolytic capacitor
R1	100 kΩ ¼ W 5% resistor
R2	1 kΩ ¼ W 5% resistor
R3	680 kΩ ¼ W 5% resistor
R4	25 kΩ potentiometer
R5	500 Ω potentiometer
R6	47 Ω ¼ W 5% resistor

McGRAW-HILL   *Fig. 40-16*

This receiver is designed to demodulate amplitude-modulated (AM) IR light beams and will drive a loudspeaker. R5 is an auxiliary volume control and it could be omitted. Q1 should be suitably mounted and shielded from stray light pickup.

41

Instrumentation Amplifier Circuits

The sources of the following circuits are contained in the Sources section, which begins on page 707. The figure number in the box of each circuit correlates to the entry in the Sources section.

×100 INSTRUMENTATION AMPLIFIER

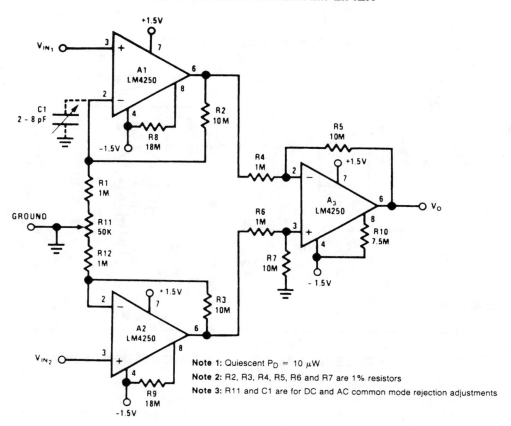

Note 1: Quiescent $P_D = 10 \mu W$

Note 2: R2, R3, R4, R5, R6 and R7 are 1% resistors

Note 3: R11 and C1 are for DC and AC common mode rejection adjustments

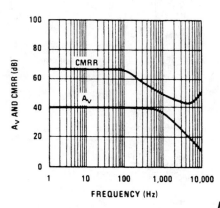

NATIONAL SEMICONDUCTOR

Fig. 41-1

CMRR vs. frequency.

INSTRUMENTATION AMPLIFIER

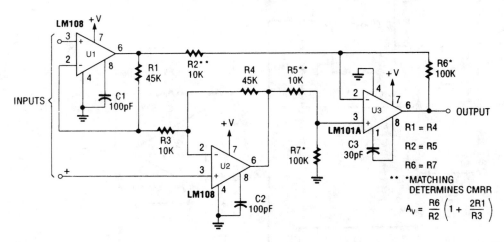

Fig. 41-2

VARIABLE-GAIN DIFFERENTIAL INPUT INSTRUMENTATION AMPLIFIER

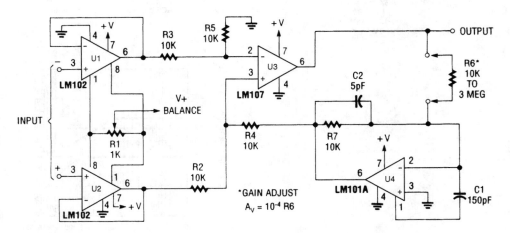

Fig. 41-3

PROGRAMMABLE GAIN INSTRUMENTATION AMPLIFIER
FOR SINGLE-SUPPLY APPLICATIONS

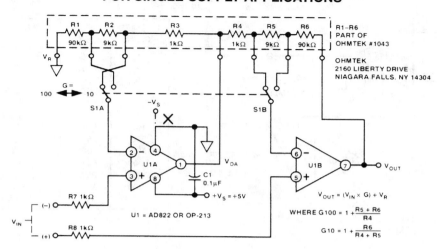

ANALOG DEVICES

Fig. 41-4

This is a two-op-amp programmable-gain instrumentation amplifier for single-supply applications. U1A and U1B are Analog Devices AD822 or OP-213 ICs.

DIFFERENTIAL-INPUT INSTRUMENTATION AMPLIFIER

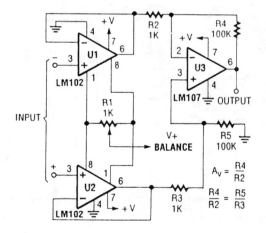

POPULAR ELECTRONICS

Fig. 41-5

HIGH INPUT-IMPEDANCE INSTRUMENTATION AMPLIFIER

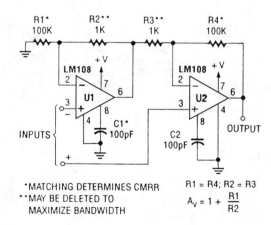

POPULAR ELECTRONICS

Fig. 41-6

ac-COUPLED INSTRUMENTATION AMPLIFIER

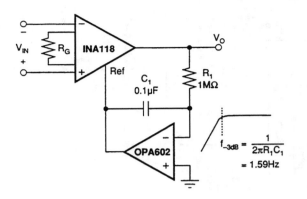

DESIRED GAIN	R_G (Ω)	NEAREST 1% R_G (Ω)
1	NC	NC
2	50.00k	49.9k
5	12.50k	12.4k
10	5.556k	5.62k
20	2.632k	2.61k
50	1.02k	1.02k
100	505.1	511
200	251.3	249
500	100.2	100
1000	50.05	49.9
2000	25.01	24.9
5000	10.00	10
10000	5.001	4.99

BURR-BROWN

Fig. 41-7

LOW-NOISE INSTRUMENTATION AMPLIFIER

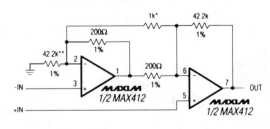

* TRIM FOR GAIN
** TRIM FOR COMMON-MODE REJECTION

MAXIM

Fig. 41-8

A Maxim MAX412 IC amplifier is used in this circuit. The supply current is ±5 V at 5 mA.

LOW-POWER INSTRUMENTATION AMPLIFIER

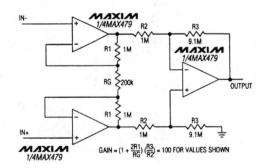

$$\text{GAIN} = \left(1 + \frac{2R1}{RG}\right)\left(\frac{R3}{R2}\right) = 100 \text{ FOR VALUES SHOWN}$$

MAXIM

Fig. 41-9

This amplifier requires less than 20 mA from a ±15-V supply.

ULTRA-LOW-NOISE SINGLE-SUPPLY INSTRUMENTATION AMPLIFIER

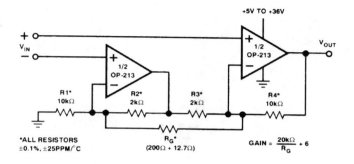

$$\text{GAIN} = \frac{20k\Omega}{R_G} + 6$$

ANALOG DEVICES

Fig. 41-10

INSTRUMENTATION AMPLIFIER

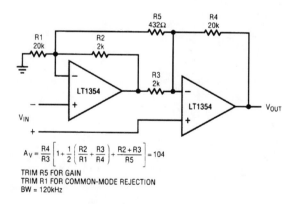

$$A_V = \frac{R4}{R3}\left[1 + \frac{1}{2}\left(\frac{R2}{R1} + \frac{R3}{R4}\right) + \frac{R2+R3}{R5}\right] = 104$$

TRIM R5 FOR GAIN
TRIM R1 FOR COMMON-MODE REJECTION
BW = 120kHz

LINEAR TECHNOLOGY

Fig. 41-11

42

Integrator Circuits

The sources of the following circuits are contained in the Sources section, which begins on page 707. The figure number in the box of each circuit correlates to the entry in the Sources section.

INTEGRATOR WITH BIAS-CURRENT COMPENSATION

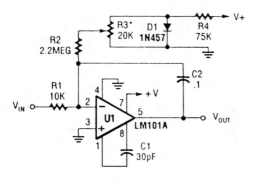

*ADJUST FOR ZERO INTEGRATOR DRIFT.
CURRENT DRIFT TYPICALLY 0.1 n/A°C
OVER −55°C TO 125°C
TEMPERATURE RANGE.

POPULAR ELECTRONICS   **Fig. 42-1**

SIMPLE INTEGRATOR

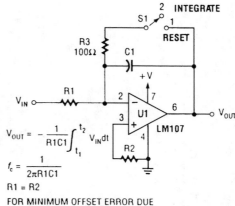

$$V_{OUT} = -\frac{1}{R1C1}\int_{t_1}^{t_2} V_{IN}\,dt$$

$$f_c = \frac{1}{2\pi R1C1}$$

R1 = R2

FOR MINIMUM OFFSET ERROR DUE
TO INPUT BIAS CURRENT

POPULAR ELECTRONICS   **Fig. 42-2**

ac INTEGRATOR

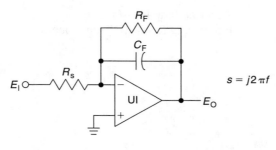

$$s = j2\pi f$$

$$\frac{E_O}{E_I}(s) = \frac{R_F}{R_S(1 + sR_FC_F)} \approx \frac{1}{R_FC_F} \text{ for } s \gg \frac{1}{R_FC_F}$$

WILLIAM SHEETS   **Fig. 42-3**

This op-amp circuit can be used with a wide variety of op amps. The values of R_f and R_i depend on gain, but will be 1 kΩ to 1 MΩ in most cases. C_f depends on the pole frequency needed. U1 is a 741-type op amp, etc.

43

Interface Circuits

The sources of the following circuits are contained in the Sources section, which begins on page 707. The figure number in the box of each circuit correlates to the entry in the Sources section.

Timer/ac Line Interface
Interfacing Resistive Transducers

TIMER/ac LINE INTERFACE

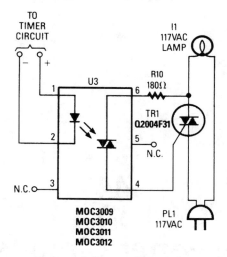

POPULAR ELECTRONICS

Fig. 43-1

This circuit illustrates the use of an optoisolator to enable the control of a triac connected to the ac line and load, while maintaining dc and ac isolation between the ac line and the timer circuit. A 555 or other timer circuit can be used.

INTERFACING RESISTIVE TRANSDUCERS

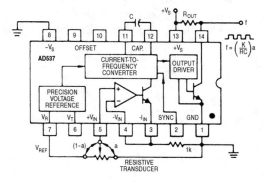

ANALOG DEVICES

Fig. 43-2

All types of resistive-element transducers, such as servo-pots, level indicators, thermistors, photosensors, strain gages, and so on, can be directly connected to the AD537. The scale-correction factor, K, is a function of resistance, varying from 0.65 to 0.98 for values from 3 to 100 kΩ.

44

Inverter Circuits

The sources of the following circuits are contained in the Sources section, which begins on page 707. The figure number in the box of each circuit correlates to the entry in the Sources section.

SCR Inverter and Trigger Circuit
Simple Inverter
Vehicle Audio Amplifier Inverter
Positive-to-Negative dc/dc Inverter
1-kW 10-kHz Sine-Wave Inverter

SCR INVERTER AND TRIGGER CIRCUIT

An SCR inverter and trigger circuit. This inverter operates well over a wide load range and can be used to power inductive devices, such as motors. General Electric Semiconductor Products Dept

R10, R11	1 Ω, 20 W
R12, R13	12 Ω, 1 W
C5	2-4 μfd units in parallel, G.E. #61F254
D4, D5	G.E. 1N2157
SCR1, SCR2	G.E. C40A
T2	G.E. #9T33Y267
L1	G.E. #9⅝33¾ 266

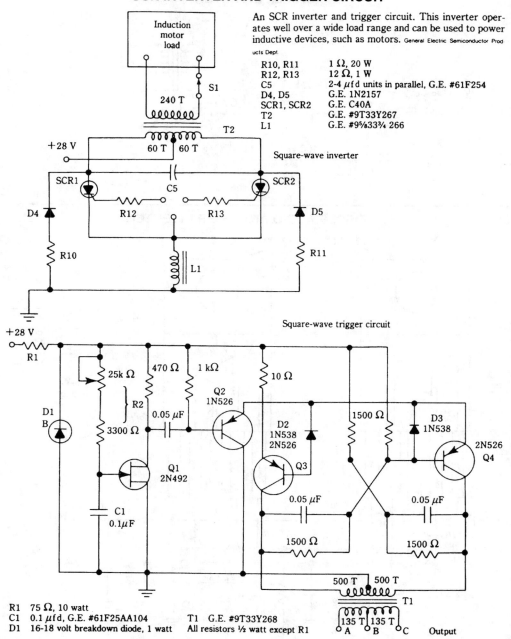

R1 75 Ω, 10 watt
C1 0.1 μfd, G.E. #61F25AA104   T1 G.E. #9T33Y268
D1 16-18 volt breakdown diode, 1 watt   All resistors ½ watt except R1

McGRAW-HILL

Fig. 44-1

In this circuit, L1 and C5 are used as commutating elements. L1 resonates with C5 at the frequency corresponding to the half period of the waveform.

SIMPLE INVERTER

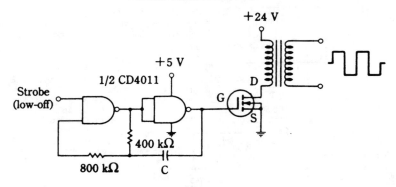

McGRAW-HILL

Fig. 44-2

VEHICLE AUDIO AMPLIFIER INVERTER

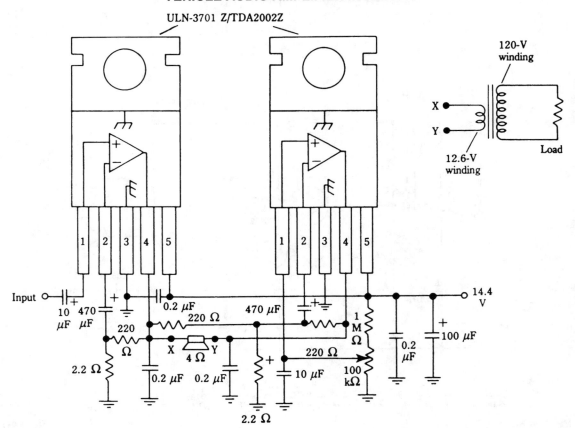

McGRAW-HILL

Fig. 44-3

An audio amplifier can drive a step-up transformer to obtain 120 Vac.

POSITIVE-TO-NEGATIVE dc/dc INVERTER

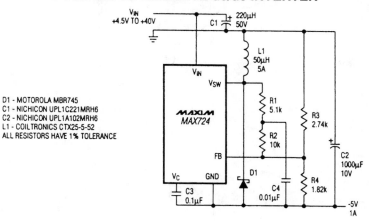

D1 - MOTOROLA MBR745
C1 - NICHICON UPL1C221MRH6
C2 - NICHICON UPL1A102MRH6
L1 - COILTRONICS CTX25-5-52
ALL RESISTORS HAVE 1% TOLERANCE

MAXIM   **Fig. 44-4**

If a source of negative 5 Vdc is needed and only a positive supply is available, this circuit can be used.

1-kW 10-kHz SINE-WAVE INVERTER

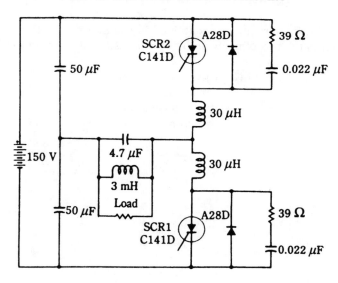

McGRAW-HILL   **Fig. 44-5**

SCRs can produce considerable power at frequencies up to 30 kHz or more. This circuit can supply 1 kW at 10 kHz. The load is shown as an equivalent load, and practically this will be the primary of the transformer for isolation purposes. The power supply can be a 120-V bridge rectifier and filter combination.

45

Ion Circuits

The sources of the following circuits are contained in the Sources section, which begins on page 707. The figure number in the box of each circuit correlates to the entry in the Sources section.

Negative Ion Generator
Ion-Sensing Electroscope
Negative Ion Generator
Ion Detector

NEGATIVE ION GENERATOR

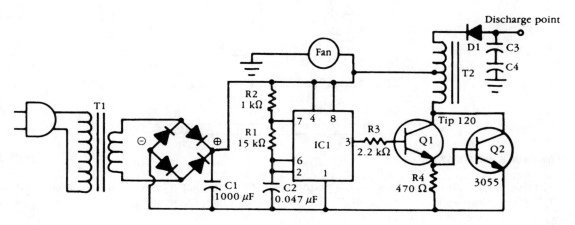

McGRAW-HILL   **Fig. 45-1**

An NE555 drives a Darlington connected pair of transistors. T1 is a small high-voltage transformer or auto ignition coil, B/W TV flyback, etc. C3, C4, and D1 must be rated for 10 to 15 kV. The fan blows air across the discharge point.

ION-SENSING ELECTROSCOPE

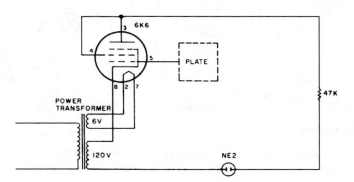

73 AMATEUR RADIO TODAY   **Fig. 45-2**

Negative ions are sensed by a plate antenna. A negative charge induced on the plate cuts off a vacuum tube, causing the neon indicator to go out.

NEGATIVE ION GENERATOR

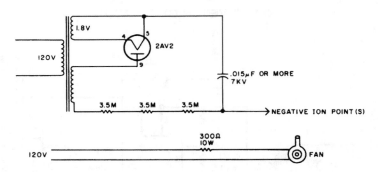

73 AMATEUR RADIO TODAY

Fig. 45-3

In this circuit, air is circulated past a pointed electrode that has a high negative voltage applied to it. The transformer is a small 4- to 6-kV output type with a filament winding. A good source of parts is a discarded electronic bug catcher.

ION DETECTOR

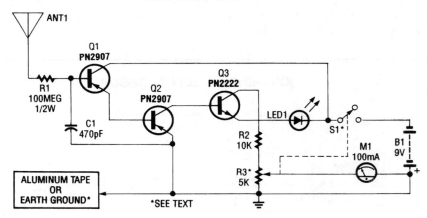

ELECTRONIC HOBBYISTS HANDBOOK

Fig. 45-4

This circuit detects static charges and free ions in the air. It can be used to indicate the presence of ion emissions, high-voltage leakage, static electricity, electrostatic fields, etc. The ground connection is made by either an earth ground or by touching the aluminum foil electrode with your hand. M1 is a 100-μA meter. R3 is a sensitivity control.

46

Laser Circuits

The sources of the following circuits are contained in the Sources section, which begins on page 707. The figure number in the box of each circuit correlates to the entry in the Sources section.

HIGH-CURRENT DRIVE CIRCUIT FOR SINGLE HETEROSTRUCTURE LASER DIODES

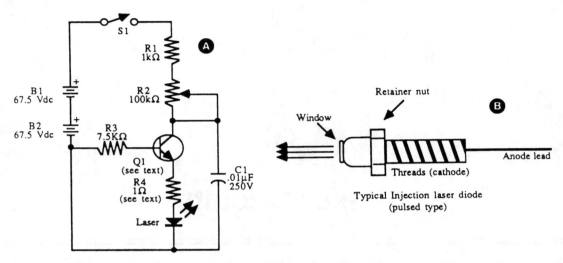

(A) High-current drive circuit for a single heterostructure laser diode. (B) Power leads for the typical sh laser diode, showing single lead for the anode.

R1	1 kilohm resistor
R2	100 kilohm potentiometer
R3	7.5 kilohm resistor
R4	1 ohm resistor, carbon composition, 5 watts
C1	0.01 μF capacitor, 250 V or higher
Q1	2N2222 or equivalent; see text
B1,B2	67.5 Vdc batteries
Misc.	Single heterostructure laser diode, heatsink

All resistors are 5 to 10 percent tolerance, ¼ watt, unless otherwise indicated.

McGRAW-HILL   *Fig. 46-1*

The transistor is operated in the avalanche mode. You might need to try several 2N2222 devices before finding one that oscillates. R2 is adjusted for optimum oscillation. This supply provides pulse of 10 to 20 amps at about 50 ns.

12-V HIGH-VOLTAGE SUPPLY FOR HE-NE LASER

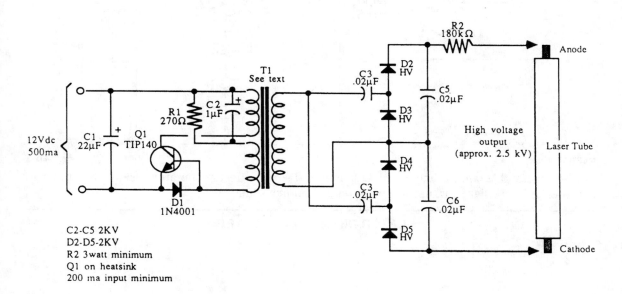

McGRAW-HILL

Fig. 46-2

T1 is a 6-V:330-V dc/dc inverter transformer with a 57.4:1 turn ratio, rated at 7 W.

R1	270-Ω resistor
R2	180-kΩ resistor, 3 to 5 W
C1	22-μF electrolytic capacitor
C2	1-μF electrolytic capacitor
C3-C6	0.02-μF capacitor, 1 kV or more
D1	1N4001 diode
D2-D5	High-voltage diode (3 kV or more)
Q1	TIP 140 power transistor
T1	High-voltage dc-to-dc converter transformer; see text for specifications

All resistors are 5 to 10% tolerance, ¼ W, unless otherwise indicated. All capacitors are 10 to 20% tolerance, rated 35 V or more, unless otherwise indicated.

LIGHT-BEAM RECEIVER AND SOUND EFFECTS GENERATOR FOR LASER PISTOLS

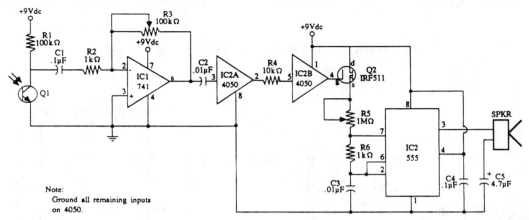

McGRAW-HILL

Fig. 46-3

Schematic diagram for light beam amplifier and sound-effects generator (using a 555 timer IC and speaker). The light striking Q1 generates a siren-like sound.

LASER DIODE TRANSMITTER

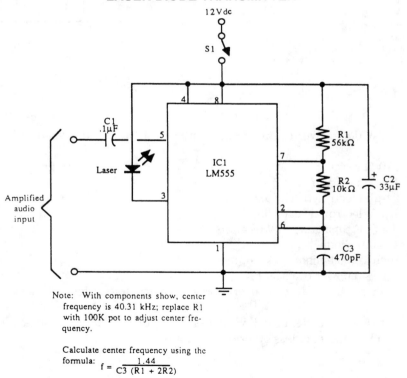

Note: With components show, center frequency is 40.31 kHz; replace R1 with 100K pot to adjust center frequency.

Calculate center frequency using the formula:

$$f = \frac{1.44}{C3\ (R1 + 2R2)}$$

McGRAW-HILL

Fig. 46-4

IR LASER LIGHT DETECTOR

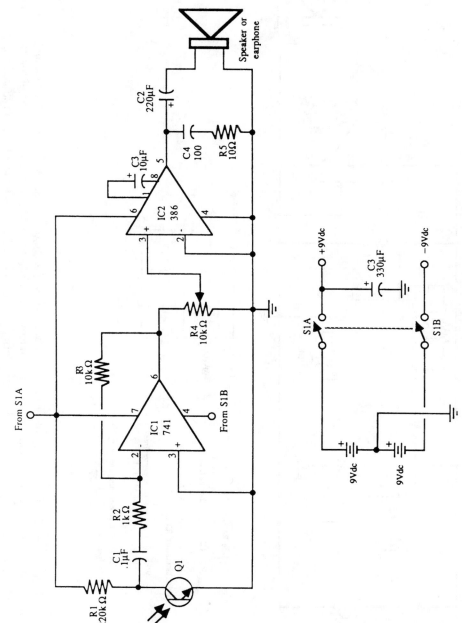

Fig. 46-5

The universal laser light detector. The output of the LM386 audio amplifier can be connected to a small 8-Ω speaker or ear-phone. Two 9-V batteries provide power. Decrease R1 to lower sensitivity; increase R3 to increase gain of the op amp (avoid very high gain or the op amp might oscillate). Q1 is an infrared phototransistor.

PLL IR LASER LIGHT RECEIVER

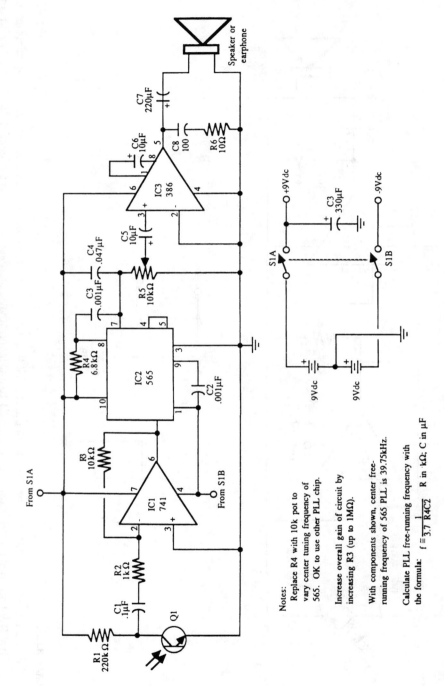

Notes:

Replace R4 with 10k pot to vary center tuning frequency of 565. OK to use other PLL chip.

Increase overall gain of circuit by increasing R3 (up to 1MΩ).

With components shown, center free-running frequency of 565 PLL is 39.75kHz.

Calculate PLL free-running frequency with the formula: $f = \dfrac{1}{3.7\ R4C2}$ R in kΩ; C in μF

McGRAW-HILL

Fig. 46-6

Circuit schematics for the 555-based PLL laser light PFM receiver. Although R4 is shown as a resistor, you might want to substitute it with a 10-kΩ precision potentiometer so that you can "dial in" the center frequency of the transmitter. Experiment with the value of C_1 for the best high-frequency response. Notice that circuit is functionally identical to the laser light detector/receiver shown in the figure, but with the addition of the 565.

OP-AMP DIODE LASER DRIVER

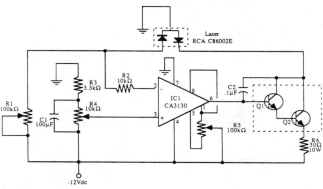

McGRAW-HILL

Fig. 46-7

This circuit is one way to automatically adjust drive current using a discrete op amp. Use the transistors specified or replace them with a suitable Darlington power transistor (such as TIP 120).

IC1	RCA CA 313 operational amplifier
R1, R5	100-kΩ potentiometer
R2	10-kΩ resistor
R3	3.3-kΩ potentiometer
R4	10-kΩ potentiometer
R6	30-Ω, 10-W resistor
C1	100-μF electrolytic capacitor
C2	0.1-μF disc capacitor
Q1	2N2101 transistor
Q2	2N3585 transistor
Laser	RCA C86002 (or equivalent laser diode)

LASER dc SUPPLY

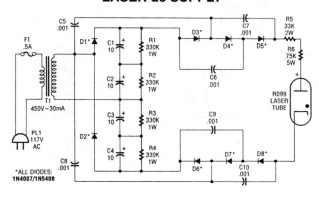

POPULAR ELECTRONICS

Fig. 46-8

The supply provides about 6 kVdc when open circuited, dropping to around 1375 Vdc when loaded. The R099 is a laser tube.

IC LASER DIODE DRIVER

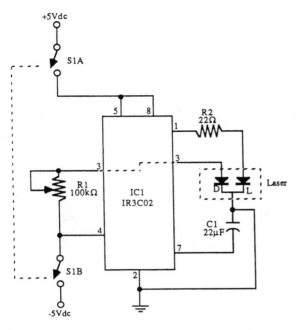

IC	Sharp IR3C02 laser diode driver IC
R1	100-kΩ resistor
R2	22-Ω resistor
C1	22-μF electrolytic capacitor
S1	DPDT switch
Misc.	Double heterostructure laser diode (such as Sharp LT020), heatsink

Fig. 46-9

PULSED DOUBLE-HETEROSTRUCTURE LASER DRIVER

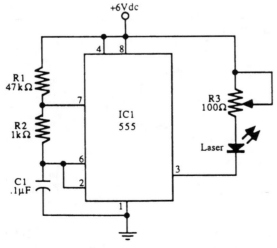

IC1	555 timer IC
R1	47-kΩ resistor
R2	1-kΩ resistor
R3	100-kΩ potentiometer
C1	0.1-μF disc capacitor
Misc.	Double heterostructure laser diode, heatsink

All resistors are 5 to 10% tolerance, ¼ W. All capacitors are 10 to 20% tolerance, rated 35 V or more.

Fig. 46-10

47

Light-Controlled Circuits

The sources of the following circuits are contained in the Sources section, which begins on page 707. The figure number in the box of each circuit correlates to the entry in the Sources section.

Traffic Light-Sequencer Circuit
Tachometer Adapter
Sun-Tracking Circuit for Solar Arrays
Optical Fringe Counter
Low-Noise Light Sensor with dc Servo
Photodiode Amplifier
Light-Switched LED Blinker
Single-Supply Photodiode Amplifier
Light-Controlled Monostable
Darkness Monitor
Programmable Light-Activated Relay
Traffic Light Controller
Colorimeter
Eight Decade Light Meter
LED Lightwave Communications Transmitter
LED Lightwave Receiver
Solar Power Supply
Solar Power Supply with Linear Regulator
Photodiode Log Converter/Transmitter
Rechargeable Solar Power for Sun Tracker

TRAFFIC LIGHT-SEQUENCER CIRCUIT

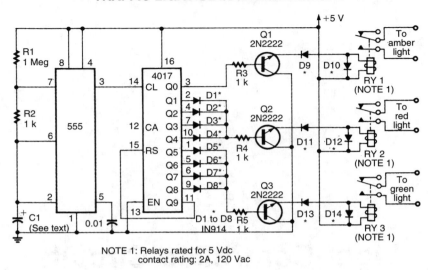

ELECTRONICS NOW

NOTE 1: Relays rated for 5 Vdc
contact rating: 2A, 120 Vac

Fig. 47-1

This circuit uses a 555 timer to drive a 4017 counter. The counter outputs drive transistor relay drivers. Time lights "on" can be proportioned by changing connections of outputs of counter.

TACHOMETER ADAPTER

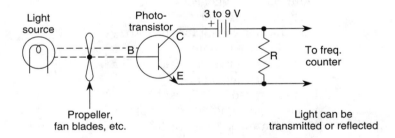

WILLIAM SHEETS

Fig. 47-2

Use of a phototransistor and light source can enable a frequency counter to act as a tachometer:

$$RPM = 60 \times \frac{Frequency\ Counter\ Reading\ Hz}{\#\ Blades\ or\ Spokes}$$

The light source is interrupted by the number of propeller blades, fan blades, spokes, or other marking. R can be anywhere from 1 to 100 kΩ. Try several values for best results.

SUN-TRACKING CIRCUIT FOR SOLAR ARRAYS

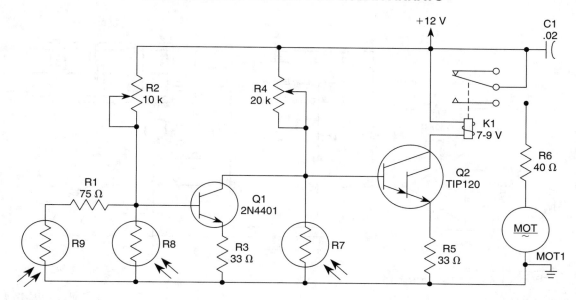

POPULAR ELECTRONICS

Fig. 47-3

The sun tracker uses a combination of three photoresistors. R7, R8, and R9, to ensure that the circuit will follow the sun during the day, but not look for it at night. Photoresistor cells, R7, R8, and R9 have a value of 160 Ω in full sunlight and 4880 Ω in the shade, that is not absolutely critical. R7 is mounted in a "well" with a narrow slit so that sunlight falls upon it only when the photoresistor is pointed directly at the sun. When that occurs, the resistance of R7 drops. That photoresistor and potentiometer R4 form a voltage divider at the base of the Darlington transistor, Q1. When R7's resistance is low, Q1 will be kept off.

When the sun swings a little westward, R7 will no longer be in sunlight, causing its resistance to go up, which raises the base voltage of Q1 and turns that Darlington on. That, in turn closes the relay, K1, providing current to the drive motor, MOT1, which is a 1.5-Vdc, low-torque hobby motor. The motor then turns slowly (resistor R6 limits the maximum current to the motor and keeps it from running too fast), putting R7 in direct sunlight again; Q1's base voltage then drops and the tracker stops. That is repeated again and again as the sun moves across the sky. Photoresistor R8 is mounted on the outside of the well so that it receives a wide angle of full sunlight. When the sun is shining, R8's resistance is low, keeping Q2 turned off, and allowing the tracker to act as described, without interference. But if the sun "slips" behind a cloud, R8's resistance goes high, producing a forward bias on the base of Q2. That turns that transistor on and sinks the base of Q1 to near ground so that Q1 then remains off. That immobilizes the tracker drive; that also keeps the drive shut down in the dark of night.

Photoresistor, R9, is the dawn sensor. It is mounted on the back of the sun tracker. When the tracker stops at sunset, pointing toward the west, R9 is pointing toward the east. When the sun rises the following morning and shines on R9, its resistance goes low, turning Q2 off and allowing Q1's base to go high. That presents current to the relay and therefore to the drive motor, causing the tracker to swing around to the east.

OPTICAL FRINGE COUNTER

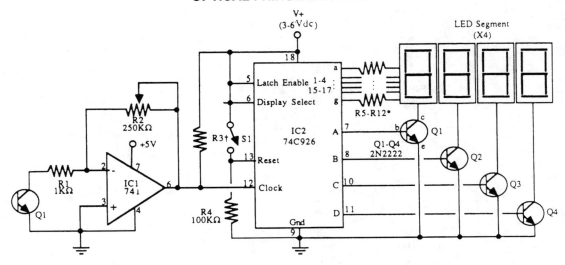

* Not required if +V is 4V or less
† Optional, 10KΩ to 10MΩ, for sensitivity

McGRAW-HILL

Fig. 47-4

For work with interferometer and optical experiments, this fringe counter can be useful. Photo transistor Q1 provides light and dark sensing. As the sensor is moved across the fringe pattern alternate light and dark areas translate to an electrical waveform. This is amplified by IC1 and counted by IC2. A Schmitt trigger circuit can be added, if desired.

LOW-NOISE LIGHT SENSOR WITH dc SERVO

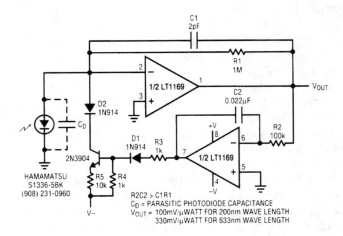

$R2C2 > C1R1$
C_D = PARASITIC PHOTODIODE CAPACITANCE
V_{OUT} = 100mV/µWATT FOR 200nm WAVE LENGTH
330mV/µWATT FOR 633nm WAVE LENGTH

LINEAR TECHNOLOGY

Fig. 47-5

PHOTODIODE AMPLIFIER

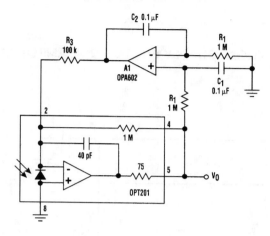

ELECTRONIC DESIGN

Fig. 47-6

A photodiode amplifier combined with a dc-restoration circuit will reject low-frequency ambient background light, easing measurement of a light signal.

LIGHT-SWITCHED LED BLINKER

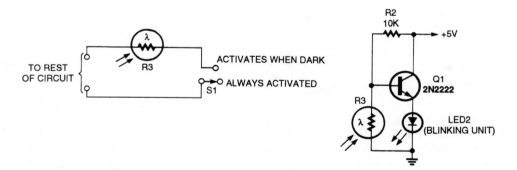

POPULAR ELECTRONICS

Fig. 47-7

This circuit can be used to flash an LED during periods of darkness. Use it for burglar alarm simulators for boats, docks, autos, etc.

SINGLE-SUPPLY PHOTODIODE AMPLIFIER

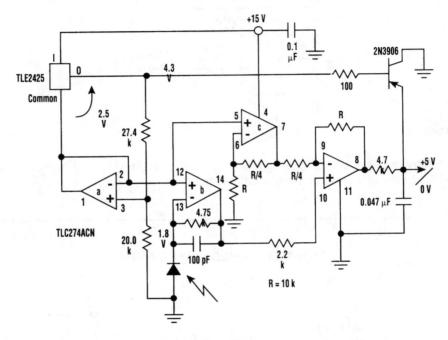

Fig. 47-8

This circuit provides a reverse-bias operating point and output voltage offset and uses a single-polarity power supply. The floating reference voltage from TLE2425 serves to bias the diode in a reverse-polarity mode. It also provides a clamping level at the output. Consequently, linear response to illumination is maintained for a 5-V range from dark current to full sunlight conditions.

LIGHT-CONTROLLED MONOSTABLE

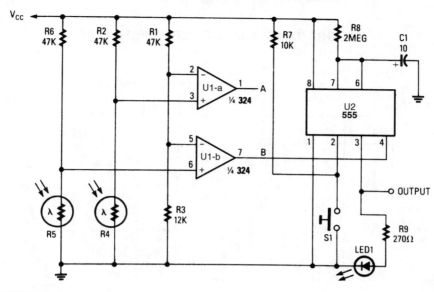

POPULAR ELECTRONICS

Fig. 47-9

The light-controlled monostable was produced by combining a 555 monostable multivibrator with a pair of light-controlled comparators. The circuit can be used to enable the operation of the load device, depending on the time of day. During the daylight hours, the timer U2, is disabled, and so produces no output. However, during the nighttime hours, U2 is enabled by the output of U1-b so that pressing S1 initiates a timing cycle, which activates LED1 for a time determined by R8 and C1.

DARKNESS MONITOR

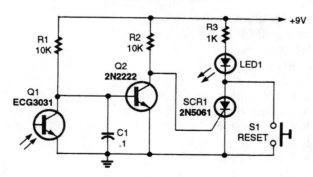

POPULAR ELECTRONICS

Fig. 47-10

When light strikes detector Q1, Q2 is cut off, allowing bias to reach SCR1, triggering SCR1 and lighting LED1. S1 resets the circuit.

PROGRAMMABLE LIGHT-ACTIVATED RELAY

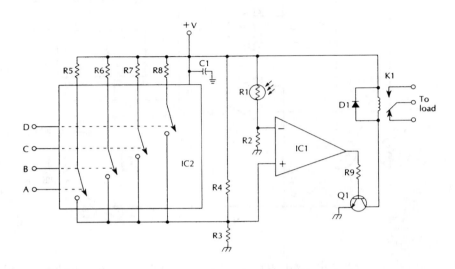

McGRAW-HILL

Fig. 47-11

Digital inputs A, B, C, D select different light levels by varying the value of bridge circuit resistance R_4.

IC1	741 op amp
IC2	CD4066 quad bilateral switch
Q1	NPN transistor (2N2222, 2N3904, or similar)
D1	diode (1N4002, or similar)
C1	0.1-μF capacitor
R1	photoresistor
R2, R3	390-kΩ, ¼-W 5% resistor
R4, R5	1-MΩ, ¼-W 5% resistor
R6	820-kΩ, ¼-W 5% resistor
R7	470-kΩ, ¼-W 5% resistor
R8	270-kΩ, ¼-W 5% resistor
R9	100-kΩ, ¼-W 5% resistor
K1	relay to suit load

TRAFFIC LIGHT CONTROLLER

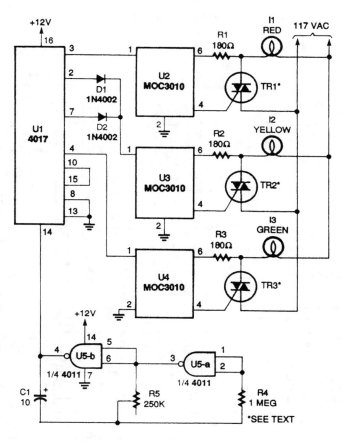

POPULAR ELECTRONICS

Fig. 47-12

Oscillator U5A-B drives a 4017 divide-by-10 counter. The first output of U1 appears at pin 3, which supplies a positive voltage to U2, a MOC 3010 optocoupler/triac-driver IC, turning it and triac TR1 on. That lights I1, the red lamp. The second output appears at pin 2 and passes through D1 to the second MOC 3010, U3, thereby lighting the yellow lamp, I2. The third output at pin 4 turns on UJ4 and the green lamp, I3. The fourth output at pin 7 travels through D2 and into U3 to light the yellow lamp, I2, again.

If you would like the traffic-light system to follow the normal sequence of green, yellow, and red, make the following circuit changes: Disconnect pins 10 and 15 of U1 from each other. Remove D1 and D2 and connect pin 2 of U1 to pin 1 of U3. Then connect pins 7 and pins 15 of U1 together. Use U2 to drive I3 (the green light) and U4 for I1 (the red light).

COLORIMETER

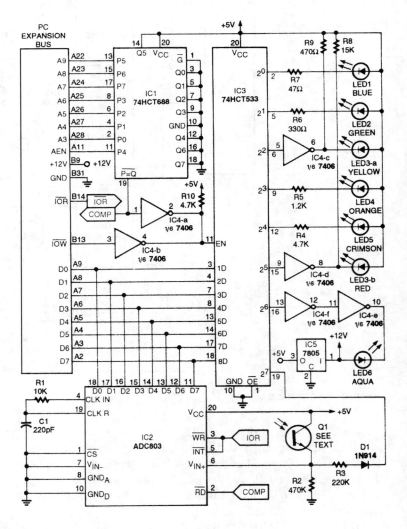

TABLE 1—LED COLORS AND CODES

LED	Wavelength	Color	Activation Value
LED1	470 nm	Blue	$2^0=1$
LED2	560 nm	Green	$2^1=2$
LED3-a	590 nm	Yellow	$2^2=4$
LED3-b	700 nm	Red	$2^3=8$
LED4	630 nm	Orange	$2^4=16$
LED5	665 nm	Crimson	$2^5=32$
LED6	482 nm	Aqua	$2^6=64$

LISTING 1—CALIBRATION PROGRAM

```
10 'CALIBRAT.BAS calibration program
20 CLS:KEY OFF:N=0:ADR=512:OPEN"R",1,"CAL1",16:OPEN"r",2,"cal2",24
30 FIELD 1,2AS B$,2AS G$,2AS Y$,2AS O$,2AS C$,2AS R$,2AS A$,2AS AG$
40 FIELD 2,24AS ID$
50 PRINT "reference number",N+1:OUT ADR,255:BEEP:INPUT "Enter Name of Standard
or 'E' To End";TEMPID$
60 IF TEMPID$="E" OR TEMPID$="e" THEN N=0: GOTO 200
70 IF TEMPID$="n" THEN INPUT"enter n to redo ",N:N=N-1:GOTO 50
80 N=N+1:FOR H=0 TO 7:K=0:IF H<7 THEN Z=2^H ELSE Z=194
90 OUT ADR,Z:FOR I=1 TO 500:NEXT I
100 FOR J=1 TO 50:K=K+INP(ADR):NEXT J
110 IF H=0 THEN LSET B$=MKI$(K)
120 IF H=1 THEN LSET G$=MKI$(K)
130 IF H=2 THEN LSET Y$=MKI$(K)
140 IF H=3 THEN LSET O$=MKI$(K)
150 IF H=4 THEN LSET C$=MKI$(K)
160 IF H=5 THEN LSET R$=MKI$(K)
170 IF H=6 THEN LSET A$=MKI$(K)
180 IF H=7 THEN LSET AG$=MKI$(K)
190 NEXT H:LSET ID$=TEMPID$:PUT 1,N:PUT 2,N:CLS:GOTO 50
200 N=N+1:GET #1,N:GET #2,N:IF N>(LOF(1)/16) THEN END
210 B=CVI(B$):G=CVI(G$):Y=CVI(Y$):O=CVI(O$):C=CVI(C$):R=CVI(R$):A=CVI(A$):
AG=CVI(AG$)
220 PRINT N,ID$:GOTO 200
```

LISTING 2—IDENTIFICATION PROGRAM

```
1 'IDENTIFY.BAS  identification program
10 ADR=512:OUT ADR,255:PRINT:INPUT "Hit Enter To Scan/Identify Unknown
Color";A
20 IF A=9 THEN RUN"fcal"
30 ERP=1E+20:OPEN"R",1,"cal1",16
40 FOR H=0 TO 7:K=0:IF H<7 THEN Z=2^H ELSE Z=194
50 OUT ADR,Z: FOR I=1 TO 500:NEXT I
60 FOR J=1 TO 50:K=K+INP(ADR):NEXT J
70 IF H=0 THEN BU=K ELSE IF H=1 THEN GU=K ELSE IF H=2 THEN YU=K
80 IF H=3 THEN OU=K ELSE IF H=4 THEN CU=K ELSE IF H=5 THEN RU=K
90 IF H=6 THEN AU=K ELSE IF H=7 THEN AGU=K
100 NEXT H:BEEP
110 OUT 512,255:OPEN"r",2,"cal2",24:FIELD 1,2AS B$,2AS G$,2AS Y$,2AS O$,2AS
C$,2    AS R$,2AS A$,2AS AG$:B=LOF(1)/16
120 FOR N=1 TO B:GET #1,N:IF ABS( CVI(B$)-BU)>400 THEN 140
130 ER=(CVI(B$)-BU)^2+(CVI(G$)-GU)^2+(CVI(Y$)-YU)^2+(CVI(O$)-OU)^2+(CVI(C$)-
CU)^    2+(CVI(R$)-RU)^2+1*((CVI(A$)-AU)^2)+2*((CVI(AG$)-AGU)^2):IF ER<ERP
THEN ERP=      ER:NN=N
140 NEXT N
150 FIELD 2, 24AS ID$: GET #2,NN
160 CLS:PRINT "Best Color Match",ID$:PRINT"Relative Error",ERP:PRINT"reference
number",NN:RUN
```

A hardware/software combination activates. In turn, one of several LEDs emits a portion of the visible spectrum. A phototransistor measures the light reflected by the surface being measured, and an 8-bit analog-to-digital converter (ADC) translates the phototransistor's output into a digital format that the computer can interpret. Seven LEDs (blue, aqua, green, yellow, orange, crimson, and red) provide a range of readings across the visible spectrum. Lack of spectral continuity among adjacent LED colors could skew results, so the circuit provides built-in compensation for this error.

Two simple BASIC programs control the circuit's operation. One allows you to define a set of standards by measuring known color samples and recording the values with an associated name. The other program measures unknown samples and provides the best match with the defined standards, as well as a relative error factor.

EIGHT-DECADE LIGHT METER

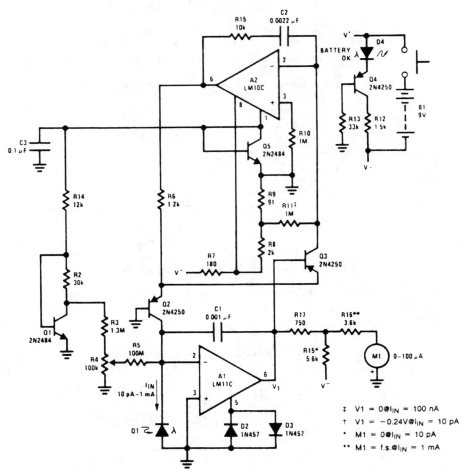

NATIONAL SEMICONDUCTOR

Fig. 47-14

A logarithmic amplifier is adapted to a battery-powered light meter. An LM10, combined op amp and reference, is used for the second amplifier and to provide the regulated voltage for offsetting the logging circuit and powering the bias-current compensation. This can provide input current resolution of better than ±2 pA over 15 to 55° C. Because a meter is the output indicator, there is no need to optimize frequency compensation. Low-cost single transistors are used for logging because the temperature range is limited. The meter is protected from overloads by clamp diodes D2 and D3.

Silicon photodiodes are more sensitive to infrared than visible light, so an appropriate filter must be used for photography. Alternately, gallium-arsenide-phosphide diodes with suppressed IR response are becoming available.

LED LIGHTWAVE COMMUNICATIONS TRANSMITTER

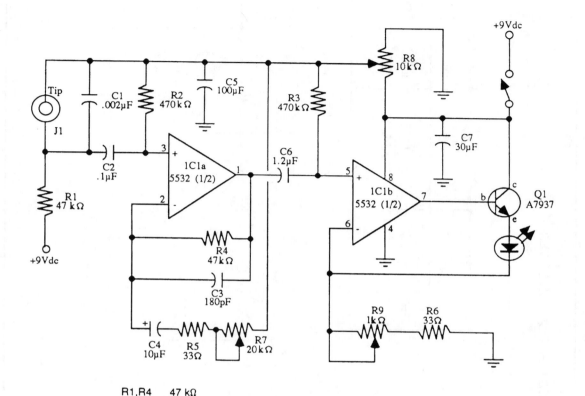

R1,R4   47 kΩ
R2,R3   470 kΩ
R5,R6   33 Ω
R7,R9   1 kΩ potentiometer
R8   10 kΩ potentiometer
C1   0.002 μF disc
C2   0.1 μF disc
C3   180 pF disc
C4   10 μF polarized electrolytic
C5   100 μF polarized electrolytic
C6   1.2 μF polarized electrolytic
C7   30 μF polarized electrolytic
IC1   5532 low-noise amplifier IC
Q1   A7937 transistor
LED1   High-output LED (see text)
J1   Miniature phone jack (for electret condenser microphone)
S1   SPST switch

All resistors are 5 to 10 percent tolerance, 1/4 watt. All capacitors are 10 to 20 percent tolerance, rated at 35 volts or more.

LED LIGHTWAVE RECEIVER

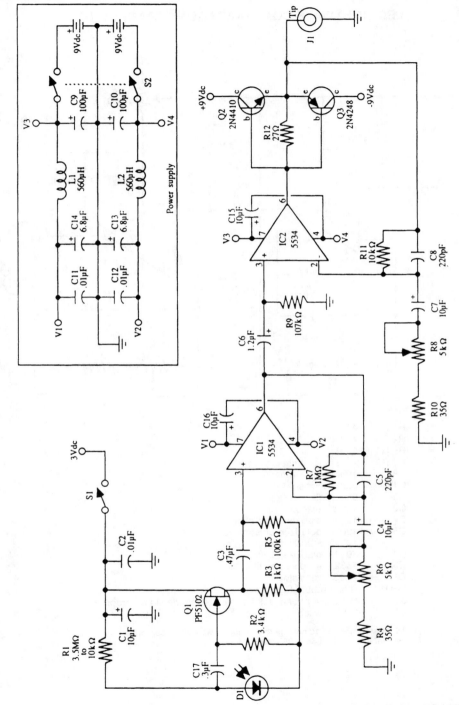

Fig. 47-16

McGRAW-HILL

310

SOLAR POWER SUPPLY

ELECTRONICS HOBBYIST HANDBOOK

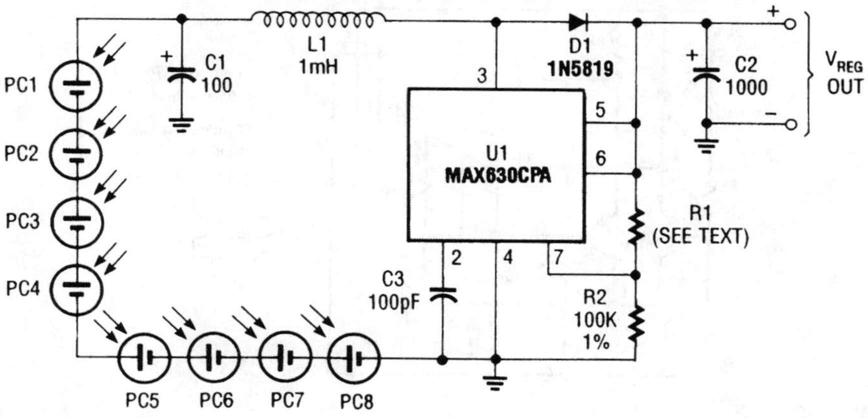

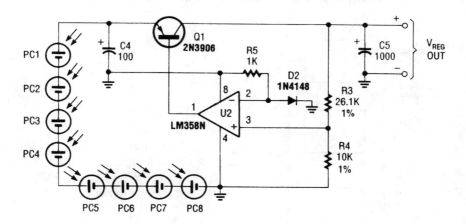

Fig. 47-17

This circuit delivers either 4.8 or 7.2 V regulated at 15 mA with a 3-V input from a bank of photocells. R1 should be 453 kΩ for a 7.2-V output and 274 kΩ for a 4.8-Vdc output. Regulator efficiency is around 70%. This should be considered when selecting suitable solar cells.

SOLAR POWER SUPPLY WITH LINEAR REGULATOR

ELECTRONICS HOBBYIST HANDBOOK

Fig. 47-18

This regulator delivers a constant 2.4 Vdc for powering small devices that run on two AA cells, such as cassettes and small radios. Regulator drop is about 0.3 volt. This should be considered when choosing solar cells. Load current is typically 125 mA.

PHOTODIODE LOG CONVERTER/TRANSMITTER

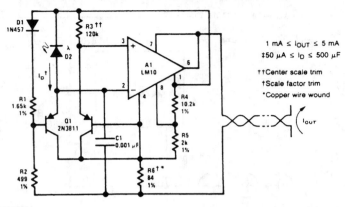

1 mA ≤ I_{OUT} ≤ 5 mA

‡50 μA ≤ I_D ≤ 500 μF

††Center scale trim

†Scale factor trim

*Copper wire wound

NATIONAL SEMICONDUCTOR

Fig. 47-19

A logarithmic conversion is made on the output current of a photodiode to compress a four-decade, light-intensity variation into a standard transmission range. The circuit is balanced at mid-range, where R3 should be chosen so that the current through it equals the photodiode current. The log-conversion slope is temperature compensated with R6. Setting the reference output to 1.22 V gives a current through R2 that is proportional to absolute temperature because of D1 so that this level-shift voltage matches the temperature coefficient of R6. C1 has been added so that large-area photodiodes with high capacitance do not cause frequency instabilities.

RECHARGEABLE SOLAR POWER FOR SUN TRACKER

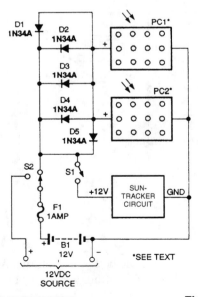

This application circuit provides rechargeable solar power for the sun tracker, as well as for another 12-volt device. PC1 and PC2 can be mounted on sun tracker assembly.

POPULAR ELECTRONICS

Fig. 47-20

48

Logic Circuits

The sources of the following circuits are contained in the Sources section, which begins on page 707. The figure number in the box of each circuit correlates to the entry in the Sources section.

Logic State Change Indicator
Combinatorial Logic Multiplexer
AND Gate
Relay "AND" Circuit
Relay "OR" Circuit

LOGIC STATE CHANGE INDICATOR

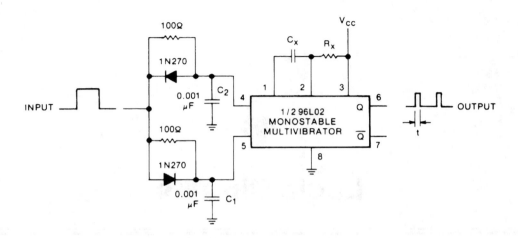

NASA TECH BRIEFS   *Fig. 48-1*

A circuit consisting of a one-shot multivibrator IC, a pair of diodes, and some resistors and capacitors delivers an output pulse when the logic state at its input terminal changes—either from high to low or from low to high. Thus, this circuit can serve as a state-change indicator or as a frequency doubler for a square-wave input.

Any monostable can be used; the arrangement in the figure achieves low power dissipation (80 milliwatts) by using half of a Fairchild 96L02 transistor-transistor-logic dual multivibrator. The 96L02 is triggered when pins 3 and 5 are high and pin 4 changes state from low to high. It also triggers if pin 3 is high, pin 4 is low.

The circuit shown here allows these conditions to be satisfied with a single input terminal, plus the fixed bias on pin 3; the arrangement of resistors, capacitors, and diodes automatically biases pin 5 high when transmitting a rising transition to pin 4 and biases pin 4 low when applying a falling transition to pin 5.

For example, if the input terminal has been low and then goes high, C1 charges through a forward-biased diode that shunts its 100-Ω resistor; therefore pin 5 goes high immediately. C2 charges through 100 Ω; however, because its diode is back-biased, the rising level is not applied to pin 4 until 5 is already high. Therefore, the conditions for triggering an output pulse are satisfied.

The output pulse duration, t, is set by the value of time constant $R_x C_x$.

COMBINATORIAL LOGIC MULTIPLEXER

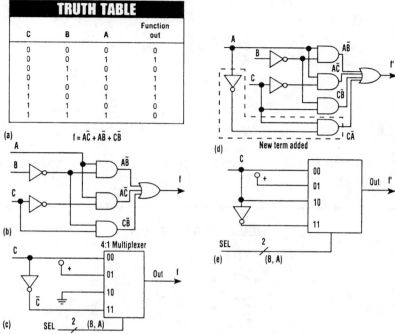

TRUTH TABLE

C	B	A	Function out
0	0	0	0
0	0	1	1
0	1	0	0
0	1	1	1
1	0	0	1
1	0	1	1
1	1	0	0
1	1	1	0

(a) $f = A\bar{C} + A\bar{B} + C\bar{B}$

(b)

(c)

(d)   New term added

(e)   (B, A)

ELECTRONIC DESIGN   *Fig. 48-2*

Combinatorial logic can be implemented simply by using a multiplexer instead of logic gates. Shown are the truth table (A), its logic circuit (B), and the multiplexer connections (C). If the logic circuitry is changed (D), the multiplexer would be reconnected (E).

AND GATE

POPULAR ELECTRONICS   *Fig. 48-3*

An LM139 is configured as an AND gate (TTL or CMOS is usually used). With this idea, you can use leftover IC sections and save an extra package in some instances.

RELAY "AND" CIRCUIT

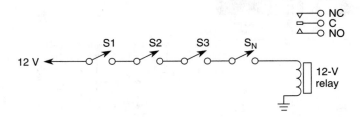

WILLIAM SHEETS

Fig. 48-4

All switches S1 through S_n must be closed to operate the relay. If one opens, the relay drops out. Use this circuit for burglar alarms, etc.

RELAY "OR" CIRCUIT

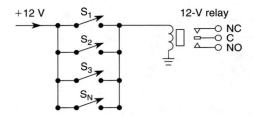

WILLIAM SHEETS

Fig. 48-5

Closing any switch S1, S2, S3, or S_n will actuate the relay (N = any number). Use this circuit for burglar alarms, etc.

49

Mathematical Circuits

The sources of the following circuits are contained in the Sources section, which begins on page 707. The figure number in the box of each circuit correlates to the entry in the Sources section.

Low-Cost One-Quadrant Multiplier/Divider
Low-Cost Accurate Square-Root Circuit
Low-Cost Accurate Squaring Circuit
Bridge Linearizing Function
Square Rooter
Analog Variable Multiplier/Divider
Difference of Squares
Approximation for Sin ϕ
Simple Analog Averaging Circuit
Simple Analog Multiplier
Δ% Ratio Computer

LOW-COST ONE-QUADRANT MULTIPLIER/DIVIDER

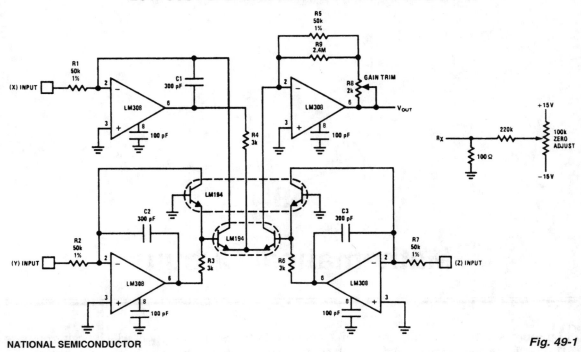

NATIONAL SEMICONDUCTOR

Fig. 49-1

This circuit will produce an output that is proportional to the product of the (X) and (Y) inputs divided by the Z input. All inputs must be positive, limiting operation to one quadrant. For very low level inputs, the offset voltage in the LM308s might create large percentage errors referred to input. A simple scheme for offsetting any of the LM308s to zero is shown in dotted line; the positive input of the appropriate LM308 is simply tied to R_x instead of ground for zeroing. The summing mode of operation on all inputs allows easy scaling on any or all inputs. Simply set the input resistor equal to $(V_{in(max)}/(200\ \mu A)$. V_{out} is equal to:

$$V_{out} = \frac{\left(\dfrac{X}{R_1}\right)\left(\dfrac{Y}{R_2}\right)(R_5)}{\dfrac{Z}{R_7}}$$

Input voltages above the supply voltage are allowed because of the summing mode of operation. Several inputs can be summed at X, Y, and Z.

For a simple (X) • (Y) or (X)/Z function, the unused input must be tied to the reference voltage. Perturbations in this reference will be seen at the output as scale factor changes, so a stable reference is necessary for precision work. For less critical applications, the unused input can be tied to the positive supply, with

$$R = \frac{V+}{200\ \mu A}$$

LOW-COST ACCURATE SQUARE-ROOT CIRCUIT

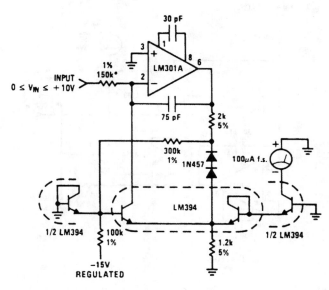

*Trim for full scale accuracy.

NATIONAL SEMICONDUCTOR

Fig. 49-2

The circuit will generate a square-root function, accurately and inexpensively. The output is a current that can be used to drive a meter directly or be converted to a voltage with a summing junction current-to-voltage converter. The −15-V supply is used as a reference, so it must be stable. A 1% change in the −15-V supply will give a ½% shift in output reading. No positive supply is required when an LM301A is used because its inputs can be used at the same voltage as the positive supply (ground). The two 1N457 diodes and the 300-kΩ resistor are used to temperature-compensate the current through the diode-connected ½LM394.

LOW-COST ACCURATE SQUARING CIRCUIT

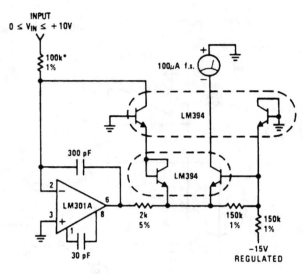

INPUT
$0 \leq V_{IN} \leq +10V$

100k*
1%

100μA f.s.

LM394

300 pF

2

LM301A

6

3

8

2k
5%

150k
1%

150k
1%

LM394

1

30 pF

−15V
REGULATED

*Trim for full scale accuracy.

NATIONAL SEMICONDUCTOR   *Fig. 49-3*

The circuit shown will square the input signal and deliver the result as an output current. Full-scale input is 10 V, but this can be changed simply by changing the value of the 100-kΩ input resistor. As in the square root circuit, the −15-V supply is used as the reference. In this case, however, a 1% shift in supply voltage produces a 1% shift in the output signal. The 150-kΩ resistor across the base-emitter of ½LM394 provides slight temperature compensation of the reference current from the −15-V supply. For improved accuracy at low input signal levels, the offset voltage of the LM301A should be zeroed out, and a 100-kΩ resistor should be inserted in the positive input to provide input to provide optimum dc balance.

BRIDGE LINEARIZING FUNCTION

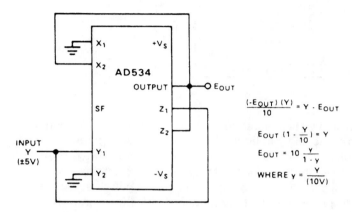

$$\frac{(-E_{OUT})\,(Y)}{10} = Y \cdot E_{OUT}$$

$$E_{OUT}\left(1 - \frac{Y}{10}\right) = Y$$

$$E_{OUT} = 10\,\frac{y}{1-y}$$

$$\text{WHERE } y = \frac{Y}{(10V)}$$

ANALOG DEVICES   *Fig. 49-4*

If one arm of a Wheatstone Bridge varies from its nominal value by a factor, $(1 + 2x)$, the voltage or current output of the bridge will be (with appropriate polarities):

$$y \approx \frac{x}{1 + x}$$

Linear response requires very small x and, usually, preamplification. The circuit shown here enables large-deviation bridges to be used without losing linearity.

The circuit computes the inverse of the bridge function, i.e.,

$$x \approx \frac{y}{1 + y}$$

Depending on which arm of the bridge varies, it might be necessary to reverse the polarity of the z connections.

SQUARE ROOTER

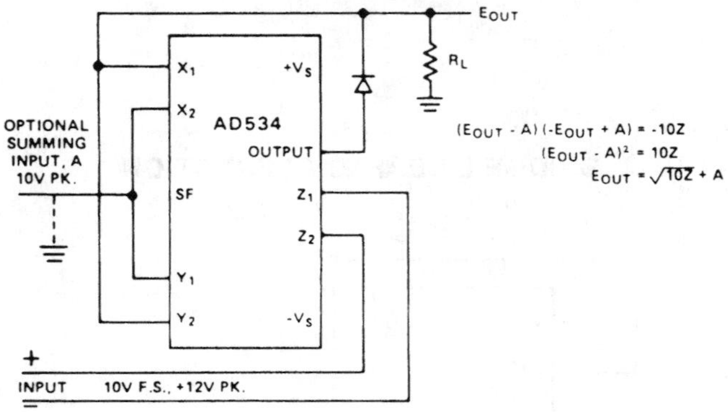

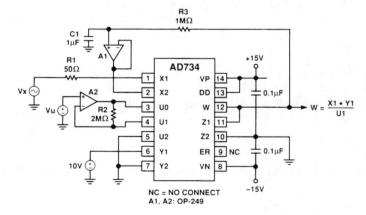

$(E_{OUT} - A)(-E_{OUT} + A) = -10Z$

$(E_{OUT} - A)^2 = 10Z$

$E_{OUT} = \sqrt{10Z} + A$

Fig. 49-5

This illustration shows the connection of the AD534 for square-rooting, with differential inputs. The diode prevents a latching condition—common to this configuration—which would occur if the input momentarily changed polarity. As shown, the output is always positive; it can be changed to a negative output by reversing the diode polarity and interchanging the X inputs. Because the signal input is differential, all combinations of input and output polarities can be realized. If the output circuit does not provide a resistive load to ground, one should be connected to maintain diode conduction. For critical applications, the Z offset can be adjusted for greater accuracy below 1 V.

ANALOG VARIABLE MULTIPLIER/DIVIDER

$W = \dfrac{X_1 \cdot Y_1}{U_1}$

Fig. 49-6

An output voltage ($W = X_1 \times Y_1/U_1$) is produced by this multiplier circuit. The AD734 is a four-quadrant multiplier.

DIFFERENCE OF SQUARES

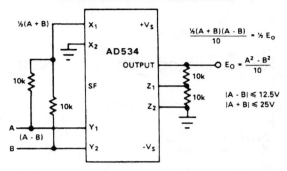

ANALOG DEVICES

Fig. 49-7

A single AD534 can be used to compute the difference of the squares of two input signals. The function can be useful in vector computations, and in weighting the difference of two magnitudes to emphasize the greater nonlinearity.

APPROXIMATION FOR SIN φ

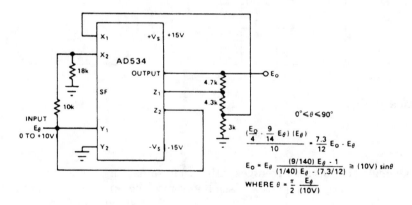

$$0° < \theta < 90°$$

$$\frac{(\frac{E_O}{4} - \frac{9}{14} E_\theta)(E_\theta)}{10} = \frac{7.3}{12} E_O - E_\theta$$

$$E_O = E_\theta \frac{(9/140) E_\theta - 1}{(1/40) E_\theta - (7.3/12)} \cong (10V) \sin\theta$$

$$\text{WHERE } \theta = \frac{\pi}{2} \frac{E_\theta}{(10V)}$$

ANALOG DEVICES

Fig. 49-8

The AD534 is remarkably easy to use in the implementation of the approximation formulas described in Chapter 2-1 of the *Nonlinear Circuits Handbook*. Many of these involve implicit loops to generate the function and previously required several additional op amps for the addition and subtraction of the various terms. This circuit is an example of what can be done with external resistors only. For φ between 0° and 90°, the approximation maintains a theoretical accuracy to within 0.5% of full-scale; 0.75% is practical with AD534L and 0.1% resistances were used.

SIMPLE ANALOG AVERAGING CIRCUIT

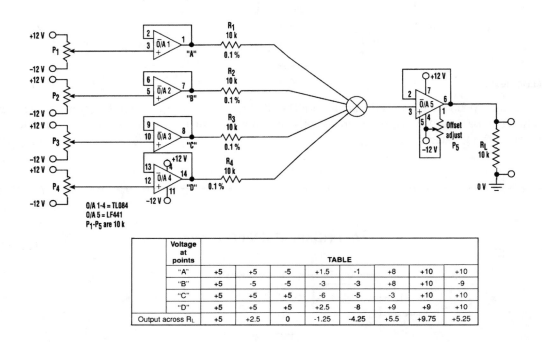

Voltage at points				TABLE				
"A"	+5	+5	-5	+1.5	-1	+8	+10	+10
"B"	+5	-5	-5	-3	-3	+8	+10	-9
"C"	+5	+5	+5	-6	-5	-3	+10	+10
"D"	+5	+5	+5	+2.5	-8	+9	+9	+10
Output across R$_L$	+5	+2.5	0	-1.25	-4.25	+5.5	+9.75	+5.25

ELECTRONIC DESIGN

Fig. 49-9

At times, an analog circuit that averages rather than sums can be quite handy. You won't usually find this type of circuit in op-amp books, possibly because the op amp is used only as a buffer. For best accuracy, an FET should be used with the offset adjusted out. In addition, the "averaging" resistors (R1 through R4) should be of close tolerance.

Looking at the test circuit, op amps 1 through 4 are used to alleviate interaction between adjustment potentiometers P1 through P4 and so that R1 through R4 see the same low impedance.

The table shows some arbitrarily set voltages and the resulting output voltage across R_L.

SIMPLE ANALOG MULTIPLIER

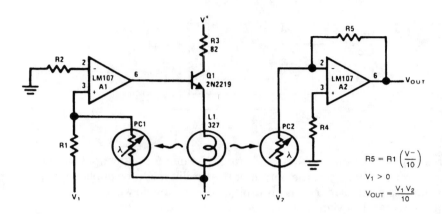

$$R5 = R1 \left(\frac{V-}{10}\right)$$

$$V_1 > 0$$

$$V_{OUT} = \frac{V_1 V_2}{10}$$

NATIONAL SEMICONDUCTOR

Fig. 49-10

Circuit operation can be understood by considering A2 as a controlled-gain amplifier, amplifying V_2, whose gain depends on the ratio of the resistance of PC2 to R5 and by considering A1 as a control amplifier, which establishes the resistance of PC2 as a function of V1. In this way, V_{OUT} is a function of both V_1 and V_2.

A1, the control amplifier, provides drive for the lamp, L1. When an input voltage, V_1, is present, L1 is driven by A1 until the current to the summing junction from the negative supply through PC1 is equal to the current to the summing junction from V1 through R1. Because the negative supply voltage is fixed, this forces the resistance of PC1 to a value that is proportional to R1 and to the ratio of V_1 to V–. L1 also illuminates PC2 and, if the photoconductors are matched, causes PC2 to have a resistance equal to PC1.

A2, the controlled gain amplifier, acts as an inverting amplifier whose gain is equal to the ratio of the resistance of PC2 to R5. If R5 is chosen equal to the product of R_1 and V–, then V_{OUT} becomes simply the product of V_1 and V_2. R5 can be scaled in powers of 10 to provide any required output scale factor.

Δ% RATIO COMPUTER

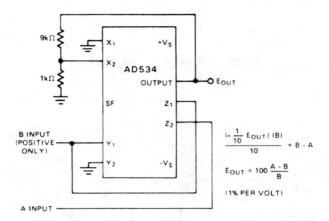

The equations shown in the figure:

$$\frac{(-\frac{1}{10} E_{OUT})(B)}{10} = B - A$$

$$E_{OUT} = 100 \frac{A - B}{B}$$

(1% PER VOLT)

ANALOG DEVICES

Fig. 49-11

The percentage-deviation function is of practical value for many applications in measurement, testing, and control. For example, the output of this circuit might be applied to a pair of biased comparators to stimulate particular actions or displays, depending on whether the gain of a circuit under test were within limits, or deviating by a preset amount in either direction.

The indicated scale factor, 1%/V, is convenient. However, other sensitivities, from 10%/V to 0.1%/V, as required by the application, can be obtained by altering the feedback attenuation ratio, from 1 to $\frac{1}{100}$. Gain or attenuation is easily applied to the A signal externally for calibration to the normalized form.

50

Measuring and Test Circuits

The sources of the following circuits are contained in the Sources section, which begins on page 707. The figure number in the box of each circuit correlates to the entry in the Sources section.

Electronic Level
Single-Chip Digital Voltmeter
Inductance with DVM Measuring Circuit
Negative Reference Voltage Circuit
Precision Current Source
1-kW Power Meter
Logic Chip Tester
Power Supply for 10-MHz Frequency Standard
Three-Terminal Regulator Current Source
Four-Wire Resistance Measurement Hookup
Audio Frequency Meter
ELF Monitor
Strain-Gage Sensor
Minute Marker
Digital Barometer
Reference Circuit
Transistor Matching Circuit
Auto-Ranging Digital Capacitance Meter
Frequency Divider for 10-MHz Frequency Standard
Electroscope
Optical Isolator Wattmeter
Digital Three-Phase Wave Generator
Simple Test Audio Amplifier
Gate Dip Oscillator I
Accelerometer (G Meter) Circuit

Gate Dip Oscillator II
Two Remote Meters
Novel RF Power Meter
Nanoammeter
1.5-V Logarithmic Light Level Meter
ac Power Monitor
100-W Variable Resistor Simulator
IMD Test Circuit for Pin Diodes
VCO and Input Frequency Comparer
ECG Amplifier with Right Leg Drive
Power Transformer Tester
4- to 20-mA Process Controller
Simple High-Current Measurer
Analog Circuitry Calibrator
Simple Signal Generator for Signal Tracing
Simple Harmonic Distortion Analyzer
Sound Subcarrier Generator
Inductance and Capacitance Determiner
  with SWR Bridge
Motorcycle Tune-Up Aid
50-MHz Frequency Counter
10-MHz Frequency Standard
Programmable Capacitor Circuit
Programmable Resistor Circuit

ELECTRONIC LEVEL

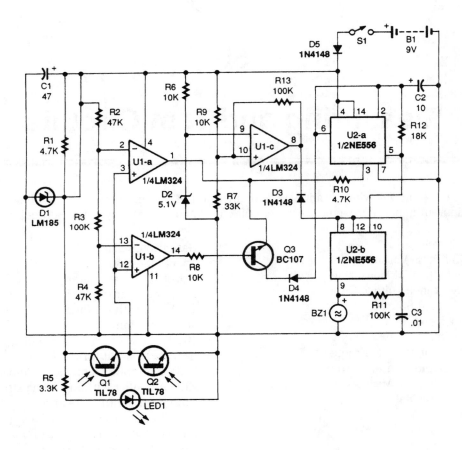

POPULAR ELECTRONICS

Fig. 50-1

The electronic level uses a pair of phototransistors and an infrared LED to sense bubble position. In this circuit, the amounts of infrared radiation received by phototransistors Q1 and Q2 are translated by op-amp U1 and dual-timer U2 into either a steady tone, or a fast- or slow-pulsing one.

SINGLE-CHIP DIGITAL VOLTMETER

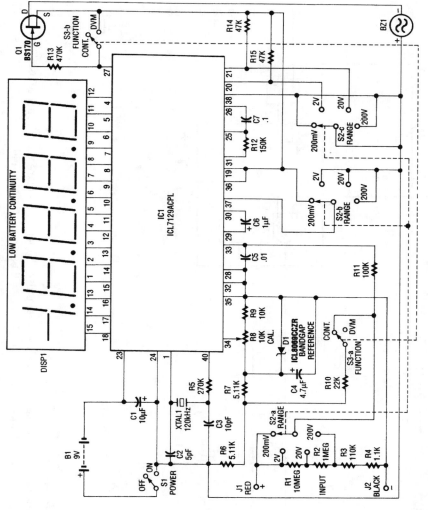

Fig. 50-2

ELECTRONIC EXPERIMENTERS HANDBOOK

This 4½-digit DVM circuit is built around a Maxim ICL7129ACPL A/D converter and LCD driver. An ICL8069 CCZR 1.2-V band-gap reference diode is used for a voltage reference. S2a-b-c select one of four ranges up to 200 V (maximum). The meter also has a piezoelectric buzzer for continuity testing. S3 selects either DVM or continuity. Crystal 1 can be changed to 100 kHz if maximum rejection of 50 Hz is desired. The crystal normally provides 120 kHz for best 60-Hz rejection. This is caused by the dual-slope conversion technique used in IC1.

329

INDUCTANCE WITH DVM MEASURING CIRCUIT

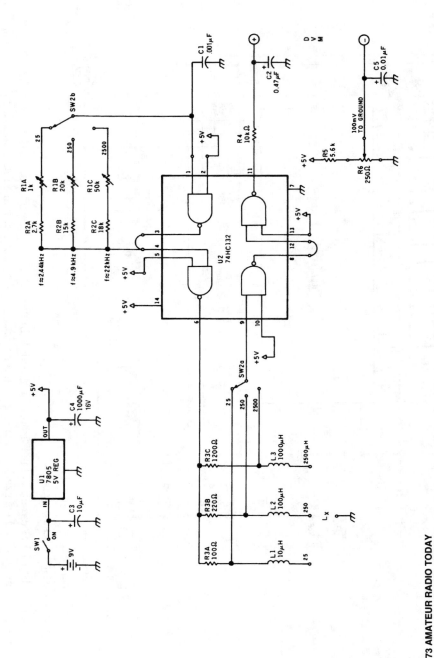

Fig. 50-3

73 AMATEUR RADIO TODAY

The inductance meter comprises an oscillator and pulse shaper. The square wave from the oscillator is differentiated and the differentiator output is shaped and read on a DVM. This will be proportional to inductance. R1A, B, C set calibration and R6 is an offset control.

NEGATIVE REFERENCE VOLTAGE CIRCUIT

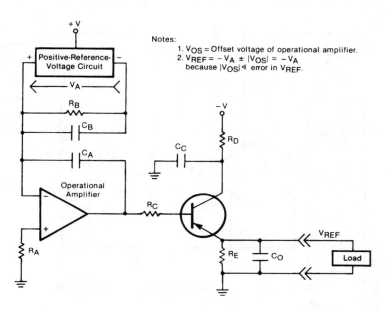

Notes:
1. V_{OS} = Offset voltage of operational amplifier.
2. $V_{REF} = -V_A \pm |V_{OS}| \approx -V_A$
   because $|V_{OS}| \lll$ error in V_{REF}.

Fig. 50-4

This figure illustrates a circuit that supplies a precise negative reference voltage. To meet requirements of accuracy and stability, it incorporates a highly precise positive reference voltage by use of a high-gain, stable feedback booster circuit.

The booster circuit includes an operational amplifier and a transistor, which handles the load current. Typically, a positive-reference-voltage circuit can handle only relatively small load currents. This consideration does not apply in the present circuit because the positive-reference-voltage unit is placed in the voltage feedback loop of the booster circuit in parallel with resistor R_B. Thus, from the perspective of the positive-reference-voltage unit, R_B is a constant load. This feature enhances the stability of the circuit by removing the load regulation factor.

Provided that the offset voltage of the operational amplifier is low, the accuracy of the overall circuit depends only on the accuracy of the positive-reference-voltage unit. The overall circuit draws very little power for its own operation. It can handle unexpectedly heavy loads; the feedback configuration and the high gain provided by the combination of the operational amplifier and the transistor give the circuit a very low output impedance. The capacitors reduce the noise voltage and help stabilize the circuit. In the event that the load becomes a short circuit, R_D protects the transistor by limiting the load current.

PRECISION CURRENT SOURCE

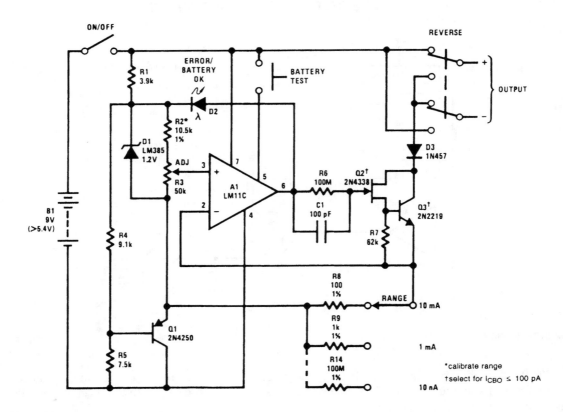

NATIONAL SEMICONDUCTOR

Fig. 50-5

A complete, battery-powered current source suitable for laboratory use is shown. The op amp regulates the voltage across the range resistors at a level determined by the voltage on the arm of the calibrated potentiometer, R3. The voltage on the range resistors is established by the current through Q2 and Q3, which is delivered to the output.

The reference diode, D1, determines basic accuracy. Q1 is included to ensure that the LM11 inputs are kept within the common-mode range with diminishing battery voltage. A light-emitting diode, D2, is used to indicate output saturation. However, this indication cannot be relied upon for output-current settings below about 20 nA, unless the value of R6 is increased. The reason is that very low currents can be supplied to the range resistors through R6 without developing enough voltage drop to turn on the diode.

If the LED illuminates with the output open, there is sufficient battery voltage to operate the circuit. But a battery test switch is also provided. It is connected to the base of the op-amp output stage and forces the output toward V+.

1-kW POWER METER

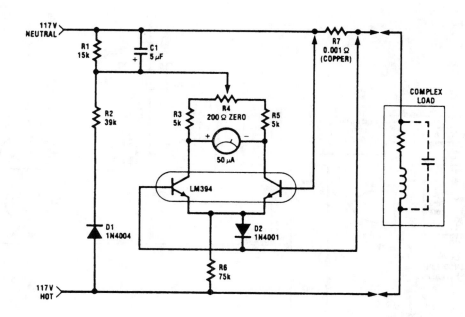

NATIONAL SEMICONDUCTOR

Fig. 50-6

The power meter shown uses only one transistor pair to provide the complete (X) (Y) function. The circuit is intended for 117 Vac ± 50 Vac operation, but can be easily modified for higher or lower voltages. It measures true (nonreactive) power being delivered to the load and requires no external power supply. Idling power drain is only 0.5 W. The load current-sensing voltage is only 10 mV, keeping load voltage loss to 0.01%. Rejection of reactive load currents is better than 100:1 for linear loads. Nonlinearity is about 1% full scale when using a 50-μA meter movement. The temperature correction for gain is accomplished by using a copper shunt (+0.32%/°C) for load-current sensing. This circuit measures power on negative cycles only, so it cannot be used on rectifying loads.

LOGIC CHIP TESTER

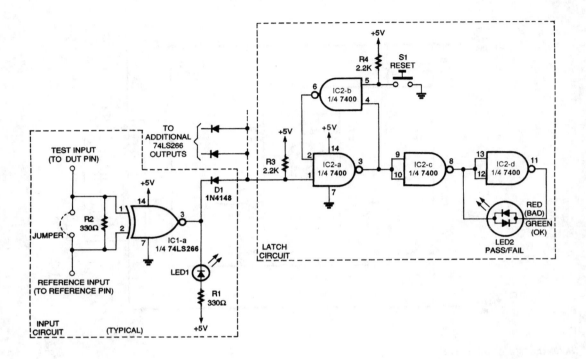

ELECTRONICS NOW

Fig. 50-7

This device compares two logic signals and indicates whether the two logic signals are the same or different. To use the tester, first connect the test input to the suspect pin of the DUT. Next, connect the reference input to the same pin of an identical reference chip that is known to be good. Push the reset button to begin the test; the green section of the bicolor LED will be illuminated. Any signal on the test device that differs from the one on the reference device will then momentarily light the LED lamp that corresponds to that pin, and also latch on the red section of the bi-color LED. That indicates that the device under test is faulty. If the reference and DUT signals are the same, the DUT is OK, and the green LED will remain lit.

POWER SUPPLY FOR 10-MHz FREQUENCY STANDARD

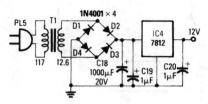

RADIO-ELECTRONICS   **Fig. 50-8**

This simple power supply can be used in place of battery B1 of the 10-MHz frequency standard.

THREE-TERMINAL REGULATOR CURRENT SOURCE

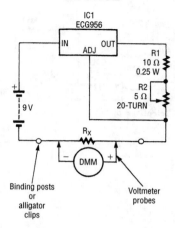

ELECTRONICS NOW   **Fig. 50-9**

A three-terminal voltage regulator acts as a current source in this circuit. A resistor is being calibrated using a DMM and the current source.

FOUR-WIRE RESISTANCE MEASUREMENT HOOKUP

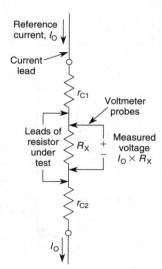

ELECTRONICS NOW   **Fig. 50-10**

A true four-wire resistance measurement hookup.

AUDIO FREQUENCY METER

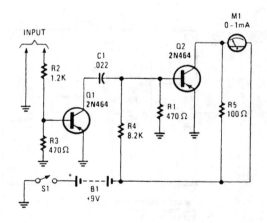

POPULAR ELECTRONICS   **Fig. 50-11**

A pulse-shaper is used in a tachometer circuit to drive a meter.

ELF MONITOR

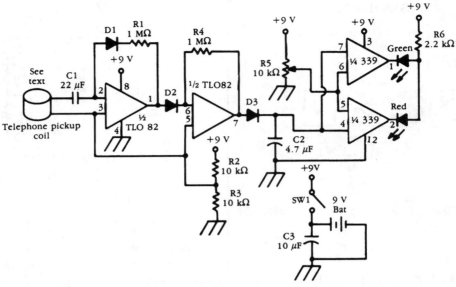

McGRAW-HILL

Fig. 50-12

A telephone pick-up coil is used as a sensor for low-frequency magnetic fields. The signal is amplified and detected, then used to drive a comparator.

STRAIN-GAGE SENSOR

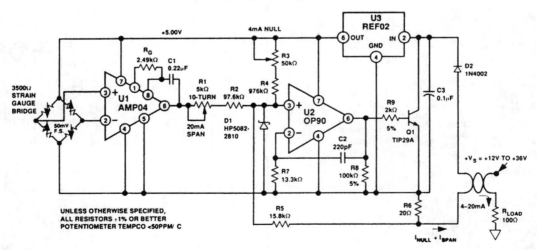

ANALOG DEVICES

Fig. 50-13

In this loop-powered strain-gage sensor application, a 50-mV full-scale (FS) bridge output is amplified and calibrated for a 4–20-mA transmitter output. Power is furnished by the remote loop supply of 12 to 36 V.

MINUTE MARKER

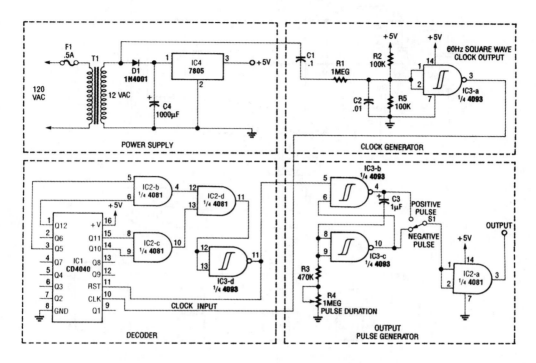

ELECTRONICS NOW

Fig. 50-14

The figure shows the schematic of a minute marker. The output of transformer T1 is 12.6 Vac at 60 Hz, which is rectified by D1 and regulated by IC4, and LM7805 regulator, to provide 5 Vdc for the circuit. The unrectified ac is bandpass-filtered by R1, R2, R5, C1, and C2. Resistors R2 and R5 also form a dc-voltage divider, which biases the input of Schmitt trigger IC3-a to 2.5 V. The Schmitt trigger generates a 60-Hz square wave, which is fed to the input of IC1, a CE4040 12-stage binary counter.

The outputs of the counter are a 4081 quad AND gate (IC2), and the decoded output is fed back to the reset input of the counter, which resets the counter when the desired count is reached.

The pulse from IC2-d is inverted by Schmitt trigger IC3-d, and passed along to the output pulse generator. The output pulse is generated by two Schmitt triggers cross-connected as an RS flip-flop (IC3-b and IC3-c). The output of the flip-flop is fed to 3, R4, and C3, whose values set the output pulse duration. The output pulse duration (T) can be approximated by the formula $T = 1.2 \times C_3 \times (R_3 + R_4)$. A positive or negative-going pulse is selected by S1, and buffered by the remaining AND gate (IC2-a).

DIGITAL BAROMETER

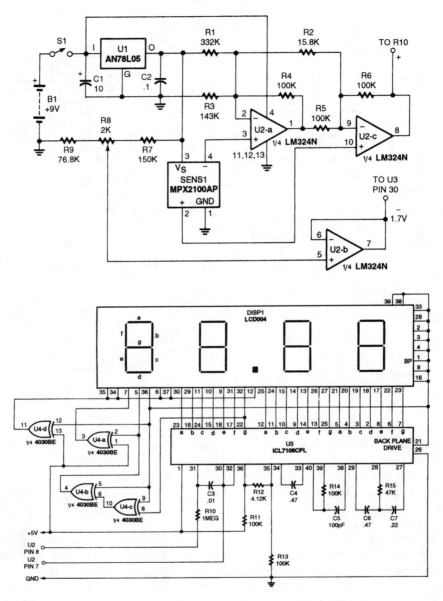

Fig. 50-15

A pressure sensor is used in this application. This outputs a voltage to amplifier U2, and a 3½ digit A/D converter module. It is calibrated to read barometric pressure in inches of mercury.

REFERENCE CIRCUIT

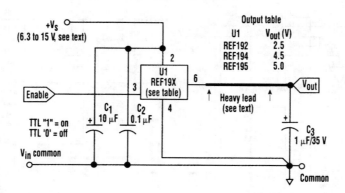

Output table		
U1	V_{out} (V)	
REF192	2.5	
REF194	4.5	
REF195	5.0	

ELECTRONICS DESIGN

Fig. 50-16

In this high-performance reference circuit, U1 is a device from the REF190 series producing device-selectable outputs of 2.5, 4.5, and 5 V with simple, noncritical external circuitry. An Analog Devices REF 19 X (see the table in the figure) is used to derive a reference voltage.

TRANSISTOR MATCHING CIRCUIT

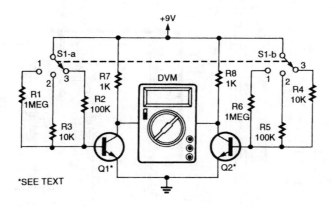

POPULAR ELECTRONICS

Fig. 50-17

In order to match two transistors, select Q1 and Q2 so that they give zero readings (or as close as possible) on a DVM. The DVM acts as a null detector. An analog meter can be substituted. S1 should be set for an appropriate level of base current (approximately 8, 80, or 800 μA).

AUTO-RANGING DIGITAL CAPACITANCE METER

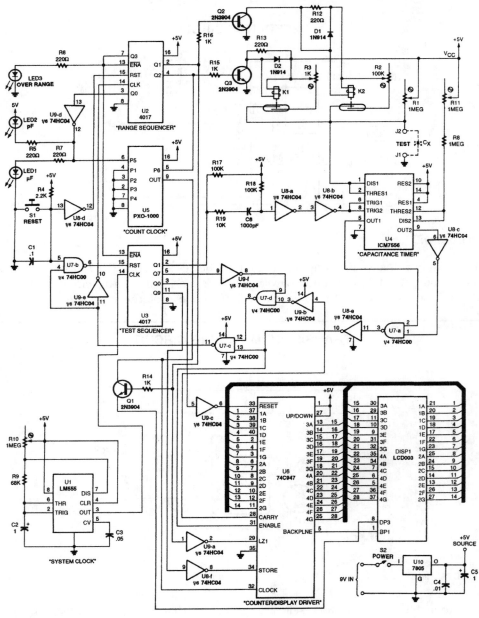

Fig. 50-18

This digital capacitance meter reads from 1 pF to 1000 µF. Basically, a timer (U4) uses the unknown capacitance to generate a pulse of duration, depending on the value of unknown capacitance, and the pulse duration is measured. The display is an LCD 0003 driven by a 74C947 counter/display driver.

FREQUENCY DIVIDER FOR 10-MHz FREQUENCY STANDARD

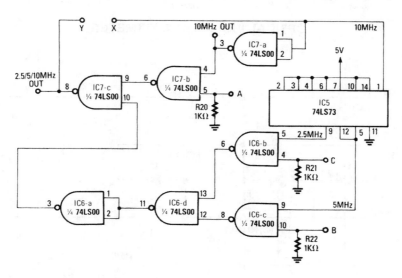

Fig. 50-19

ELECTROSCOPE

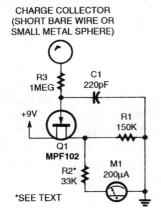

CHARGE COLLECTOR
(SHORT BARE WIRE OR
SMALL METAL SPHERE)

This circuit is useful for detecting electrostatic charges. In operation, C1 reduces ac noise, but lowers the sensitivity a bit. The MPF102 and R1 form a voltage divider. When the FET's gate is earth-grounded, the divider's output will be about 4.5 V giving a half-scale reading on M1, a 200-µA meter. A positively charged object (like cotton-rubbed glass) will give a positive deflection from half-scale, and a negatively charged object (a plastic comb, for example) will give a negative meter deflection.

The whole circuit (including the 9-V battery supply) should be in a metal enclosure, and a short piece of bare wire makes a fine charge collector.

Fig. 50-20

OPTICAL ISOLATOR WATTMETER

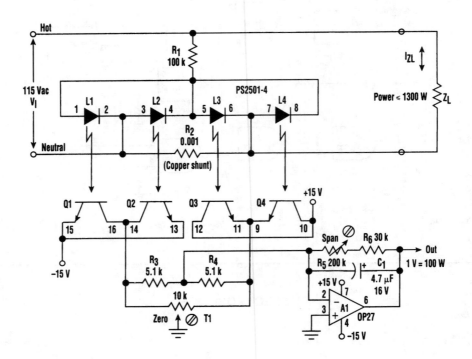

ELECTRONIC DESIGN

Fig. 50-21

The quad-channel optical isolator, consisting of LED L1 through L4 and phototransistors Q1 through Q4, is connected in a double bridge configuration. The arrangement serves to compute the four-quadrant product of ac line voltage and Z1 load current. The result is an accurate representation of the true instantaneous power delivered to the load—even if the line voltage wanders and the load is reactive and nonlinear. This wattmeter function is, of course, optically isolated from the ac line, has full-scale limit of 1300 W, and is output with scale factor of 1 V/100 W.

DIGITAL THREE-PHASE WAVE GENERATOR

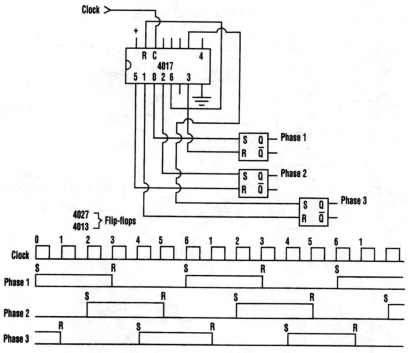

ELECTRONIC DESIGN   *Fig. 50-22*

With a simple digital circuit, three-phase square waves can be produced from a single-phase square-wave signal source. The timing diagram shows that the second and third phases are 120° and 240° behind the first phase, respectively.

The frequency range over which the three-phase outputs will occur is limited only by the capability of the logic used. The output frequency is ⅙ of the input frequency.

SIMPLE TEST AUDIO AMPLIFIER

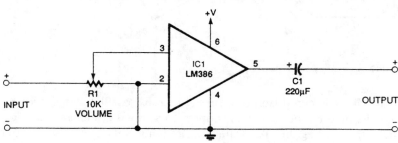

ELECTRONICS NOW   *Fig. 50-23*

This circuit has a gain of about 20. A suitable power supply voltage is 5 to 12 V, depending on the desired audio output power level.

GATE DIP OSCILLATOR I

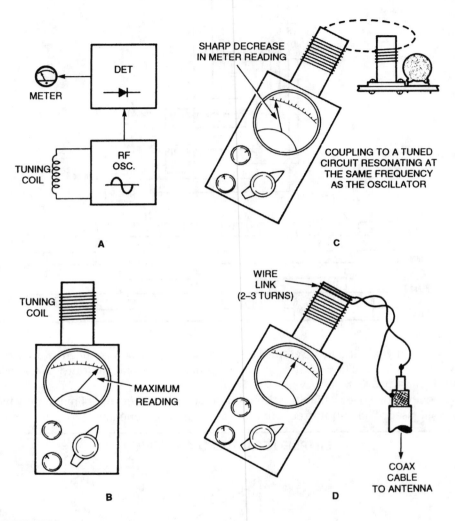

POPULAR ELECTRONICS

Fig. 50-24

The typical dip meter is comprised of a tuning coil, RF oscillator, a detector, and a meter as shown in A. When the meter's tuning coil is coupled to a tuned circuit resonating at the same frequency as the GDM, the reading dips (C). The GDM's tuning coil can be coupled to the coaxial feed line of an antenna through a few (perhaps 2 to 3) turns of wire, and used to determine the antenna's resonant frequency (D).

ACCELEROMETER (G METER) CIRCUIT

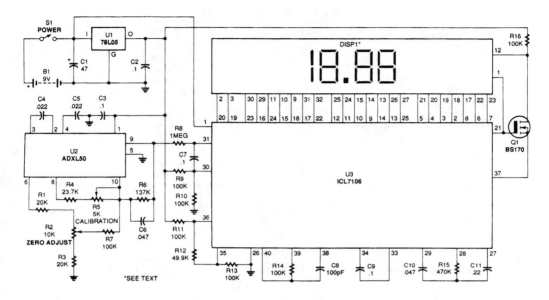

POPULAR ELECTRONICS

Fig. 50-25

As this schematic shows, the ADXL50 accelerometer, U2, interfaces with an A/D converter, U3, to drive a 3½-digit LCD module, DISP1. Because that module displays any number from −19.99 to +19.99, the circuit is designed to measure g's within that range. The heart of the circuit is U2, the ADXL50 accelerometer. The sensitivity of that chip is set to ±20 g's in order to accommodate the full scale capability of LCD module DISP1 (19.99). Circuit gain is determined by the values of R4, R5, and R6, potentiometer R2, and R3 provides a way to manually set the zero-g voltage-output level at pin 9 of U2 to half the supply voltage—2.5 V. That output voltage will vary linearly by 0.1-V/g of acceleration.

In order to achieve good circuit performance at low g levels, the bandwidth of the amplifier is limited to about 30 Hz by C6. The digital-display section of the circuit is composed of DISP1 and U3. Included in U3 are the A/D converter, clock oscillator, storage resistors and latches, 3½-digit seven-segment decoders, and backplane generator.

The differential analog input of U3 is applied between pins 30 and 31. The positive input, pin 31, is driven by output-pin 9 of U2 through R8, a buffer resistor, and the negative input, pin 30, is biased at a fixed voltage of 2.5 V by a voltage-divider string composed of R9 and R10.

A reference voltage is required by U3. Full-scale display, 19.99, occurs when the differential, analog input voltage applied between pins 31 and 30 is equal to twice the reference voltage. The decimal point of the LCD has to be illuminated to display readings from 0.00 to 19.99. That is done by inverting the backplane square-wave drive signal appearing at pin 21 of U3, through MOSFET Q1, and applying the 180-degree out-of-phase signal to pin 12 of DISP1.

GATE DIP OSCILLATOR II

TABLE 1—COIL WINDING DATA

Band (MHz)	Turns	Wire Size/Type
3.5 6.5	45	32-enameled
6.5 11	32	26-enameled
11 19	14	20-enameled
15 24	10	20-enameled
21 36	7	insulated connection wire
32 56	4	insulated connection wire
60 110	U-shaped*	16 enameled

*1.8-inches long

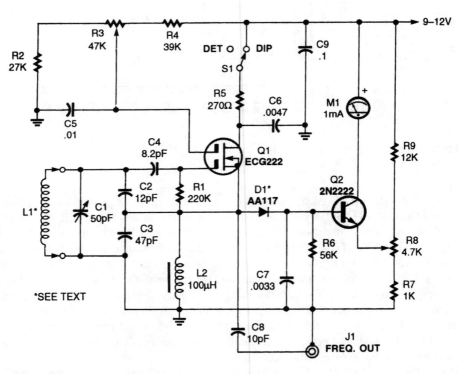

POPULAR ELECTRONICS

Fig. 50-26

Useful for measuring the resonant frequencies of antennas, tuned circuits, and also as a tuned detector, this circuit is a modern variation of the classic vacuum tube grid dip oscillator. It coupled the G.D.O. to a tuned circuit and caused RF energy to be absorbed by the unknown tuned circuit when the G.D.O. frequency was the same as that of the tuned circuit in question. This showed as a "dip" in the meter reading.

TWO REMOTE METERS

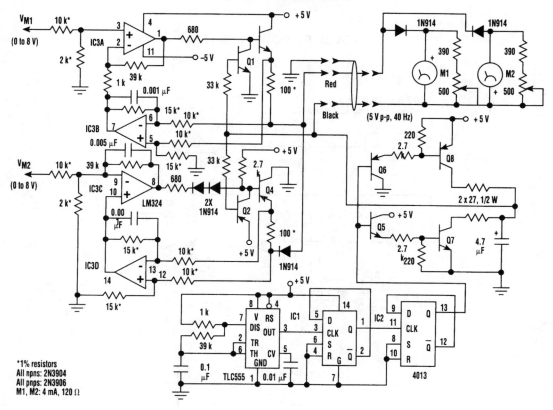

*1% resistors
All npns: 2N3904
All pnps: 2N3906
M1, M2: 4 mA, 120 Ω

ELECTRONIC DESIGN

Fig. 50-27

Two remote meters can be driven independently using just one wire pair. This "constant current" design eliminates the effects of wire-pair resistance up to 200 Ω. Driving two remote meters independently usually requires two wire pairs (one pair for each meter).

In the circuit, IC1 and IC2 generate a 40-Hz symmetrical square wave (the frequency isn't critical). Q5 through Q8 amplify the square wave to 5 V p-p, which is applied to the "return" (black wire) for the remote meters.

Amplifier IC3A buffers the input signal voltage V_{mi}, intended for meter M1 (0 to 8 V), and sends it through emitter-follower Q3 to a 100-Ω current-sense resistor. The other end of this resistor is tied to the "supply" (red wire) of the remote meters. IC3B amplifies the voltage across the sense resistor, which corresponds to the current sent to remote meter M1, and closes the feedback loop to IC3A.

This results in a voltage of 0 to 8 V at the M1 input, generating a current of 0 to 10 mA to M1. Transistor Q1 gates this current on and off synchronous to the 40-Hz square wave so that meter M1 actually sees a 50% 0-to-+10-mA peak (0 to 5 mA average) current.

Similarly, IC3C, IC3D, Q2, and Q4 provide a 0- to –10-mA peak current for M2. M1 and M2 are isolated by the two-reverse-connected 1N914 diodes in the remote-meter box. Variable resistors across M1 and M2 permit calibration. The extra 1N914 diode in the M2 drive circuit prevents interference between M1 and M2.

NOVEL RF POWER METER

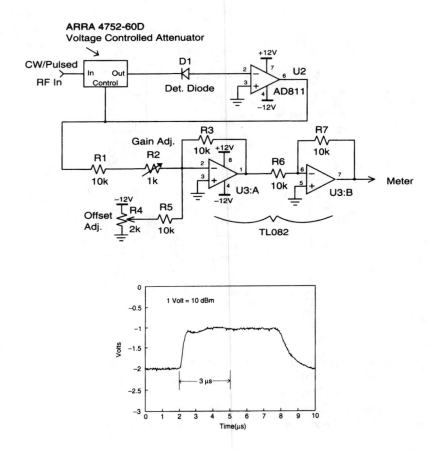

Fig. 50-28

The circuit matches the diode with a voltage-variable attenuator that has a logarithmic response. By varying the attenuation until the diode output is zero, the resulting attenuation value then corresponds to the input power level. Because the voltage-variable attenuator's output is logarithmic, diode nonlinearities become negligible.

NANOAMMETER

**Resistance Values for
DC Nano and Micro Ammeter**

I FULL SCALE	R_f [Ω]	R'_f [Ω]
100 nA	1.5M	1.5M
500 nA	300k	300k
1 μA	300k	0
5 μA	60k	0
10 μA	30k	0
50 μA	6k	0
100 μA	3k	0

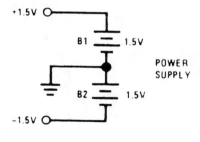

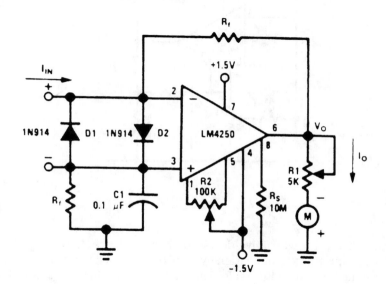

NATIONAL SEMICONDUCTOR   *Fig. 50-29*

Potentiometer R2 provides an electrical meter zero by forcing input offset voltage V_{os} to zero. Full-scale meter deflection is set by R1. Both R1 and R2 only need to be set once for each op amp and meter combination. For a 50-μA 2-kΩ meter movement, R1 should be about 4 kΩ to give full-scale meter deflection in response to a 300-mV output voltage. Diodes D1 and D2 provide full input protection for overcurrents up to 75 mA.

With an R_f resistor value of 1.5 MΩ, the circuit becomes a nanoammeter with a full-scale reading capability of 100 nA. Reducing R_f to 3 kΩ in steps, as shown in the figure increases the full-scale deflection to 100 μA, the maximum for this circuit configuration. The voltage drop across the two input terminals is equal to the output voltage (V_o) divided by the open loop gain. Assume that an open loop gain of 10,000 gives an input voltage drop of 30 μV or less.

1.5-V LOGARITHMIC LIGHT LEVEL METER

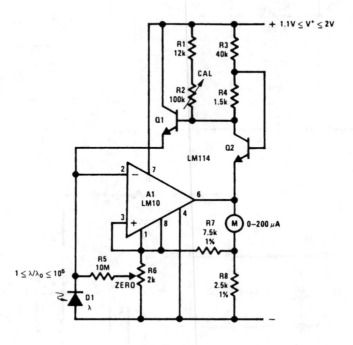

NATIONAL SEMICONDUCTOR

Fig. 50-30

A portable light-level meter with a five-decade dynamic range is shown. The circuit is calibrated at mid-range with the appropriate illumination by adjusting R2 such that the amplifier output equals the reference and the meter is at center scale. The emitter-base voltage of Q22 will vary with supply voltage; so R4 is included to minimize the effect on circuit balance. If photocurrents less than 50 nA are to be measured, it is necessary to compensate the bias current of the op amp.

The logging slope is not temperature compensated. With a five-decade response, the error at the scale extremes will be about 40% (a half stop in photography) for a ±18°C temperature change.

If temperature compensation is desired, it is best to use a center-zero meter to introduce the offset, rather than the reference compensation. It can be obtained by making the resistor in series with the meter a copper wire-wound unit.

If this design is to be used for photography, it is important to remember that silicon photodiodes are sensitive to near-infrared light, whereas ordinary film is not. Therefore, an infrared-stop filter is called for. A blue-enhanced photodiode or an appropriate correction filter would also produce best results.

ac POWER MONITOR

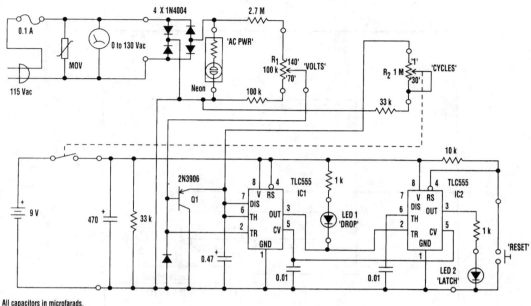

ELECTRONIC DESIGN   *Fig. 50-31*

The 0- to 130-V voltmeter and neon "AC PWR" lamp provide an average indication of the ac power. The fuse and metal oxide varistor (MOV) protect the monitor against overvoltage spikes.

Four 1N4004 diodes rectify the ac voltage, generating negative-going pulses twice per cycle (every 8.33 ms for 60-Hz power). Variable-resistor F1 supplies a reduced amplitude sample of these pulses to a missing pulse detector consisting of Q1, IC1, and associated circuitry. As long as the pulse amplitude exceeds the threshold value set by R1, IC1 continually triggers, keeping its output high. When the pulse amplitude drops below the threshold value, IC1 times out with a time constant set by variable resistor R2 and the 0.47-μF capacitor. R2 is calibrated to read the number of cycles required for a dropout indication. It can be set between 1 cycle (about 17 ms) and 30 cycles (0.5 second).

When IC1 times out, its output goes low. This turns on LED1. The low output also triggers IC2, which is configured as a set-reset flip-flop. This turns on LED2. When the voltage returns to normal, IC1 again starts triggering and its output returns high, turning off LED1. LED2, however, remains on until the manual reset button is pressed.

The circuit is powered by a 9-V battery and is assembled in a plastic or grounded metal case. Notice that there's no isolation between the ac power line and the monitor circuitry. Be careful to avoid electrical shock when testing the circuitry.

100-W VARIABLE RESISTOR SIMULATOR

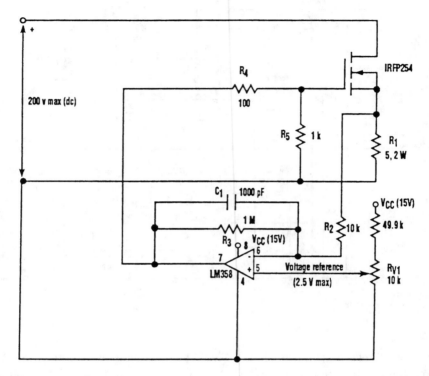

Fig. 50-32

Variable resistive loads with precise load steps often are required in automatic testers used to test and calibrate power supplies. The action of a high-power (100 W) variable resistor can be simulated with the circuit shown.

The voltage drop across R1, which is proportional to the FET current, is compared against a variable input voltage reference using a high-gain op amp. Error voltage developed by the amplifier drives the gate, controlling the transconductance of the FET.

Power dissipation is limited by the safe-operating-area curve of the selected FET. The FET should be mounted onto a properly sized heatsink or a heatsink-fan combination to maintain its case temperature within safe limits. The circuit is designed to dissipate a maximum power of 100 W if the FET-case temperature is maintained below 50°C. The potentiometer (RV1) can be replaced by a digital-to-analog converter so that it can adapt to the computer control for use in automatic testers.

IMD TEST CIRCUIT FOR PIN DIODES

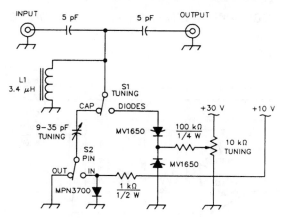

QST

Fig. 50-33

A loosely coupled tuned circuit for testing IMD production by PIN and tuning diodes in a narrow-band preselector, S1, TUNING, selects whether C1 or a pair of back-to-back MV1650 tuning diodes resonate L1. S2, PIN, adds or removes an MPN3700 PIN diode in series with C1. L1 consists of 33 turns of #28 enameled wire on a t-37-6 toroidal powdered-iron core. The MV1650, a "20-V" tuning diode, exhibits a nominal capacitance of 100 pF at a tuning voltage of 4 V.

VCO AND INPUT FREQUENCY COMPARER

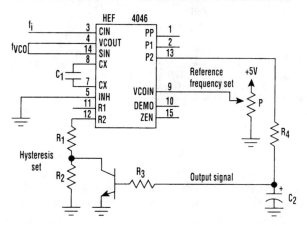

ELECTRONIC DESIGN

Fig. 50-34

Comparison of an input signal's frequency (f_i) with that of voltage-controlled oscillator (f_{VCO}) can be accomplished with just one CMOS phase-locked loop IC and a transistor (see figure). The phase and the frequency can be compared with a phase comparator, which, along with the VCO, is part of the HEF4046 PLL IC. The transistor helps introduce hysteresis, enabling the circuit to be used as a switch driver.

ECG AMPLIFIER WITH RIGHT LEG DRIVE

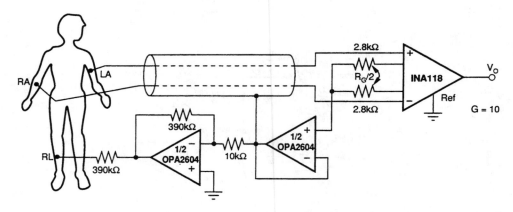

BURR-BROWN

Fig. 50-35

POWER TRANSFORMER TESTER

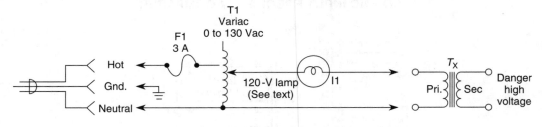

Warning: observe existing shock hazard

WILLIAM SHEETS

Fig. 50-36

Very often a power transformer is suspect and connecting a shorted transformer to an ac source can be hazardous. This test method will detect a defective or shorted transformer. The primary of the power transformer is energized through a Variac (0 to 130 Vac) and a lamp equal in wattage to about half that of the transformer under test. Connect the transformer, set Variac at zero, then energize circuit. Apply voltage to suspected transformer (Tx) as shown. The lamp should not light. If it does, Tx is shorted. Next, short the secondary of suspected Tx. This time, the lamp should light. For multiple winding transformers, repeat for each secondary winding. Beware of the shock hazard as the open windings of Tx can develop full-rated voltage.

4- TO 20-mA PROCESS CONTROLLER

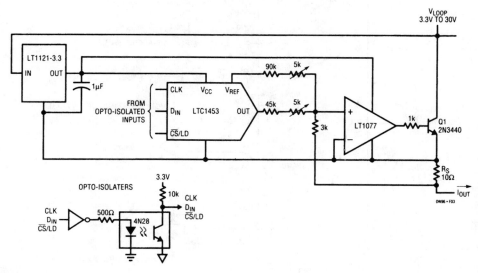

LINEAR TECHNOLOGY

Fig. 50-37

The figure shows how to use an LTC1453 to make an optoisolated digitally controlled 4- to 20-mA process controller. The controller circuitry, including the optoisolator, is powered by the loop voltage that can have a wide range of 3.3 V to 30 V. The 1.22-V reference output of the LTC1453 is used for the 4-mA offset current and V_{OUT} is used for the digitally controlled 0- to 16-mA current. R_S is a sense resistor and the LT1077 op amp modulates the transistor Q1 to provide the 4- to 20-mA current through this resistor. The control circuitry consumes well under the 4-mA budget at zero scale.

SIMPLE HIGH-CURRENT MEASURER

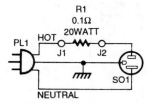

POPULAR ELECTRONICS

Fig. 50-38

Testing heavy-load devices with a ten-amp maximum meter can be accomplished with this straightforward meter add-on. If done right, it could be made from a high-current extension cord. J1 and J2 are well-insulated jacks to accept meter probe tips.

ANALOG CIRCUITRY CALIBRATOR

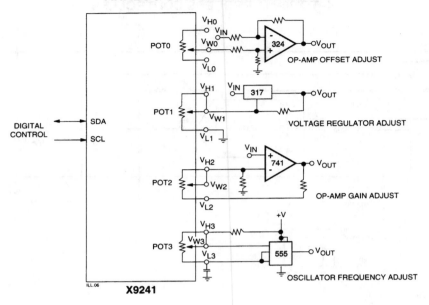

XICOR

Fig. 50-39

An XICOR X9241 Quad POT IC can be used to digitally adjust four analog circuits, as shown in the example schematic.

SIMPLE SIGNAL GENERATOR FOR SIGNAL TRACING

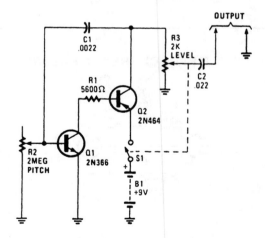

POPULAR ELECTRONICS

Fig. 50-40

A simple R-C oscillator generates a harmonic-rich waveform for signal injection.

SIMPLE HARMONIC DISTORTION ANALYZER

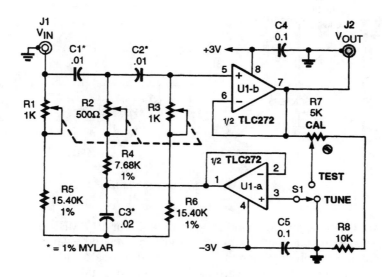

POPULAR ELECTRONICS

Fig. 50-41

This simple circuit lets you accurately measure the total harmonic distortion (THD) using your true-RMS voltmeter.

This THD circuit is somewhat different from the usual types: it can operate at the standard frequency of 1000 Hz, but it also is tunable from 970 Hz to 1030 Hz, and has an adjustable Q factor of 0.3 to over 50. Op-amp U1, a TLC272 CMOS unit, contains the two voltage-followers required to buffer the input to the bootstrapped twin-T notch filter. Tuning is accomplished by R1, R2, and R3, which are standard linear-taper slide pots "ganged" together by mounting them side-by-side and gluing their sliders together. The only other important construction hint is to use twisted pair at the circuit's input and output.

To calibrate the circuit, input a 1000-V RMS signal at 2000 Hz, set S1 to TEST, and adjust R7 for a reading of 0.99-V RMS on a true-RMS voltmeter at the output.

To use the circuit, set S1 to TUNE, input a 1000-Hz sine-wave signal to the amplifier under test, and set the amplifier's output to the THD adapter and tune R1/R2/R3 for the lowest output signal. Then, set S1 to TEST and read the RMS voltage. To calculate the percent THD use:

$$THD = \left(\frac{V_{out}}{V_{in}} \right) \times 100$$

SOUND SUBCARRIER GENERATOR

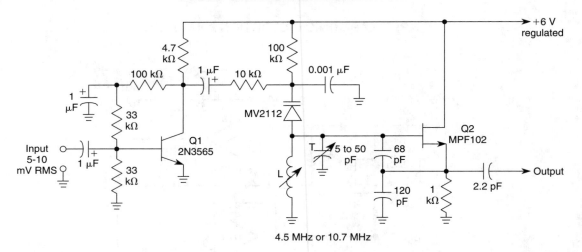

WILLIAM SHEETS

Fig. 50-42

This circuit will generate an FM sound subcarrier at 4.5 or 10.7 MHz for FM and TV IF testing and alignment. Q1 is an audio amplifier and Q2 is a VCO modulated by an MV2112 varactor. Deviation up to 1% of frequency can be obtained. L is chosen to resonate with the circuit capacitance to either 4.5 or 10.7 MHz. The values will be around 2 to 10 μH, depending on the frequency.

INDUCTANCE AND CAPACITANCE DETERMINER WITH SWR BRIDGE

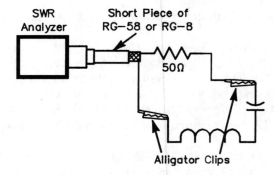

QST

Fig. 50-43

At resonance, the SWR will be 1:1 with a 50-Ω resistance, as reactance is zero. If either L or C is known:

$$|Xc| = |X1| = \frac{|X_L|}{2\pi fC} = 2\pi fL$$

$$L_{\text{unknown}} = \frac{1}{(2\pi f)^2 C} \qquad C_{\text{unknown}} = \frac{1}{(2\pi f)^2 L}$$

MOTORCYCLE TUNE-UP AID

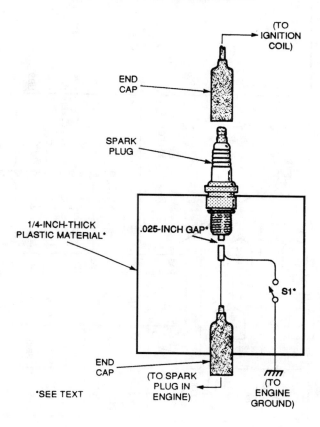

Fig. 50-44

Performing a tune-up on a newer bike is made a lot easier with this helpful circuit. Because of the high voltages present, make sure that S1 has an insulated handle and that the fixture is grounded. With the ignition turned off, remove one of the spark plug wires and connect it to the spark plug on the fixture. Slip the fixture's end cap over the spark plug on the cycle and you're ready to go. Open S1 and start the engine. Then, close S1; the cylinder with the fixture should not fire and a spark should be seen at the fixed gap. Be sure that the fixture is connected to the engine ground before closing S1.

50-MHz FREQUENCY COUNTER

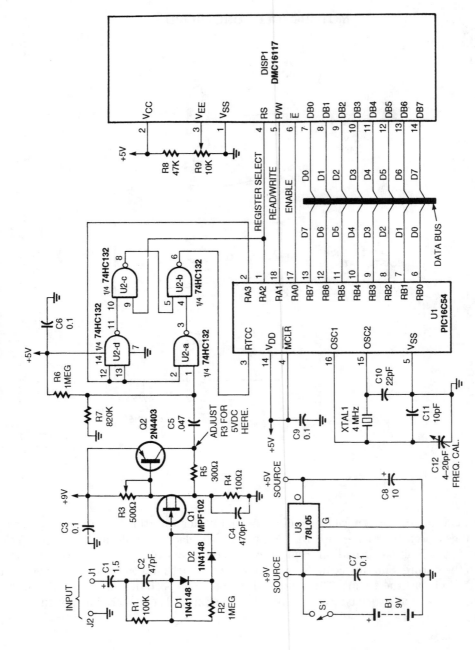

Fig. 50-45

POPULAR ELECTRONICS

This inexpensive frequency counter uses a microcontroller as the counter. The microcontroller feeds an LCD display module that accepts standard ASCII code. The frequency is displayed as Hz, kHz, or MHz and the counter is autoranging.

10-MHz FREQUENCY STANDARD

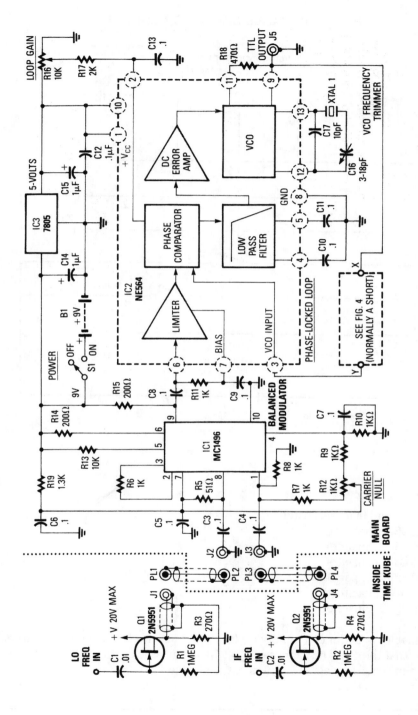

Fig. 50-46

RADIO-ELECTRONICS

A Radio Shack 10-MHz receiver is used as a basis for this circuit. The L.O. and IF frequencies are added. When the receiver is tuned to 10 MHz (WWV), the sum of the L.O. and IF are used to phase lock a VCO to the 10-MHz signal. By using a divider in the loop, 2.5 or 5 MHz can be used as well.

361

PROGRAMMABLE CAPACITOR CIRCUIT

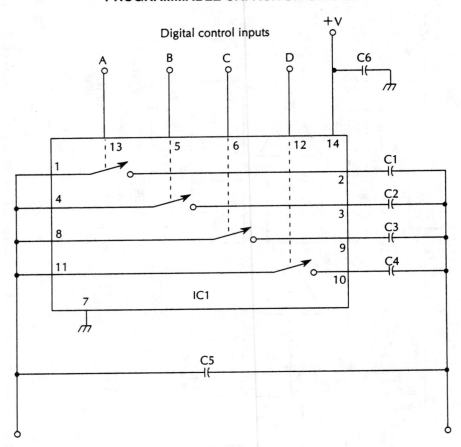

Fig. 50-47

IC1	CD4066 quad bilateral switch
C1	10-µF, 25-V capacitor
C2	22-µF, 25-V capacitor
C3	47-µF, 25-V capacitor
C4	100-µF, 25-V capacitor
C5	1-µF, 25-V capacitor
C6	0.1-µF, 25-V capacitor

The programmable capacitor can be very useful in circuits where you need to switch capacitance values. Remember that the "ON" resistance of IC1 appears in series with the capacitors and must be taken into account in some applications as it is not negligible.

PROGRAMMABLE RESISTOR CIRCUIT

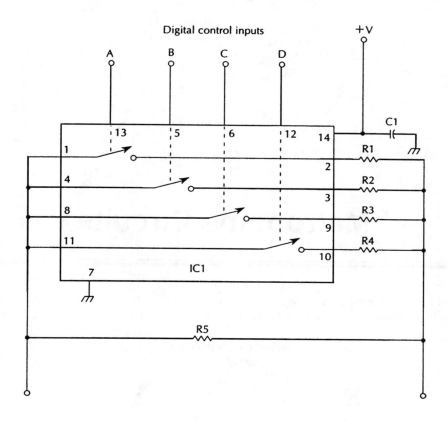

Fig. 50-48

IC1	CD4066 quad bilateral switch
C1	0.1-μF capacitor
R1	10-kΩ, ¼-W 5% resistor
R2	4.7-kΩ, ¼-W 5% resistor
R3	2.2-kΩ, ¼-W 5% resistor
R4	1-kΩ, ¼-W 5% resistor
R5	1-MΩ, ¼-W 5% resistor

A programmable resistor can replace a potentiometer or fixed resistor. Remember that the "ON" resistance of IC1 might have to be taken into account in some applications.

363

51

Metronome Circuits

The sources of the following circuits are contained in the Sources section, which begins on page 707. The figure number in the box of each circuit correlates to the entry in the Sources section.

Audible Metronome
Visual Metronome

AUDIBLE METRONOME

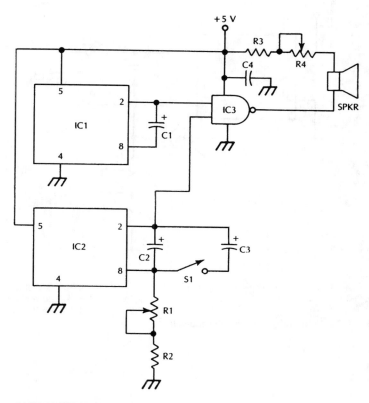

IC1, IC2	LM3909 LED flasher/oscillator	C3	47 μF 10 V electrolytic capacitor
IC3	7400 quad NAND gate	C4	0.01 μF capacitor
SPKR	small loudspeaker	R1	50 kΩ potentiometer
S1	SPST switch	R2	3.3 kΩ ¼ W 5% resistor
C1	1 μF 10 V electrolytic capacitor	R3	47 kΩ ¼ W 5% resistor
C2	100 μF 10 V electrolytic capacitor	R4	500 Ω potentiometer

McGRAW-HILL

Fig. 51-1

IC1 generates an audible frequency while a variable very low frequency is generated by IC2. R1 sets the metronome rate. The two signals are combined in IC3.

VISUAL METRONOME

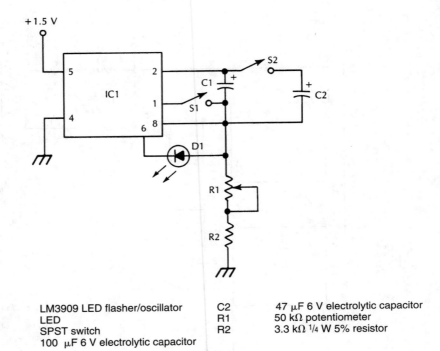

IC1	LM3909 LED flasher/oscillator	C2	47 μF 6 V electrolytic capacitor
D1	LED	R1	50 kΩ potentiometer
S1, S2	SPST switch	R2	3.3 kΩ ¼ W 5% resistor
C1	100 μF 6 V electrolytic capacitor		

Fig. 51-2

52

Miscellaneous Treasures

The sources of the following circuits are contained in the Sources section, which begins on page 707. The figure number in the box of each circuit correlates to the entry in the Sources section.

MOSFET Drive Current Booster
Simple Event Counter
Frequency Doubler
Atmosphere Noise Monitor
Tachometer Derived from Brushless Shaft Angle
  Resolver
Vocal Stripper
Vocal Stripper Power Supply
Single-Chip Message System
Television Vertical Deflection Circuit
Tone Burst Generator
Audio Volume Limiter
Simple Intercom for Noisy Environments
Ditherizer
Triac Lamp Dimmer Circuit
500-ksps 8-Channel Data Acquisition Circuit
Hydrophone
Your Name in Lights

Underwater Microphone
Pulse Echo Driver
Simple Pseudorandom Voltage Source
TV Horizontal Deflection Circuit
Muting Circuit
Simple Remote Gain Control
Loop Oscillator Eliminator
1-A Voltage Follower
Electronic Fish Lure
Heartbeat Transducer
Contact Debouncer
Positive Feedback Cable Terminator
×10 Frequency Multiplier
Jacob's Ladder
Master-Slave Device Error Checker
Ground Loop Preventer
Dual Tone Generator for Audio Servicing
Diodeless Peak-Hold Circuit

MOSFET DRIVE CURRENT BOOSTER

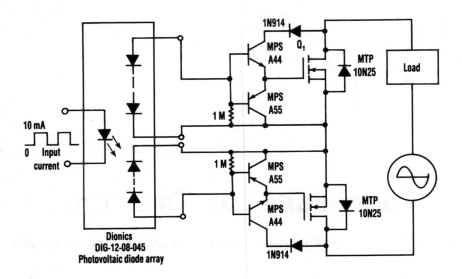

ELECTRONIC DESIGN

Fig. 52-1

A clean and inexpensive way to eliminate the floating-gate supply is to use the voltage available at the MOSFET's drain to drive its gate. Tying the collector of Q1 (a high-voltage, small-signal, 400-V NPN) to the MOSFET's drain supplies sufficient gate-drive voltage when it's needed most—when the MOSFET's drain-to-source voltage is high. Two such circuits used back-to-back form an ac relay.

Using the emitter follower attached to the drain increases gate-drive current and decreases the MOSFET's turn-on time by a factor equal to the high-voltage NPN's beta. The resulting drain-to-source voltage fall times depend on the MOSFET's size and its required gate charge. The circuit that's used gives a fall time of 200 µs for an MTP10N2f5 10-A, 250-V MOSFET. With such fall times cutting switching losses, pulse-width modulation at frequencies under 100 Hz is possible.

During tune-on, V_{DS} falls rapidly until it reaches the sum of the 1N914 diode's 0.7-V drop, the collector-emitter saturation voltage of Q1, and the gate-to-source voltage required to support the load current. At that point, the diode array completes the MOSFET's turn-on, unaided by the buffer. This slows the fall of V_{DS} considerably when it reaches about 5 to 7 V. In high-voltage, low-frequency systems, tailing of V_{DS} is tolerable because the tail's voltage magnitude constitutes a small fraction of the switching voltage. The IN914 makes it possible for V_{GS} to exceed V_{DS} as the MOSFET completes turn-on.

SIMPLE EVENT COUNTER

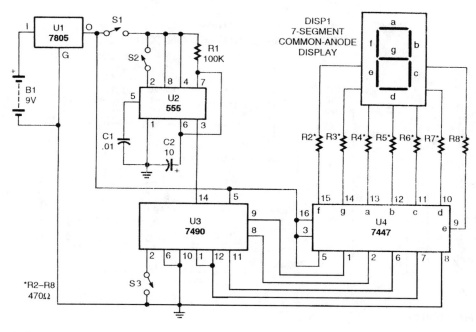

POPULAR ELECTRONICS

Fig. 52-2

S1 is a power switch. U2 drives counter U3 by producing a pulse when S2 is depressed. U4 and DISP1 read the count of counter IC U3. S3 is a reset to zero switch. The counter is a basic one-digit circuit useful as a holding block or by itself.

FREQUENCY DOUBLER

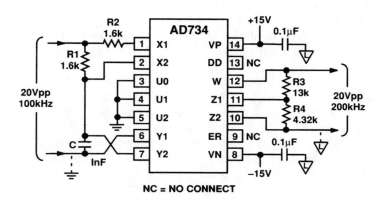

ANALOG DEVICES

Fig. 52-3

An Analog Devices AD734 four-quadrant analog multiplier is used as a frequency doubler.

ATMOSPHERE NOISE MONITOR

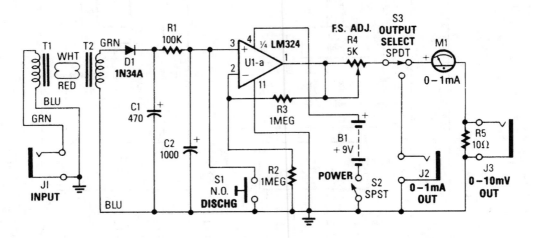

Fig. 52-4

Tune an unmodified transistor radio to an unused frequency near 540 kHz that's free of broadcast-station interference; the receiver is used to pick up sferics. The received signal is fed from the receiver's earphone jack through a patch cord to the input jack (J1) of the circuit. The back-to-back audio transformers, T1 and T2, provide a suitable impedance match and signal level when the unit is used with various receivers.

Diode D1 rectifies the audio input from the receiver to pulsating dc, which is filtered by C1, R1, and C2 to provide a time constant of several minutes. That dampens out fluctuations in most cases, unless lightning flashes are very infrequent.

The voltage appearing at the output of the filter is a function of signal strength transferred by C1. Switch S1 is included to provide a convenient way to discharge the capacitors, if adjustments are required during a monitoring session.

Integrated circuit U1 (one section of an LM324 quad pro op amp) is used as a high input-resistance voltmeter. Resistors R2 and R3 determine amplifier gain, and potentiometer R4 is used to adjust full-scale meter deflection for a suitable voltage level at the input. A value of 1.5 V has been satisfactory for use with several receivers tried, but they can be changed.

If the monitor is to be used only as a meter, the milliammeter can be connected directly between R4 and chassis ground, omitting R5, S3, J2, and J3. The latter components provide suitable output for use with a chart recorder having a full-scale range of either 10 mV or 1 mA. The circuit, when powered from a 9-V battery, draws about 1 mA.

TACHOMETER DERIVED FROM BRUSHLESS SHAFT ANGLE RESOLVER

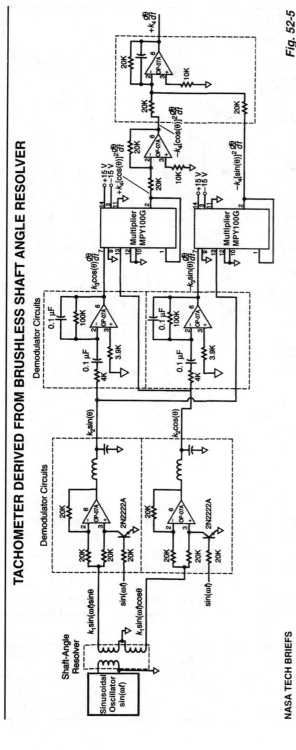

Fig. 52-5

NASA TECH BRIEFS

The tachometer circuit operates in conjunction with a brushless shaft-angle resolver. By performing a sequence of straightforward mathematical operations on the resolver signals and utilizing a simple trigonometric identity, it generates a voltage proportional to the rate of rotation of the shaft.

The figure illustrates an analog tachometer circuit that processes the input and output signals of a two-phase, brushless, transformer-type shaft-angle resolver into a signal with instantaneous amplitude proportional to the instantaneous rate of rotation of the shaft. The processing in this circuit effects a straightforward combination of mathematical operations leading to a final operation based on the well-known trigonometric identity $[\sin(x)]^2 + [\cos(x)]^2 = 1$ for any value of x.

The resolver is excited with a periodic waveform; a sinusoid is indicated in the figure, but a square, triangular, or other periodic waveform could be used instead. Thus, the two outputs of the resolver are $k_1 \sin(\omega t)\sin(\theta)$ and $k_1 \sin(\omega t)\cos(\theta)$, where k_1 is a constant proportional to the amplitude of excitation, ωt is $2\pi c x$ the frequency of excitation, t is time, and θ is the instantaneous shaft angle.

The two outputs of the resolver are then processed, along with a replica of the sinusoidal excitation, by demodulators. These signals are then differentiated with respect to time in two differentiator circuits. Notice that $d\theta/dt$ is the rate of change of the shaft angle and is the quantity that one seeks to measure.

Next, a multiplier circuit forms a product of the demodulator and differentiator outputs proportional to $\sin(\theta)$, and the product of the demodulator and differentiator outputs proportional to $\cos(\theta)$. The output of the cosine multiplier is fed to a unit-gain inverting amplifier.

371

VOCAL STRIPPER

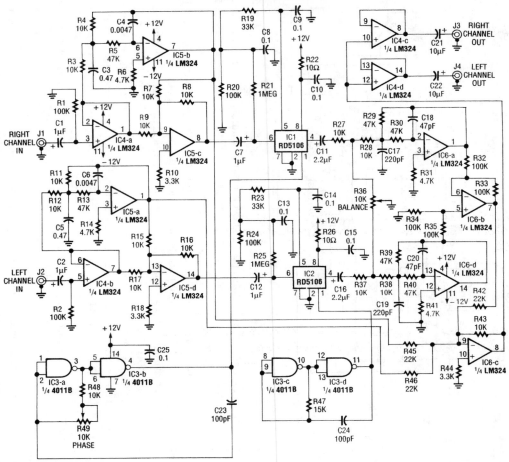

Fig. 52-6

The schematic of the lead vocal filter is shown in the figure. The left and right channel signals are coupled through C1 and C2 to buffer amps IC4-a and IC4-b. From the buffer amps, the left and right channel signals pass through active crossovers IC5-a and IC5-b, sending all low frequencies to a final mixer IC6-c, and all middle and high frequencies to analog delay lines IC1 and IC2, RD5106 256-sample bucket-brigades. Integrated circuit IC2 delays the left-channel signal by 2.4 ms, set by the fixed-frequency clock generated by ½IC3, R47, and C24. The right channel signal is delayed by IC1 with a variable-frequency clock generated by ½IC3, R48, R49, and C23. Potentiometer R49 is used for phase adjustment.

The output of each delay line from IC1 and IC2 passes through low-pass filters IC6-a and -d, and their associated parts, to filter out high-frequency sample-steps produced by IC1 and IC2. Balance control R36 is adjusted for equal amplitude of the left and right channels. IC6-b is a difference amplifier that cancels all lead vocals that are common to both channels. The resulting signal from IC6-b is remixed with low frequencies by IC6-c and is then sent to the output via buffers IC4-c and IC4-d.

VOCAL STRIPPER POWER SUPPLY

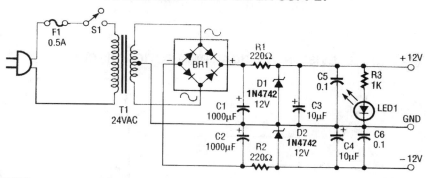

RADIO-ELECTRONICS

Fig. 52-7

The power supply schematic for the lead vocal filter circuit.

SINGLE-CHIP MESSAGE SYSTEM

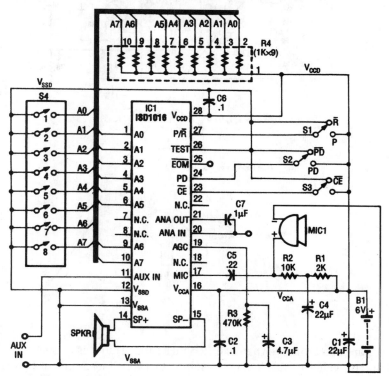

ELECTRONICS EXPERIMENTERS HANDBOOK

Fig. 52-8

The ISD1016 is a complete analog audio record/playback system on a chip. The analog signal is sampled and the samples stored in an EEPROM as analog levels. Upon playback, the analog data is read out and amplified. Up to 16 seconds of data (audio) can be stored.

TELEVISION VERTICAL DEFLECTION CIRCUIT

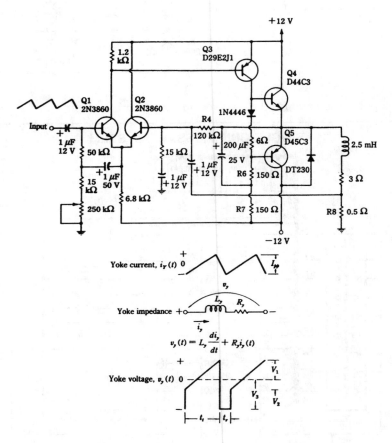

Fig. 52-9

Two transistors are used to drive the yoke (2.5 mH + 0.3 Ω) in this deflection circuit. R8 samples the yoke current and provides feedback to Q2, resulting in a very linear current ramp through the yoke.

TONE BURST GENERATOR

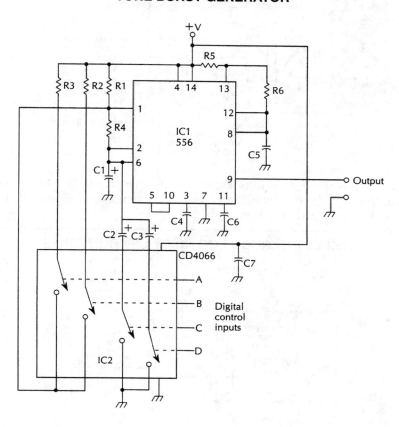

Fig. 52-10

The burst length is digitally controlled by inputs A, B, C, D. This input selects the necessary values of R and C. The circuit shown in the figure generates a burst of tone followed by a silent period, then another tone burst is sounded and so forth. The timing of the tone bursts is digitally controlled via the CD4066 (IC2). A parts list for this circuit is given in the table.

IC1	556 dual timer (or two 555 timers)	C5	0.047-μF capacitor
IC2	CD4066 quad bilateral switch	R1	100-kΩ, ¼-W 5% resistor
C1	1-μF, 25-V electrolytic capacitor	R2, R4	220-kΩ, ¼-W 5% resistor
C2	4.7-μF, 25-V electrolytic capacitor	R3	680-kΩ, ¼-W 5% resistor
C3	10-μF, 25-V electrolytic capacitor	R5	12-kΩ, ¼-W 5% resistor
C4, C6	0.01-μF capacitor	R6	4.7-kΩ, ¼-W 5% resistor

AUDIO VOLUME LIMITER

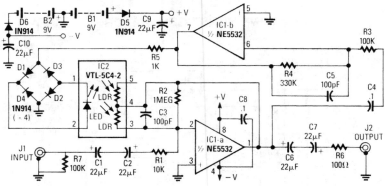

RADIO-ELECTRONICS

Fig. 52-11

In this circuit, amplifier IC1-a provides signal amplification of –40 to +40 dB depending on the value of the LDR. The LDR (light dependent resistor) is driven by rectified audio from voltage follower IC1b and bridge rectifier D1 through D4.

SIMPLE INTERCOM FOR NOISY ENVIRONMENTS

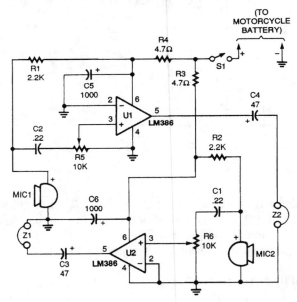

POPULAR ELECTRONICS

Fig. 52-12

This intercom was originally designed for motorcycle to passenger communications. A simple "passenger-to-pilot" intercom circuit is shown. Two LM386 ICs are connected in a low-gain amplifier circuit with the headphone output of one paired to the microphone input of the other. The microphones are electret elements and the earphones can be of the in-ear type or of the small stereo/mono type that will fit inside a helmet. Both amplifiers in the circuit operate at a minimum gain of 20 dB.

DITHERIZER

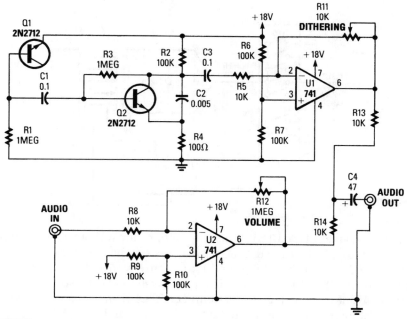

POPULAR ELECTRONICS

Fig. 52-13

In digital audio, a noise signal of amplitude less than one significant bit is often added to the audio to reduce the quantizing effect and improve the audio quality by trading digital "noise" for analog noise, which does not have the harsh sound. This circuit consists of a noise generator to add a low level of noise to an analog signal to be digitized, or an analog signal from a digital source.

TRIAC LAMP DIMMER CIRCUIT

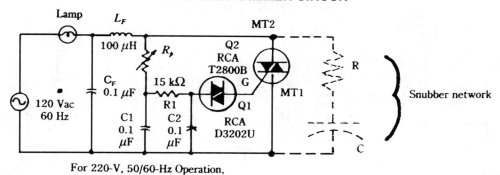

For 220-V, 50/60-Hz Operation,
replace T2800B with T2800D.

McGRAW-HILL

Fig. 52-14

The brightness of a lamp or lamps can be varied with this circuit. The snubber circuit values are typically 0.1 μF and 100-Ω. R_8 is typically 25 to 100 kΩ.

500-ksps 8-CHANNEL DATA ACQUISITION CIRCUIT

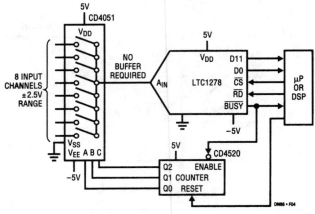

LINEAR TECHNOLOGY

Fig. 52-15

The high input impedance of the LTC1278 allows multiplexing without a buffer amplifier. Both single channel and multiplexed high-speed data acquisition systems benefit from the LTC1278/LTC1279's dynamic conversion performance. The 1.6-μs and 1.4-μs conversion and 200-ns and 180-ns S/H acquisition times enable the LTC1278/LTC1279 to convert a 500 ksps and 600 ksps, respectively. The figure shows a 500-ksps 8-channel data acquisition system. The LTC1278's high input impedance eliminates the need for a buffer amplifier between the multiplexer's output and the Adc's input.

HYDROPHONE

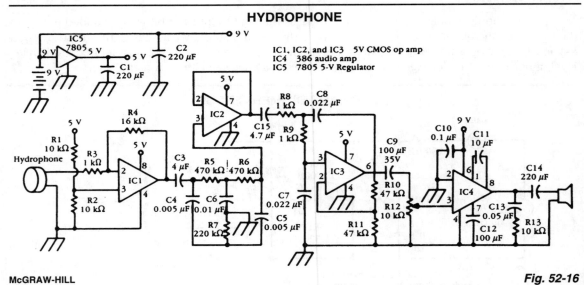

McGRAW-HILL

Fig. 52-16

A commercially available hydrophone transducer is used in this system. The transducer is connected via a cable to the amplifier, which remains out of the water. The hydrophone should be suitably mounted for intended application.

YOUR NAME IN LIGHTS

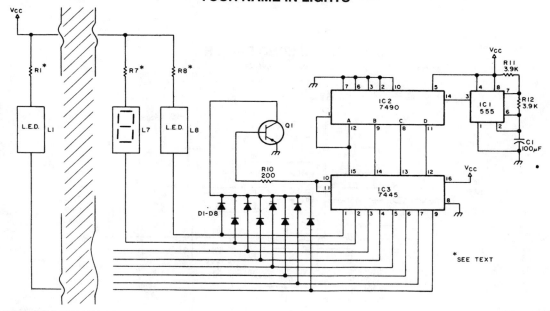

73 AMATEUR RADIO TODAY

Fig. 52-17

This circuit will enable you to put a name or callsign in lights using seven-segment LEDs. The display will spell the desired name out sequentially. Select the correct type of LED. Solder the correct leads together to form the letters you want. After mounting the appropriate current-limiting resistor, the 7445 can only sink 80 mA, so a PNP transistor is needed to handle the current required to light the letters. The heart of the circuit is a 555 oscillator into a 7490 decade counter, which is decoded by a 7445 open-collector driver chip.

UNDERWATER MICROPHONE

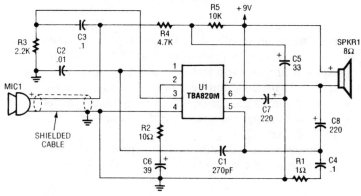

ELECTRONIC HOBBYISTS HANDBOOK

Fig. 52-18

This circuit uses a TBA820 audio IC to amplify underwater sounds. The microphone must be waterproofed. This project was originally used in a home aquarium to monitor fish sounds.

PULSE ECHO DRIVER

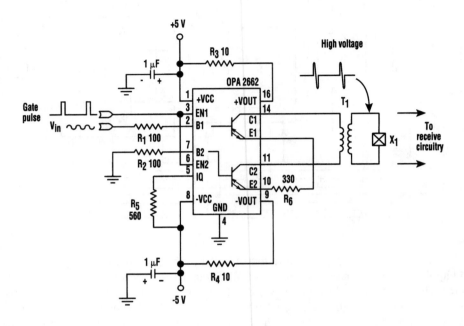

Fig. 52-19

This pulse-echo driver uses the OPA2662 dual operational transconductance amplifier (OTA) from Burr-Brown (the receive circuitry isn't shown). The OTA is preferable over an op amp for driving low impedances because it provides a current output rather than a voltage output.

Ultrasonic pulse-echo applications often incorporate a transformer-coupled crystal to obtain a high-voltage pulse because the echo can be orders of magnitude smaller in amplitude. The transformer turns ratio also provides tuning at the resonant frequency of the crystal, which usually means a relatively low-impedance primary winding.

An operational transconductance amplifier (OTA) is preferred over an op amp to drive such a low impedance. One particular application involves a pulse-echo driver circuit using the OPA2662.

SIMPLE PSEUDORANDOM VOLTAGE SOURCE

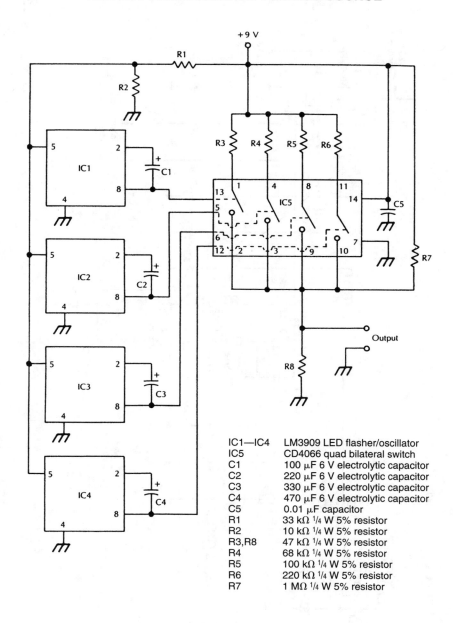

IC1—IC4	LM3909 LED flasher/oscillator
IC5	CD4066 quad bilateral switch
C1	100 μF 6 V electrolytic capacitor
C2	220 μF 6 V electrolytic capacitor
C3	330 μF 6 V electrolytic capacitor
C4	470 μF 6 V electrolytic capacitor
C5	0.01 μF capacitor
R1	33 kΩ ¼ W 5% resistor
R2	10 kΩ ¼ W 5% resistor
R3,R8	47 kΩ ¼ W 5% resistor
R4	68 kΩ ¼ W 5% resistor
R5	100 kΩ ¼ W 5% resistor
R6	220 kΩ ¼ W 5% resistor
R7	1 MΩ ¼ W 5% resistor

McGRAW-HILL

Fig. 52-20

An approximation to a pseudorandom voltage is produced by combining the outputs of four low-frequency oscillators with 0.3, 0.6, 0.9, and 1.4 Hz frequencies. The summing network is a quad bilateral switch and resistor network.

TV HORIZONTAL DEFLECTION CIRCUIT

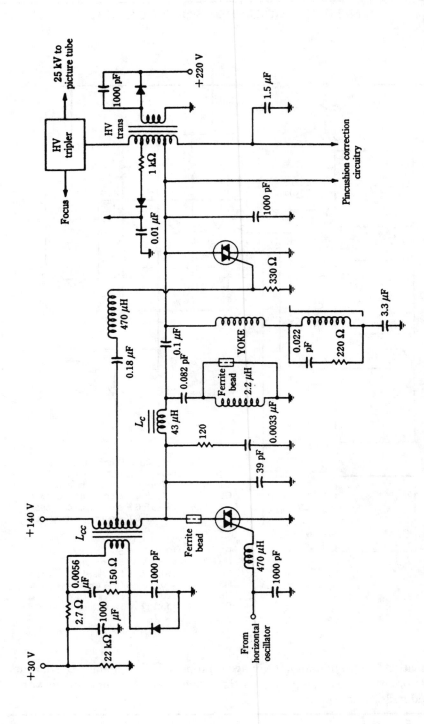

McGRAW-HILL

Fig. 52-21

The circuit illustrates the method of using two SCR devices in a TV horizontal deflection application. This circuit was widely used by certain TV manufacturers as an alternate to the vacuum tube or transistor deflection circuit.

MUTING CIRCUIT

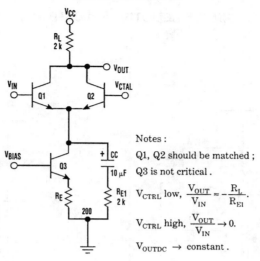

Notes :

Q1, Q2 should be matched ;

Q3 is not critical .

V_{CTRL} low, $\dfrac{V_{OUT}}{V_{IN}} \approx -\dfrac{R_L}{R_{E1}}$.

V_{CTRL} high, $\dfrac{V_{OUT}}{V_{IN}} \to 0$.

$V_{OUTDC} \to$ constant .

ELECTRONIC DESIGN   **Fig. 52-22**

The circuit operates as follows: The signal is input to V_{in} and a dc control voltage is applied to V_c. V_{bias} determines the desired bias point current. Assuming the following component and voltage values:

$$\begin{aligned}
V_{cc} &= 7.6 \text{ Vdc} \\
V_{bias} &= 1 \text{ Vdc} \\
V_{in} &= 1 \text{ Vp-p, ac signal centered about 3.8 Vdc bias} \\
R_L &= 2 \text{ k}\Omega \\
R_E &= 200 \text{ k}\Omega \\
R_{E1} &= 2 \text{ k}\Omega
\end{aligned}$$

Q3 bias current is 1 mA, and dc output voltage is about 5.8 V with an ac gain of about –1, Q1 and Q2 form a current switch and Q3 acts as a constant current source.

For unmuted operation, $V_c = 0$ Vdc, and all of the bias current flows through Q1. Consequently, the circuit operates as a normal common emitter stage, with ac gain = $-R_L/R_{E1}$. When $V_c = 5$ Vdc, all of the bias current flows through Q2, reducing the signal gain to zero. However, because the same dc current flows through R_L in both cases (unmuted and muted), the bias point at the output remains fixed. The C_c/R_{E1} network is required to bypass the Q3 current source (which is a high impedance) to achieve a low ac impedance at the emitter of the Q1 common emitter stage during unmuted operation. C_c is chosen to be a short circuit at signal frequencies of interest. The circuit works best if the Q1 and Q2 pair is matched. Typical change in the output dc voltage from unmuted to muted condition is <5 mVdc.

R_L, R_E, and V_{bias} are chosen for desired dc operating conditions and signal dynamic range. V_{bias} can be generated via a V_{cc} voltage divider. The signal at V_{in} can be ac coupled, but a bias circuit must be added to Q1's base to generate a dc component. R_{E1} is chosen for desired ac gain. V_{in} must be centered about a dc component, and, to assure proper switching action, $V_{CT}R_L$ must be higher than V_{in} by an amount greater than one V_{BE} drop.

SIMPLE REMOTE GAIN CONTROL

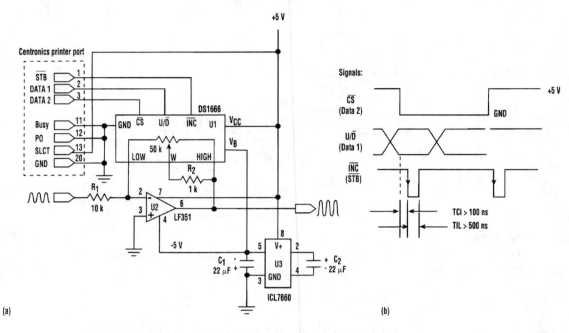

REMOTE CONTROL OF GAIN using a digital audio resistor is possible with this circuit scheme (a). It requires only three ICs and a single 5-V supply to provide gain control via a PC printer port. The input signals to U1 from the printer port are given (b).

```
100 REM LPGAIN.BAS
120 REM GAIN CONTROL FROM PC PRINTER PORT
200 OPEN "LPT1:" FOR OUTPUT AS #1
210 INPUT "GAIN UP OR DOWN (1/0): "; X
220 IF X<0 OR X>1 THEN GOTO 210
300 REM
310 INPUT "NUMBER OF COUNTS: "; C
320 REM SUBROUTINE WOULD START HERE
330 PRINT #1, CHR$(X);
340 C = C-1 : IF C>0 THEN GOTO 330
350 PRINT #1, CHR$(3)
360 GOTO 210
380 RETURN
```

ELECTRONIC DESIGN

Fig. 52-23

The listing is a test program that demonstrates circuit operation using an IBM-compatible PC. To form a subroutine for a main program, use lines 330 to 380, deleting line 360. The calling program then would pass values for X (wiper direction) and C (number of increments).

LOOP OSCILLATOR ELIMINATOR

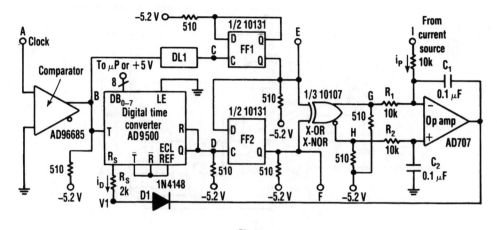

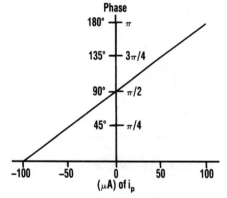

Fig. 52-24

This circuit uses negative feedback to a digital-to-time converter, and can supply a current-controlled delay to replace the oscillator in a phase-locked loop that handles input frequencies from 40 kHz to 40 MHz.

A current sourced into the inverting input of the op-amp integrator's summing node can phase shift the pulses at F in relation to those at E by up to 180°.

1-A VOLTAGE FOLLOWER

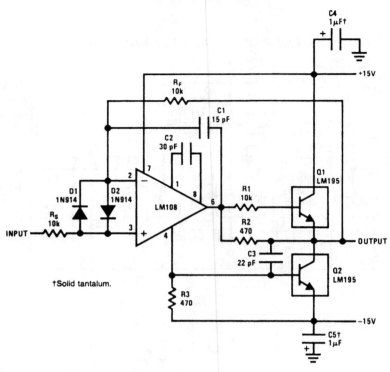

NATIONAL SEMICONDUCTOR

Fig. 52-25

This power voltage follower is good to 300 kHz.

ELECTRONIC FISH LURE

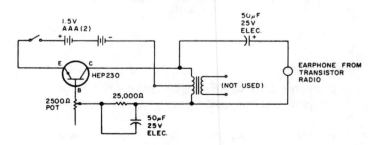

73 AMATEUR RADIO TODAY

Fig. 52-26

The click-click sound lures fish to the vicinity, where your bait or lure can do the rest. The transformer is a subminiature type with a 500-Ω, center-tapped primary and a 3.2-Ω secondary. Put the circuit in a watertight container and lower it into the water.

HEARTBEAT TRANSDUCER

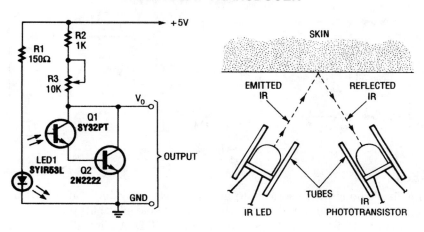

POPULAR ELECTRONICS

Fig. 52-27

A simple heart-beat transducer can be made from an infrared LED and an infrared phototransistor. It works because skin acts as a reflective surface for infrared light. The IR reflectivity of one's skin depends on the density of blood in it. Blood density rises and falls with the pumping action of the heart. So the intensity of infrared reflected by the skin (and thus transmitted to the phototransistor) rises and falls with each heartbeat.

CONTACT DEBOUNCER

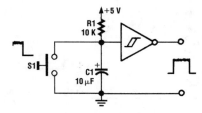

ELECTRONICS NOW

Fig. 52-28

A contact debouncer using a Schmitt trigger, such as a TTL7414, provides a "clean" pulse from a switch contact closing.

POSITIVE FEEDBACK CABLE TERMINATOR

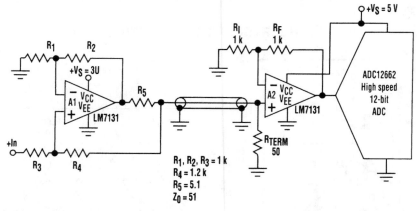

ELECTRONIC DESIGN

Fig. 52-29

Positive feedback along with a series output resistor can provide a controlled output impedance from an op-amp circuit. The circuit is useful when driving coaxial cables that must be terminated at each end in their characteristic impedance, which is often 50 Ω. Adding a 50-Ω series resistor on the op amp's output obviously reduces the available signal swing.

×10 FREQUENCY MULTIPLIER

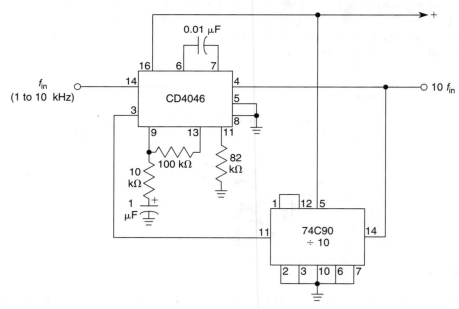

WILLIAM SHEETS

Fig. 52-30

In this circuit, the CD4046 is set up so that the V_{CO} operates at 10 to 100 kHz. The output pin (4) is fed back to a ÷10 counter. When the input frequency is ⅒ the output, lockup will occur.

JACOB'S LADDER

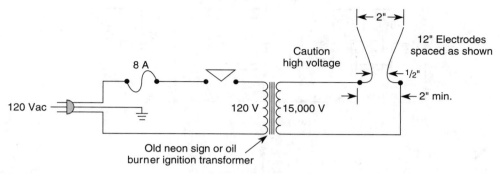

WILLIAM SHEETS

Fig. 52-31

A "Jacobs Ladder" can be made from an old neon sign or oil burner ignition transformer. A rating of 12 to 15 kV at 20 to 30 mA will be adequate. Make sure to mount the electrodes to a pair of insulators, at least 2" apart, and bent and spaced, as shown. The ladder should be enclosed in a clear plastic housing to prevent accidental contact with the high voltage and to ensure a stable arc. Vent holes should be placed top and bottom to allow gases to escape.

MASTER-SLAVE DEVICE ERROR CHECKER

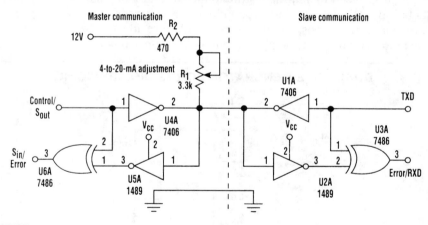

ELECTRONIC DESIGN

Fig. 52-32

An error-check mechanism introduced into master and slave communication devices can indicate mismatches when both the master and slave start sending data simultaneously. The error flag goes high and indicates a mismatch in the data.

The master is the one that can interrupt the communication from a device at the other end and force it to listen. It does this forcing a low voltage level over the communication line by raising the control line to a high level. This inhibits data flow over the lines from the slave device. As a result, the slave turns into a listen mode (not a hardware feature, but rather incorporated in the software). The slave device can transmit the data after communication from the master device ceases.

GROUND LOOP PREVENTER

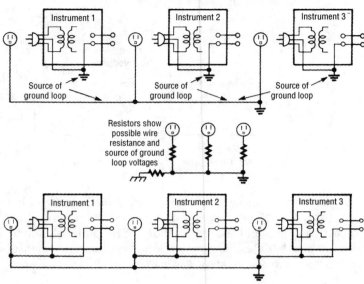

ELECTRONICS NOW

Fig. 52-33

Ground loops are caused by improper grounding. Ground-loop voltages can interfere with test measurements because the voltages in a ground loop can be larger than the signals you're trying to measure. To prevent ground loops, use two wire plugs to provide the line power to the test instruments and a separate wire to bring the input grounds of the instruments to a common ground.

DUAL TONE GENERATOR FOR AUDIO SERVICING

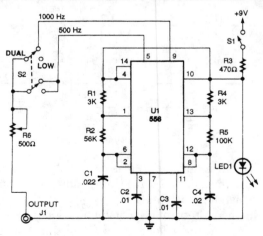

POPULAR ELECTRONICS

Fig. 52-34

This dual-tone generator can insert a distinctive tone in the audio section of a circuit under test. That way, you can work your way back from the speaker, stage-by-stage, to locate a faulty section.

DIODELESS PEAK-HOLD CIRCUIT

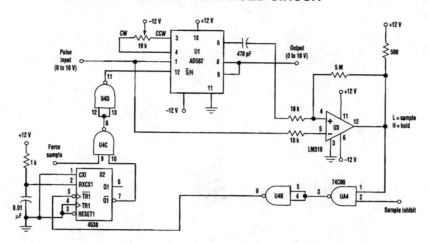

ELECTRONIC DESIGN

Fig. 52-35

The input pulse is fed into the sample-and-hold amplifier (an inexpensive AD582), as well as the comparator U3. The SHA's output also is fed into the comparator. If the input pulse is higher in amplitude than the SHA's output, the comparator output goes low and the 4538 one-shot produces a 10-µs pulse that is fed back to cause the SHA to sample and then hold the voltage. Subsequent input voltages that are less than the held value won't cause the one-shot to fire again.

Gates U4A and U4B are used to inhibit the sampling when necessary. Gates U4C and U4D, at the one-shot's output, can force the AD 582 into the sample mode. This feature is useful to reset the output to zero by forcing a sample when the input to the AD582 is zero. The polarity of the peak-hold circuit can be easily changed from positive-to-negative peak hold by reversing the inputs of the comparator.

53

Mixer Circuits

The sources of the following circuits are contained in the Sources section, which begins on page 707. The figure number in the box of each circuit correlates to the entry in the Sources section.

Simple Audio Mixer
Op-Amp Audio Mixer

SIMPLE AUDIO MIXER

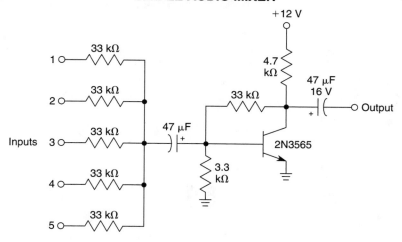

WILLIAM SHEETS

Fig. 53-1

A single transistor is used as an audio mixer, the transistor serving as a feedback amplifier.

OP-AMP AUDIO MIXER

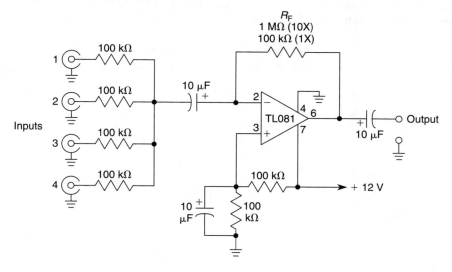

WILLIAM SHEETS

Fig. 53-2

This circuit will mix several audio signals to a common output. R_F can be made 1 MΩ for 10 × (20 dB) or 100 kΩ for unity gain.

54

Model and Hobby Circuits

The sources of the following circuits are contained in the Sources section, which begins on page 707. The figure number in the box of each circuit correlates to the entry in the Sources section.

Model Railroad Crossing Flasher
Model Railroad Track Control Signal

MODEL RAILROAD CROSSING FLASHER

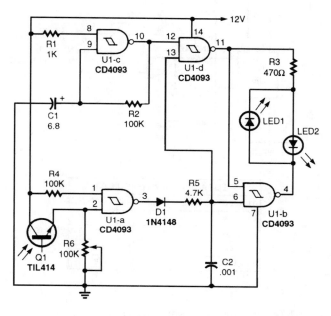

Fig. 54-1

Gate U1-c is set up as an oscillator whose frequency is determined by C1 and R1. Gates U1-b and U1-d are set up as an RS flip-flop that is gated on by U1-a. Gate U1-a in conjunction with Q1 operates as the control gate for the flip-flop. Components D1, C2, and IR5 act as a delay circuit to compensate for any light getting through the gaps between cars as they pass over the phototransistors. The light-emitting diodes are connected so that they operate alternately, depending on the outputs of U1-d and U1-b.

Basically, R6 is adjusted so that ambient room-light striking Q1 (and any other phototransistors connected in series) keeps the output of U1-a at pin 3 low. When a car passes over the phototransistor, which is installed between ties in the track, pin 3 goes high, allowing a high to be placed on pins 5 and 13. That allows the high output of U1-c at pin 10 to enable pin 12, which in turn allows pin 11 to go low. That makes a complete path for LED2 to operate. When pin 10 goes low, pin 11 goes high. That makes pin 5 high, and thus, enables pin 4 to go low and completes the circuit for LED1. That alternates the LEDs, which are installed in a railroad-crossing signal.

MODEL RAILROAD TRACK CONTROL SIGNAL

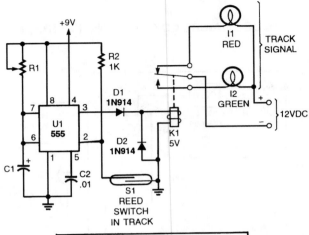

RED-LAMP ON TIME (SECONDS)		
R1 (KILOHMS)	C1 = 10µF	C1 = 100µF
100	2	16
220	3	32
470	6	70
1000	15	175

POPULAR ELECTRONICS

Fig. 54-2

When a train passes S1 (a red switch), a small magnet glued to the underside operates S1 and causes U1 to generate a pulse, activating relay K1 and changing the signal from green to red. After a time determined by R1 and C1 (see table), the relay de-energizes and the signal goes back to green.

55

Modulator Circuits

The sources of the following circuits are contained in the Sources section, which begins on page 707. The figure number in the box of each circuit correlates to the entry in the Sources section.

VARACTORLESS HF MODULATOR

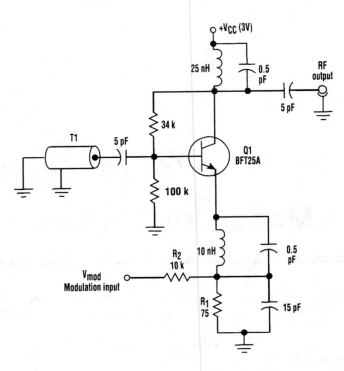

ELECTRONIC DESIGN

Fig. 55-1

Traditionally, high-frequency oscillators are frequency-modulated by using a varactor. However, varactors usually require a large voltage change to achieve a reasonable capacitance change—a problem in many battery-powered systems.

Such a problem can be overcome by employing base-charging capacitance modulation. Resistor R1 establishes Q1's current, and R2 allows control of the collector bias current by V_{mod}. The transmission line (T1) in the negative resistance-type oscillator determines the frequency of oscillation. T1 is a high-quality, low-loss, ceramic coaxial shorted quarter-wave transmission line. Under proper terminal impedances, a negative resistance is "seen" at Q1's base. T1 reacts with this negative resistance to produce sustained oscillations.

Frequency modulation is accomplished by changing Q1's collector bias current and thus changing Q1's base-charging capacitance. This effect is "seen" at Q1's base and causes a frequency shift in the resonators quarter-wave node.

MODULATOR FOR VIDEO

Sources: SAWFs

Crystal Technology, Inc.
1035 E. Meadow Circle
Palo Alto, CA 94303

Kyocera International, Inc.
8611 Balboa Ave.
San Diego, CA 92123

MuRata Corp. of America
1148 Franklin Rd. S.E.
Marietta, GA 30067

CRYSTALS

Saronix
4010 Transport at San Antonio Rd.
Palo Alto, CA 94303

COILS

Toko America, Inc.
5520 W. Touhy Ave.
Skokie, Ill. 60077

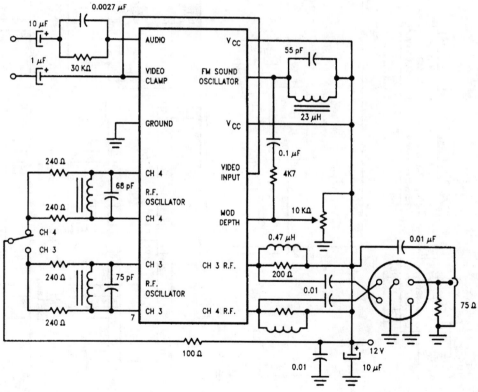

NATIONAL SEMICONDUCTOR

Fig. 55-2

This circuit uses an LM2889 and a saw filter for use as a TV modulator.

DIGITAL PULSE-WIDTH MODULATION CIRCUIT

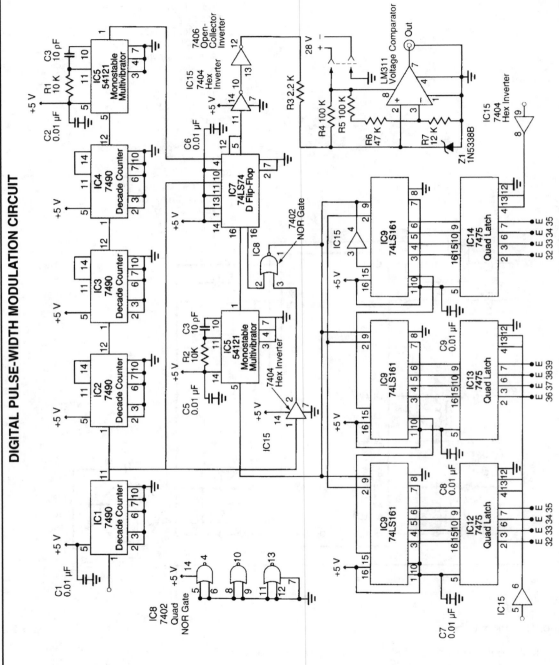

Fig. 55-3

DIGITAL PULSE-WIDTH MODULATION CIRCUIT (*Cont.*)

This circuit incorporates low-power Schottky transistor/transistor-logic (TTL) devices in critical high-speed parts. The 5-MHz clock signal is converted to a 1-MHz signal by a 7490 decade counter configured to divide by 5. The 1-MHz signal is sent, via a clock synchronizer, to a three-stage binary divider that consists of three cascaded 74LS161 binary dividers. The synchronizer consists of a 74LS74 D flip-flop, a 7404 inverter gate, and a 7402 NOR gate. The dividers are programmed from the STD bus by means of three 7475 quad latches; this makes it possible to program the frequency division from 1 to 4096 (12 bits).

The 1-MHz signal is also divided by 1000 by use of three cascaded 7490 decade counters, each configured to divide by 10; this provides a 1-kHz signal, which is sent to a 54121 monostable multivibrator configured to provide a 0.1-µs pulse, bombarded with an ion beam source in preparation for the materials about to be deposited. While the surface is bombarded with an ion beam, an electron beam source is activated so that a layer of fused silica is vapor-deposited to a total desired thickness value (typically, 1 micron or 10,000 Å). The layer of fused silica serves as a surface stabilization layer for the next step.

A metal mask with an aperture in the specified pattern of the sensor film is placed on the surface at the specified sensor location. The surface area exposed through the mask is cleaned by ion-beam bombardment for a predetermined time. Then as the bombardment continues, a metal (typically, nickel, platinum, and/or palladium) is vapor-deposited through the mask from the electron-beam source to form the sensor film. Deposition is continued until the thickness of the film reaches the value specified in the particular sensor design. A representative value for a nickel sensor film is 2500 Å.

Next, a pattern for thin film leads is defined by taping directly on the surface of the model with Kapton (or equivalent) polyimide tape. The thin film leads are fabricated by a combination of ion-beam bombardment and electron-beam vapor deposition like that used to deposit the sensor film. The metal vapor deposited in this step is typically copper, gold, or aluminum. A typical thickness for copper leads on the nickel sensor film is about 10,000 Å.

FOUR-QUADRANT MULTIPLIER AS DSB MODULATOR

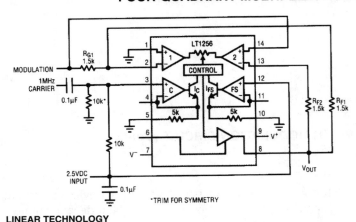

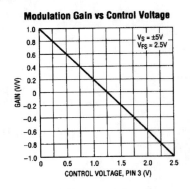

Modulation Gain vs Control Voltage

LINEAR TECHNOLOGY

Fig. 55-4

PULSE-WIDTH MODULATOR

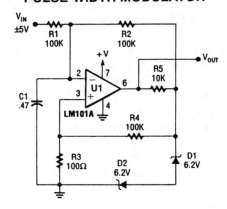

POPULAR ELECTRONICS

Fig. 55-5

LINEAR (AM) AMPLITUDE MODULATOR

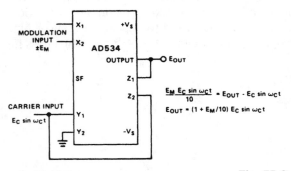

$$\frac{E_M \ E_C \sin \omega_C t}{10} = E_{OUT} - E_C \sin \omega_C t$$

$$E_{OUT} = (1 + E_M/10) \ E_C \sin \omega_C t$$

This is a very simple amplitude modulator. It makes use of the Z2 terminal to add the carrier directly to the output, thus bypassing the multiplier for zero modulation input. It has the advantage of allowing operation from a differential modulation input.

ANALOG DEVICES

Fig. 55-6

VIDEO MODULATOR HOOKUP

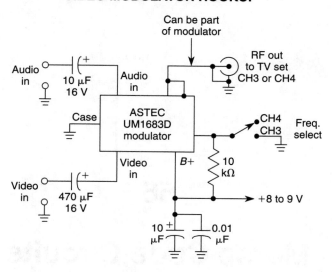

WILLIAM SHEETS

Fig. 55-7

This circuit uses an ASTEC UM1683D, but it is typical of many RF video modulators used in VCRs and satellite receivers.

56

Morse Code Circuits

The sources of the following circuits are contained in the Sources section, which begins on page 707. The figure number in the box of each circuit correlates to the entry in the Sources section.

Active CW Audio Filter
Morse Messenger
CW Identifier with Sine-Wave Audio Output
Simple Code Practice Oscillator

ACTIVE CW AUDIO FILTER

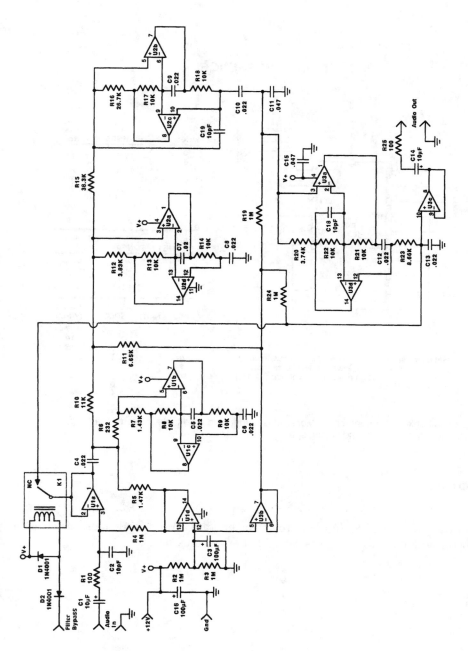

Fig. 56-1

The audio filter shown has a bandpass of 200 Hz centered on 700 Hz. Resistors are 1% tolerance and capacitors should be 5% tolerance.

MORSE MESSENGER

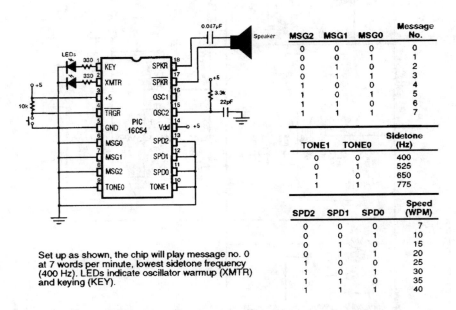

MSG2	MSG1	MSG0	Message No.
0	0	0	0
0	0	1	1
0	1	0	2
0	1	1	3
1	0	0	4
1	0	1	5
1	1	0	6
1	1	1	7

TONE1	TONE0	Sidetone (Hz)
0	0	400
0	1	525
1	0	650
1	1	775

SPD2	SPD1	SPD0	Speed (WPM)
0	0	0	7
0	0	1	10
0	1	0	15
0	1	1	20
1	0	0	25
1	0	1	30
1	1	0	35
1	1	1	40

Set up as shown, the chip will play message no. 0 at 7 words per minute, lowest sidetone frequency (400 Hz). LEDs indicate oscillator warmup (XMTR) and keying (KEY).

Simple hook-up diagram for the Morse Messenger chip. The table indicates the range of messages, sidetones, and keying speeds.

73 AMATEUR RADIO TODAY

Fig. 56-2

This keyer uses a PIC16C54 micro-controller to generate a Morse code message. The microcontroller must be programmed to suit users call IC or desired message.

MORSE MESSENGER (*Cont.*)

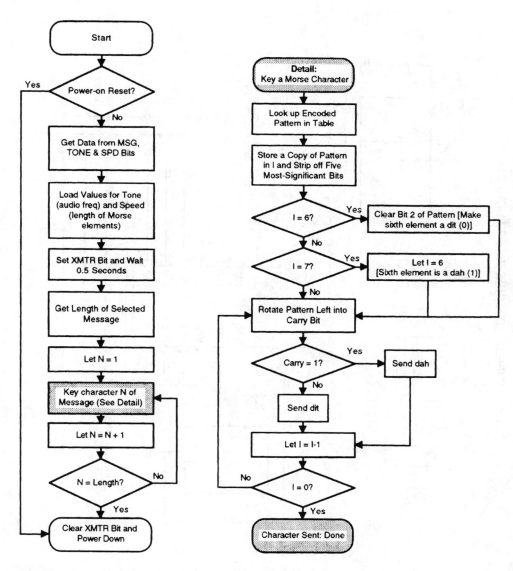

Logic of the Morse Messenger's program. This algorithm can be adapted to other devices with the help of the Morse encoding table.

CW IDENTIFIER WITH SINE-WAVE AUDIO OUTPUT

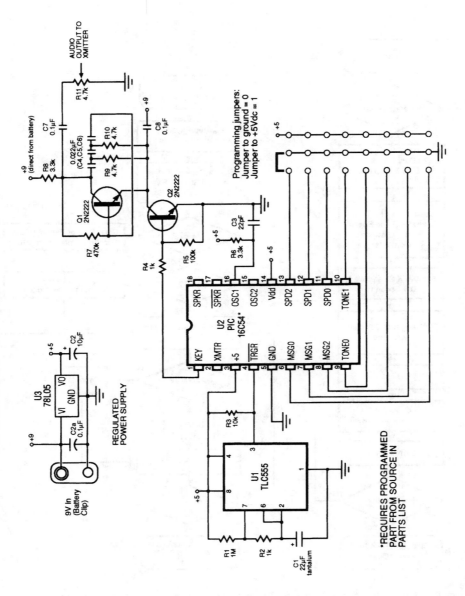

Fig. 56-3

73 AMATEUR RADIO TODAY

This identifier can be used to drive a hidden transmitter in a radio "fox hunt" activity, where the object is to locate a hidden transmitter.

SIMPLE CODE PRACTICE OSCILLATOR

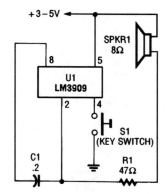

POPULAR ELECTRONICS

Fig. 56-4

With only a minor circuit change, the basic LM3909 oscillator configuration can be turned into a code-practice oscillator.

57

Motor-Control Circuits

The sources of the following circuits are contained in the Sources section, which begins on page 707. The figure number in the box of each circuit correlates to the entry in the Sources section.

PRECISE dc MOTOR SPEED CONTROLLER

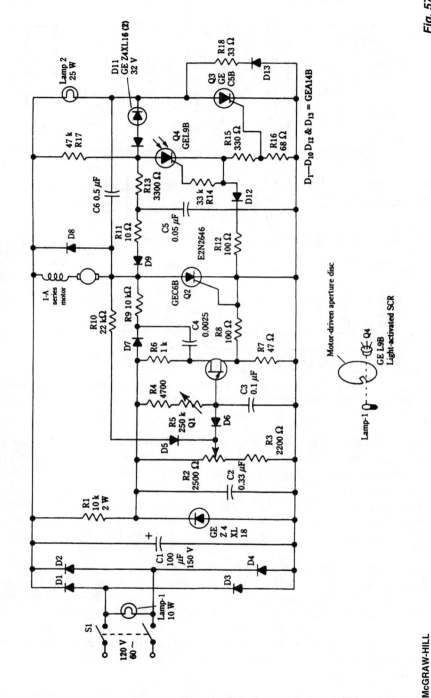

McGRAW-HILL

Fig. 57-1

A series dc motor can be made to have the same characteristics as an ac synchronous motor using this circuit. This control technique is useful where a constant motor speed is needed.

ACCURATE MOTOR SPEED CONTROL

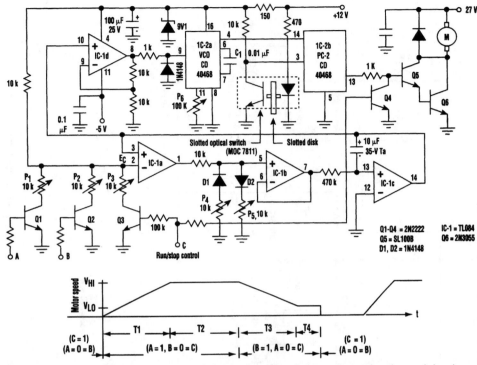

This motor velocity profile is provided by the controller. Note the smooth acceleration and deceleration profiles and the constant motor speed between the two ramps.

ELECTRONIC DESIGN

Fig. 57-2

This circuit can control dc motors used in machines that pull materials, such as wire, yarn, film, etc., from the supply rolls and rewind them onto smaller rolls. Its motor velocity profile is shown in the figure.

A lightweight disk with at least 32 slots is mounted on the motor shaft. It works in a slotted optical switch (MOC7811) to sense the motor speed. Phase comparator 2 (PC-2) of the phase-locked loop (IC-CD 4046B) compares the frequency (f_v) at the output of the VCO with the pulse rate (f_m) at the optical switch output. The PC-2 output drives the motor via the transistors when $f_v > f_m$, and removes the supply to the motor when $f_m > f_v$. The drive system quickly reaches an equilibrium condition when $f_m = f_v$.

Op amps IC-1a, 1b, and 1c form a tracking integrator whose output always smoothly reaches and remains at a voltage equal to the command voltage (E_c) presented at the inverting input of op-amp IC-1a. When the digital control inputs are set to ($A = 1, B = 0 = C$), the integrator generates a positive slope ramp that sweeps the VCO frequency and, thus, accelerates the motor. The acceleration rate and the constant speed V_{HI} can be adjusted by presets P4 and P1, respectively.

Similarly, when the inputs are set to ($A = 0 = C, B = 1$), the integrator generates a negative slope ramp that decelerates the motor. Presets P5 and P2 can be adjusted to set the required deceleration and the constant speed V_{LO}, respectively.

MOTOR DIRECTION CONTROL USING DISCRETE TRANSISTORS

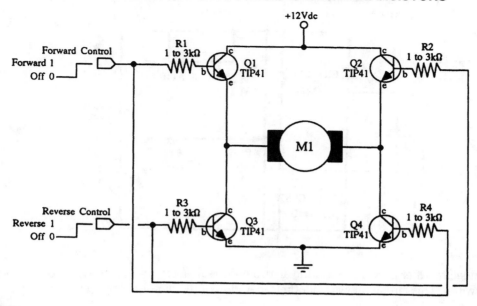

McGRAW-HILL

Fig. 57-3

For best operation, mount the transistors on heatsinks. The transistors specified are fine for small hobby motors, or up to about 6 volts dc and between 800 and 1000 mA. M1 is a small hobby motor.

LONG TIME-DELAY MOTOR-CONTROL CIRCUIT

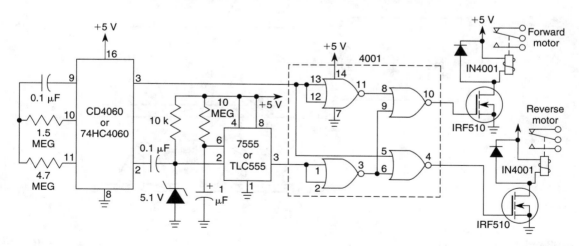

ELECTRONICS NOW

Fig. 57-4

Circuit controls forward and reverse motors. Every hour, one motor runs for 10 seconds.

FULL-WAVE SPEED CONTROL FOR MOTORS

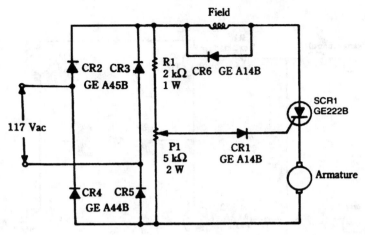

Fig. 57-5

A bridge rectifier provides pulsating dc to a universal motor, and the SCR is used as a phase-controlled switch. This circuit allows smoother operation of the motor at low speeds.

SCR MOTOR SPEED CONTROL

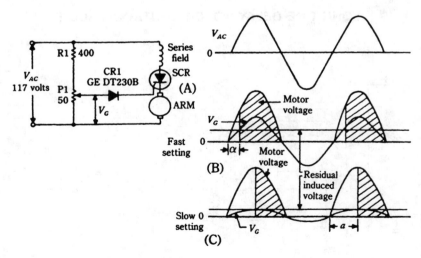

Fig. 57-6

An SCR is used in a phase-control type application to supply a variable pulsating dc voltage to a motor.

TRIAC MOTOR-CONTROL CIRCUIT

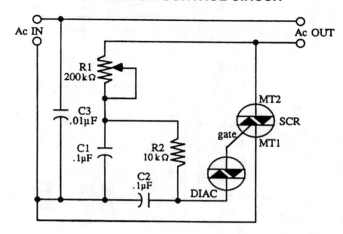

McGRAW-HILL

Fig. 57-7

An SCR-controlled ac motor control circuit. This is a full-wave circuit and is best used when the load remains constant.

LOW-VOLTAGE dc MOTOR-SPEED CONTROLLER

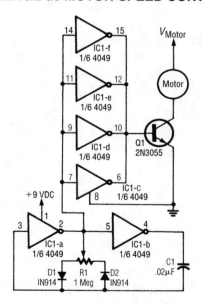

ELECTRONICS NOW

Fig. 57-8

This circuit varies the duty cycle, rather than the voltage. The two diodes control the positive and negative halves of the capacitor's charging cycle.

MOTOR DIRECTION CONTROL

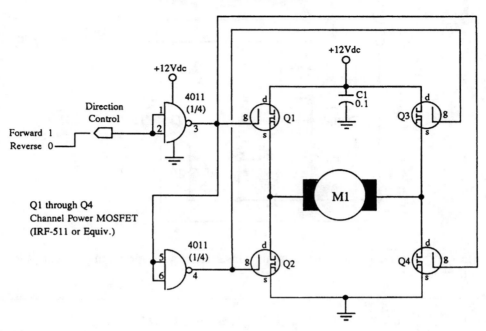

Fig. 57-9

M1 is a small hobby dc motor.

58

Multivibrator Circuits

The sources of the following circuits are contained in the Sources section, which begins on page 707. The figure number in the box of each circuit correlates to the entry in the Sources section.

Astable Multivibrator with Starting Network
Bistable Multivibrator
Astable
Astable with Variable Pulse Width
One-Shot Multivibrator
Basic 555 Astable Multivibrator
Astable Multivibrator

ASTABLE MULTIVIBRATOR WITH STARTING NETWORK

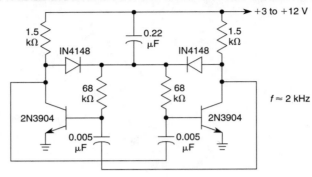

WILLIAM SHEETS

Fig. 58-1

This circuit will start with a slowly rising supply voltage waveform.

BISTABLE MULTIVIBRATOR

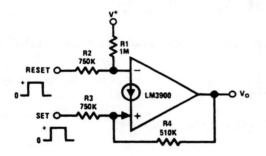

Positive feedback is provided by resistor R4, which causes the latching. A positive pulse at the "set" input causes the output to go high and a "reset" positive pulse will return the output to essentially 0 Vdc.

NATIONAL SEMICONDUCTOR

Fig. 58-2

ASTABLE

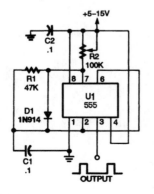

At the instant that power is applied to the 555 astable, timing capacitor C1 is initially discharged, causing the output of the chip output at pin 3 to be high. Once C1 has charged to about ⅔ of the supply voltage, its output goes low, and the discharge transistor turns on, draining the charge on C1.

POPULAR ELECTRONICS

Fig. 58-3

ASTABLE WITH VARIABLE PULSE WIDTH

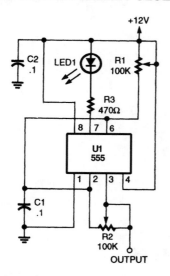

POPULAR ELECTRONICS **Fig. 58-4**

This produces a positive variable width pulse and has a symmetry control. R1 and R2 control the pulse width and symmetry.

ONE-SHOT MULTIVIBRATOR

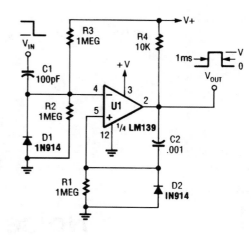

POPULAR ELECTRONICS **Fig. 58-5**

An LM139 section can be used as a one shot.

BASIC 555 ASTABLE MULTIVIBRATOR

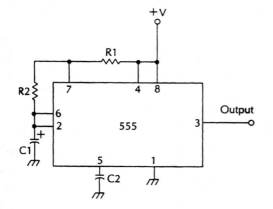

McGRAW-HILL **Fig. 58-6**

$$F = \frac{1.44}{(R_1 = 2R_2)\, C_1}$$

ASTABLE MULTIVIBRATOR

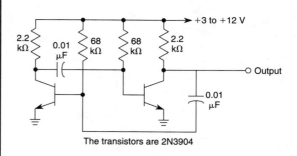

The transistors are 2N3904

WILLIAM SHEETS **Fig. 58-7**

59

Noise Circuits

The sources of the following circuits are contained in the Sources section, which begins on page 707. The figure number in the box of each circuit correlates to the entry in the Sources section.

Noise Generator
Dolby Noise-Reduction Circuit
Audio Noise-Based Voting Circuit
Adjustable Noise Clipper
Simple Noise Limiter

NOISE GENERATOR

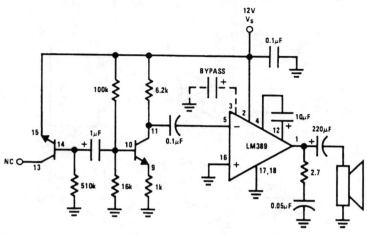

NATIONAL SEMICONDUCTOR

Fig. 59-1

This noise generator uses a Zener diode. The transistors are part of the LM389.

DOLBY NOISE-REDUCTION CIRCUIT

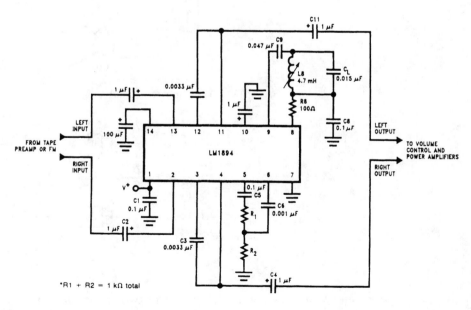

NATIONAL SEMICONDUCTOR

Fig. 59-2

AUDIO NOISE-BASED VOTING CIRCUIT

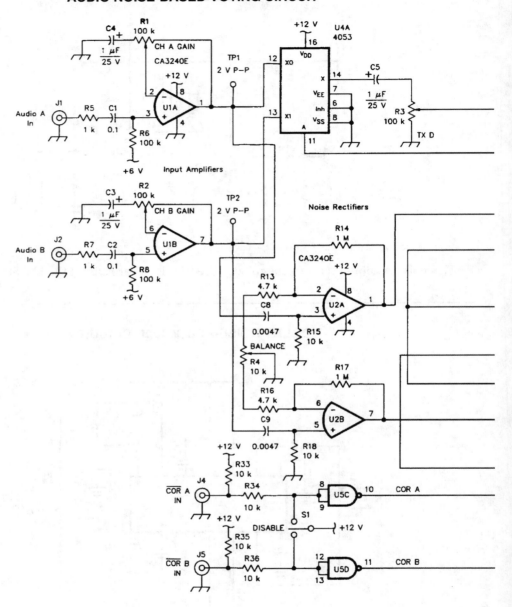

QST

The purpose of this circuit is the selection of the output of two receivers, tuned to the same channel, that has the better signal to noise ratio. This circuit compares the two noise leads from the receivers and selects the one with the lower audio noise level.

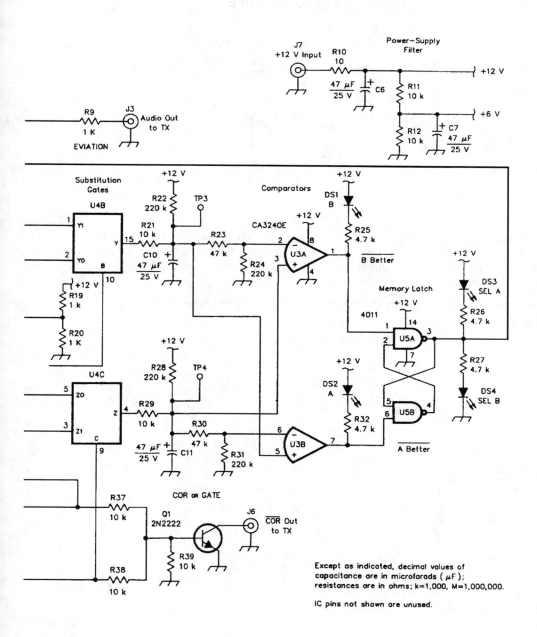

Fig. 59-3

ADJUSTABLE NOISE CLIPPER

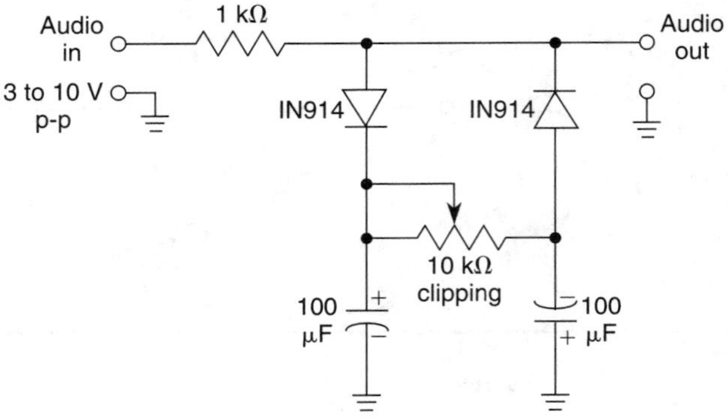

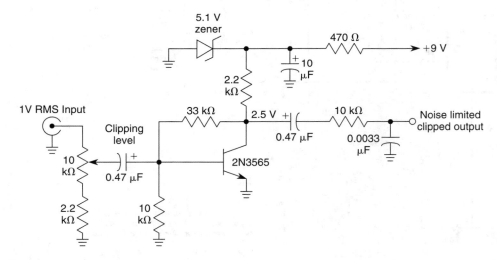

WILLIAM SHEETS

Fig. 59-4

This circuit uses two diodes and is a peak-to-peak limiter. The capacitors charge up to a dc level determined by the peak-to-peak audio signal and the clipping control. A positive or negative peak or spike is clipped if it exceeds this level plus the diode drops. The circuit should be operated at several volts level for best results.

SIMPLE NOISE LIMITER

WILLIAM SHEETS

Fig. 59-5

This circuit uses a symmetrical limiter obtained by biasing a transistor to a Q point that is half of the supply voltage and driving it into saturation and cutoff. An input of 1 to 2 V RMS is sufficient. This output will be approximately 4 V p-p into a high-impedance load.

60

Operational Amplifier Circuits

The sources of the following circuits are contained in the Sources section, which begins on page 707. The figure number in the box of each circuit correlates to the entry in the Sources section.

BASIC OP-AMP CIRCUITS

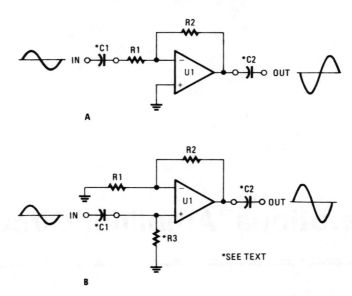

A

B

RADIO-ELECTRONICS

Fig. 60-1

The two simplest op-amp configurations are the inverting (A) and the noninverting (B). Resistor R3 is needed only if C1 is used in the noninverting circuit.

OP AMPS WITH LONG RC TIME CONSTANTS

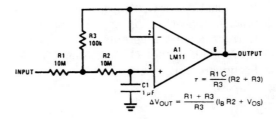

$$\tau = \frac{R1\,C}{R3}(R2 + R3)$$

$$\Delta V_{OUT} = \frac{R1 + R3}{R3}(I_B\,R2 + V_{OS})$$

NATIONAL SEMICONDUCTOR

Fig. 60-2

This circuit multiplies RC time constant to 1000 seconds and provides low output impedance. Cost is lowered because of reduced resistor and capacitor values.

OP-AMP OFFSET NULL

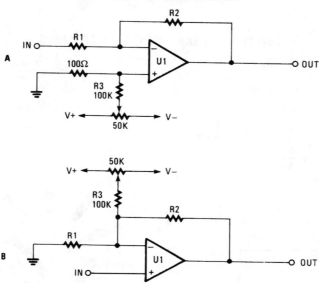

RADIO-ELECTRONICS

Fig. 60-3

Offset problems can occur in the best of circuits (and often do) without regard to whether the circuit is inverting (A), or noninverting (B). Offset-nulling potentiometers are useful in correcting the output to zero, but their effectiveness will vary under different conditions.

BASIC OP-AMP AUDIO AMPLIFIER

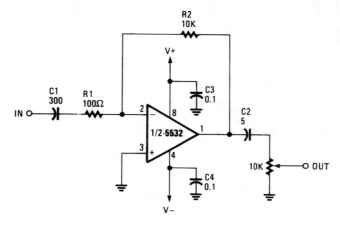

RADIO-ELECTRONICS

Fig. 60-4

Any general-purpose op amp can be used in this application.

427

INPUT GUARDING FOR HI-Z OP AMPS

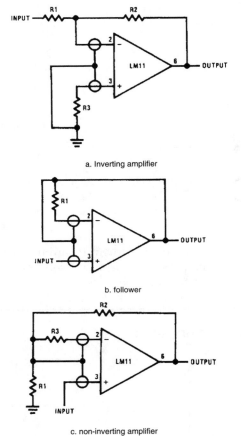

a. Inverting amplifier

b. follower

c. non-inverting amplifier

Input guarding for various op amp connections. The guard should be connected to a point at the same potential as the inputs with a low enough impedance to absorb board leakage without introducing excessive offset.

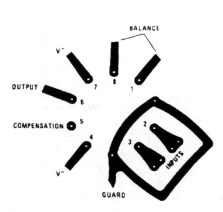

Bottom view

Input guarding can drastically reduce surface leakage. Layout for metal can is shown here. Guarding both sides of board is required. Bulk leakage reduction is less and depends on guard ring width.

ENHANCED OP-AMP BALANCED AMPLIFIER

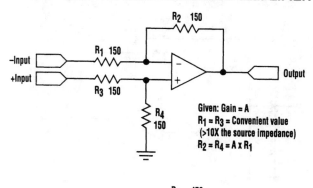

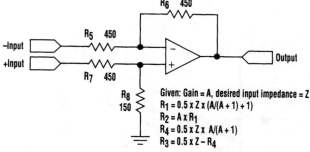

Given: Gain = A
$R_1 = R_3$ = Convenient value
(>10X the source impedance)
$R_2 = R_4 = A \times R_1$

Given: Gain = A, desired input impedance = Z
$R_1 = 0.5 \times Z \times (A/(A+1) + 1)$
$R_2 = A \times R_1$
$R_4 = 0.5 \times Z \times A/(A+1)$
$R_3 = 0.5 \times Z - R_4$

ELECTRONIC DESIGN

Fig. 60-6

The classic single op-amp balanced circuit works well in low-source-impedance configurations, but tends to struggle in higher-source-impedance applications because of the varying input impedance of the inputs referred to ground.

A modified version of the classic op-amp configuration of the figure uses a different set of formulas to determine the resistor values. It equalizes the impedance of both inputs by considering the op amp's active participation.

PARALLELED POWER OP AMPS

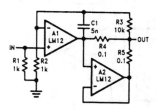

NATIONAL SEMICONDUCTOR

Fig. 60-7

Two power op amps can be paralleled using this master/slave arrangement, but high-frequency performance suffers.

SINGLE-SUPPLY OP-AMP APPLICATIONS

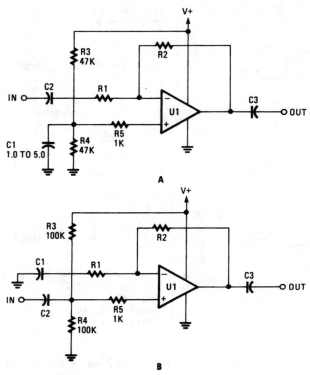

A

B

RADIO-ELECTRONICS

Fig. 60-8

An op amp that normally needs two supplies can be used when only a single supply is needed. The value of *V*+ should be twice the minimum allowable values of the positive and negative voltages normally needed. For example, a 12-V single-supply application would require an op amp capable of ±6-V operation.

CURRENT REGULATOR OP AMP

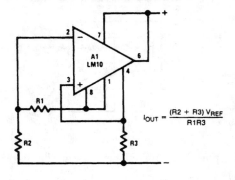

$$I_{OUT} = \frac{(R2 + R3)\,V_{REF}}{R1R3}$$

NATIONAL SEMICONDUCTOR

Fig. 60-9

OP-AMP RESISTANCE MULTIPLICATION CIRCUIT

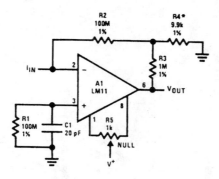

NATIONAL SEMICONDUCTOR

Fig. 60-10

Equivalent feedback resistance is 10 GΩ, but only standard resistors are used. Even though the offset voltage is multiplied by 100, output offset is actually reduced because error is dependent on offset current, rather than bias current. Voltage on summing junction is less than 5 mV.

PSEUDOGROUND

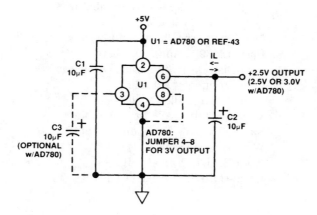

ANALOG DEVICES

Fig. 60-11

For op-amp circuits, a "pseudo ground" is often needed; a voltage reference IC can be used. The Analog Devices AD780 is used here for this application. This can sink or source current.

61

Oscillators (Audio)

The sources of the following circuits are contained in the Sources section, which begins on page 707. The figure number in the box of each circuit correlates to the entry in the Sources section.

EASILY TUNED SINE-WAVE OSCILLATOR

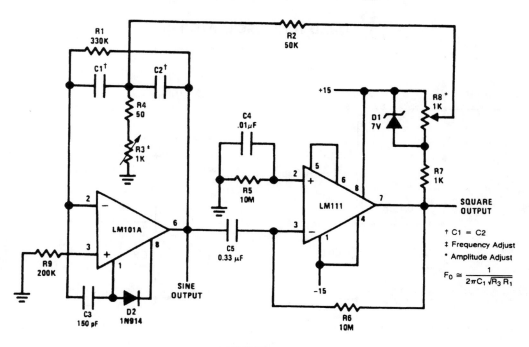

TABLE I

C_1, C_2	Min Frequency	Max Frequency
0.47 μF	18 Hz	80 Hz
0.1 μF	80 Hz	380 Hz
.022 μF	380 Hz	1.7 kHz
.0047 μF	1.7 kHz	8 kHz
.002 μF	4.4 kHz	20 kHz

NATIONAL SEMICONDUCTOR

Fig. 61-1

The circuit will provide both a sine- and square-wave output for frequencies from below 20 Hz to above 20 kHz. The frequency of oscillation is easily tuned by varying a single resistor. This is a considerable advantage over Wien bridge circuits, where two elements must be tuned simultaneously to change frequency. Also, the output amplitude is relatively stable when the frequency is changed.

An operational amplifier is used as a tuned circuit, driven by square wave from a voltage comparator. Frequency is controlled by R1, R2, C1, C2, and R3, with R3 used for tuning. Tuning the filter does not affect its gain or bandwidth, so the output amplitude does not change with frequency. A comparator is fed with the sine-wave output to obtain a square wave. The square wave is then fed back to the input of the tuned circuit to cause oscillation. Zener diode, D1, stabilizes the amplitude of the square wave fed back to the filter input. Starting is ensured by R6 and C5, which provide dc negative feedback around the comparator. This keeps the comparator in the active region.

QUAD TONE OSCILLATOR

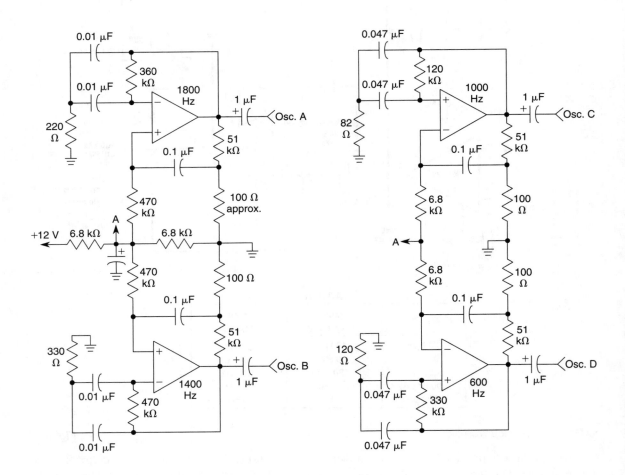

WILLIAM SHEETS

Fig. 61-2

A quad op amp (TL084, etc.) can be used to produce four audio tone generators for use in a test setup. The circuit uses a 12-V supply.

ONE-TRANSISTOR PHASE-SHIFT OSCILLATOR

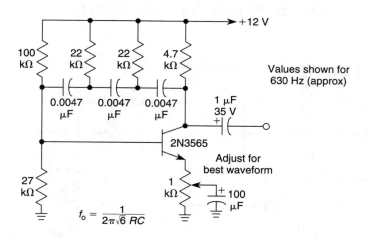

Values shown for
630 Hz (approx)

2N3565

Adjust for
best waveform

$$f_o = \frac{1}{2\pi\sqrt{6}\ RC}$$

WILLIAM SHEETS

Fig. 61-3

A single transistor is used as an active element in an RC phase shift oscillator.

BASIC LM3909 AUDIO OSCILLATOR

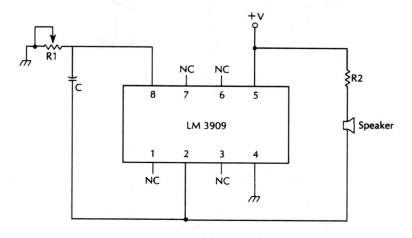

McGRAW-HILL

Fig. 61-4

The LM3909's oscillator frequency can be fine-tuned by adding a resistor to a basic circuit.

LOW-DISTORTION SINE-WAVE OSCILLATOR

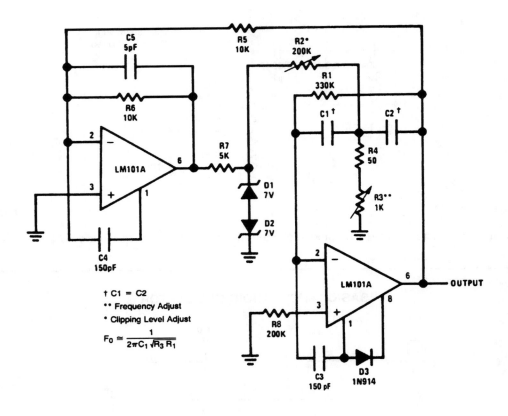

Fig. 61-5

C1, C2	Min. Frequency	Max. Frequency
0.47 μF	18 Hz	80 Hz
0.1 μF	80 Hz	380 Hz
0.022 μF	380 Hz	1.7 kHz
0.0047 μF	1.7 kHz	8 kHz
0.002 μF	4.4 kHz	20 kHz

LOW-FREQUENCY ASTABLE

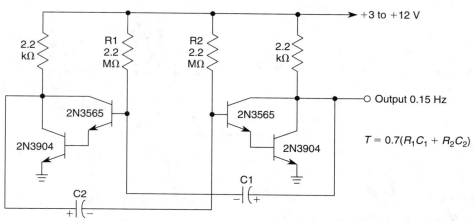

$+3$ to $+12$ V

Output 0.15 Hz

$T = 0.7(R_1 C_1 + R_2 C_2)$

C1 and C2 = 2.2 µF tantalum

WILLIAM SHEETS

Fig. 61-6

By using a high-gain low-current transistor, such as the 2N3565, a pair of Darlington-connected transistors (2N3565 and 2N3904) can be used in a high-impedance configuration.

TTL-BASED AUDIO OSCILLATOR

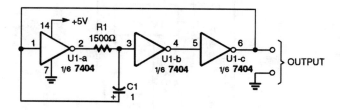

POPULAR ELECTRONICS

Fig. 61-7

Half a 7404 will produce a tone around 1000 Hz with this circuit.

VARIABLE DUTY CYCLE FROM ASTABLE

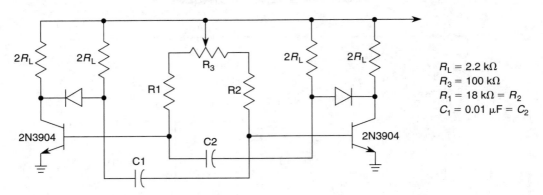

$R_L = 2.2$ kΩ
$R_3 = 100$ kΩ
$R_1 = 18$ kΩ $= R_2$
$C_1 = 0.01$ µF $= C_2$

WILLIAM SHEETS

Fig. 61-8

If $R_1 = R_2 = R_3$ and $C_1 = C_2 = C_3$

If potentiometer R3 is set at $N\%$ of rotation, then

$T_{\text{TOTAL}} \approx 0.7\ [(R + NR_3)C + [R + (1 - N)R_3C]$

$T_{\text{TOTAL}} \approx 1.4\ (R + R_3)C$ and the duty cycle can be varied without changing frequency.

SIMPLE VARIABLE-FREQUENCY OSCILLATOR

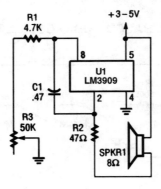

POPULAR ELECTRONICS

Fig. 61-9

In this variable audio frequency oscillator, the output of U1 at pin 2 is used to drive an 8-Ω speaker through R2 (which functions as a current-limiter).

WIEN-BRIDGE OSCILLATOR I

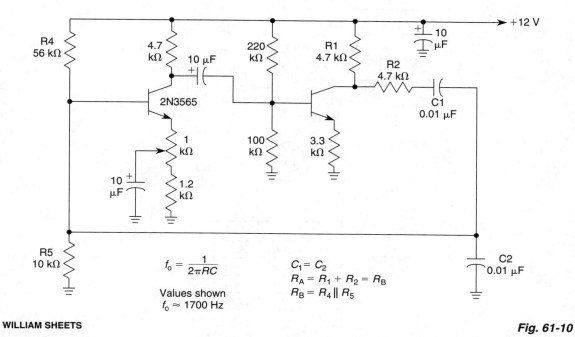

$$f_o = \frac{1}{2\pi RC}$$

Values shown
$f_o \approx 1700$ Hz

$C_1 = C_2$
$R_A = R_1 + R_2 = R_B$
$R_B = R_4 \parallel R_5$

WILLIAM SHEETS

Fig. 61-10

WIEN-BRIDGE OSCILLATOR II

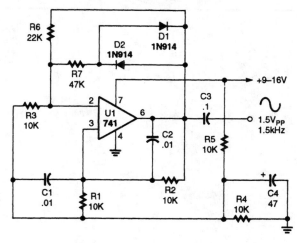

POPULAR ELECTRONICS

Fig. 61-11

The operating frequency of this Wien-bridge oscillator is determined by C1, C2, R1, and R2. It can easily be modified to act as a tunable oscillator by substituting a dual-gang linear potentiometer for R1 and R2.

LOGIC-GATE SINE-WAVE OSCILLATOR

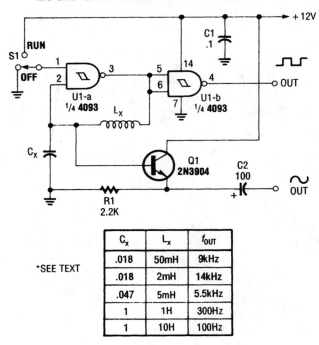

C_x	L_x	f_{OUT}
.018	50mH	9kHz
.018	2mH	14kHz
.047	5mH	5.5kHz
1	1H	300Hz
1	10H	100Hz

*SEE TEXT

POPULAR ELECTRONICS

Fig. 61-12

An inductor and capacitor are used here as frequency-determining elements in an LC oscillator.

62

Oscillators (Miscellaneous)

The sources of the following circuits are contained in the Sources section, which begins on page 707. The figure number in the box of each circuit correlates to the entry in the Sources section.

VARIABLE-FREQUENCY ASTABLE I

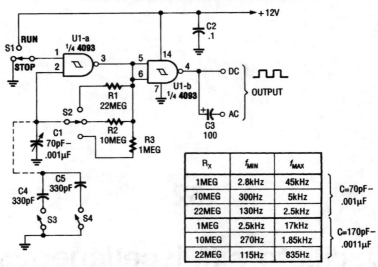

R_x	f_{MIN}	f_{MAX}	
1MEG	2.8kHz	45kHz	C=70pF–
10MEG	300Hz	5kHz	.001µF
22MEG	130Hz	2.5kHz	
1MEG	2.5kHz	17kHz	C=170pF–
10MEG	270Hz	1.85kHz	.0011µF
22MEG	115Hz	835Hz	

POPULAR ELECTRONICS   *Fig. 62-1*

This circuit is a variable-frequency oscillator using a trimmer capacitor or a three-gauge AM broadcast capacitor salvaged from an old AM radio. The three sections must be paralleled.

ASTABLE OSCILLATOR I

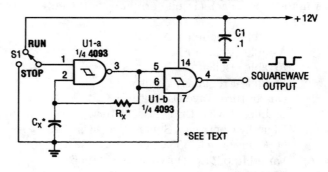

POPULAR ELECTRONICS   *Fig. 62-2*

In this circuit, two gates from the quad 4093 package are used to form a simple astable square-wave oscillator.

The values for R_x and C_x are approximately as follows:

C_x	R_x	f_o
0.001 µF	1 MΩ	3 kHz
0.1 µF	1 MΩ	30 Hz
1 µF	10 MΩ	0.03 Hz

These values can be scaled for other frequencies.

ASTABLE OSCILLATOR II

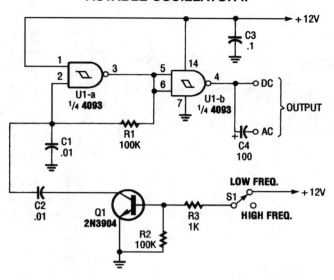

POPULAR ELECTRONICS

Fig. 62-3

By using transistor switch Q1/R2/R3, the frequency of an astable oscillator can be changed with a dc voltage or logic level.

VARIABLE-FREQUENCY ASTABLE II

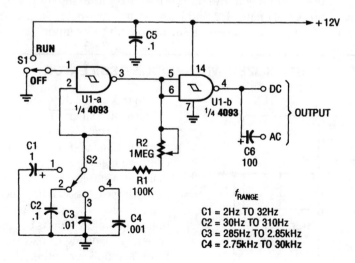

f_{RANGE}

C1 = 2Hz TO 32Hz
C2 = 30Hz TO 310Hz
C3 = 285Hz TO 2.85kHz
C4 = 2.75kHz TO 30kHz

POPULAR ELECTRONICS

Fig. 62-4

This circuit uses a single potentiometer and switched capacitors to cover 2 Hz to 30 kHz.

QUADRATURE-WAVE OSCILLATOR

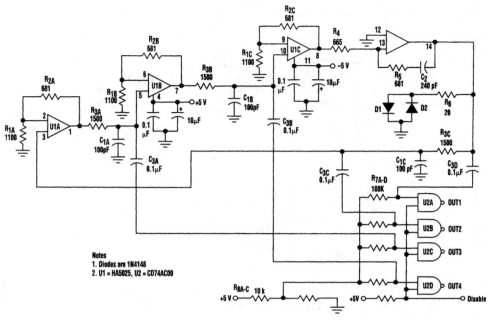

Fig. 62-5

By using a high-frequency quad current-feedback amplifier (the HA5025) as an RC oscillator, four quadrature sine waves can be generated. The HA5025's four separate amplifiers generate the sine waves, and the quad NAND gate, U2, is biased at its threshold, so it acts as a sine-wave to square-wave converter when the sine waves are ac coupled into its input.

STABILIZED WIEN-BRIDGE OSCILLATOR

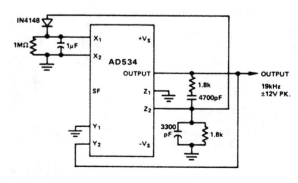

Fig. 62-6

In this application, the AD534 is used as a variable-gain amplifier for the feedback signal from the output to the Y input, via the Wien bridge. The peak-rectifier and filter combination applies sufficient voltage to the X (denominator) input to maintain a stable oscillation-amplitude (with about 0.2% ripple). At startup, because X is small (divider mode), the gain is high, and the oscillation builds up rapidly. This is but one of several possible schemes, involving no external active elements. Its forte is simplicity, rather than high performance; nevertheless, the amplitude is not greatly affected by supply and temperature variations, about 0.003 dB per volt, and 0.005 dB per degree.

DIGITALLY CONTROLLED SQUARE-WAVE OSCILLATOR

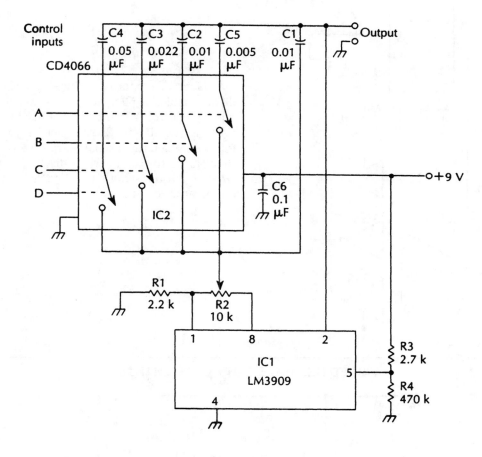

McGRAW-HILL

Fig. 62-7

IC1	LM3909 LED flasher/oscillator	C5, C6	0.1-µF capacitor
IC2	CD4066 quad bilateral switch	R1	2.2-kΩ, ¼-W 5% resistor
C1, C2	0.1-µF capacitor	R2	10-kΩ potentiometer
C3	0.033-µF capacitor	R3	2.7-kΩ, ¼-W 5% resistor
C4	0.047-µF capacitor	R4	470-Ω, ¼-W 5% resistor

50% DUTY-CYCLE 555 CIRCUIT

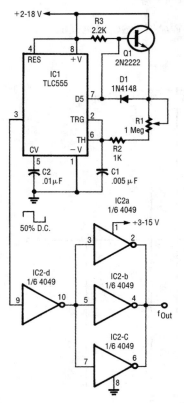

Using an external charge transistor and diode enables a 50% duty cycle and easy frequency control. When the 555's discharge transistor is cut off, the 2N2222 acts as an emitter follower. When the discharge transistor turns on, the 2N2222 turns off and C1 discharges through $(R_1 + R_2)$ at the same rate. The IN4148 provides temperature compensation.

Fig. 62-8

VARIED REP RATE, DUTY CYCLE WITH 555

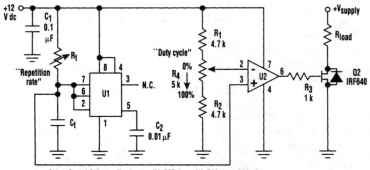

Select R_t and C_t for application   U1: 555 timer; U2: 741 or equivalent

To independently vary rpm and dwell, or duty cycle, a 555 timer is used to produce a ramp waveform, which is compared to an adjustable reference.

Fig. 62-9

DDS DIGITAL VFO

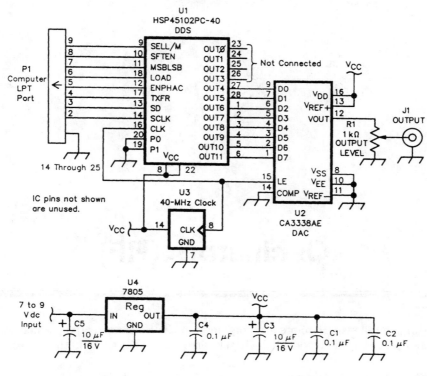

Fig. 62-10

The DDS chip (U1) generates a data stream that is converted by D/A converter U2 into a sine wave. U1 is programmed via the input from P1, from the LPT port of an IBM PC. The system uses a 40-MHz TTL output clock module.

447

63

Oscillators (RF)

The sources of the following circuits are contained in the Sources section, which begins on page 707. The figure number in the box of each circuit correlates to the entry in the Sources section.

FREQUENCY MODULATED OSCILLATOR

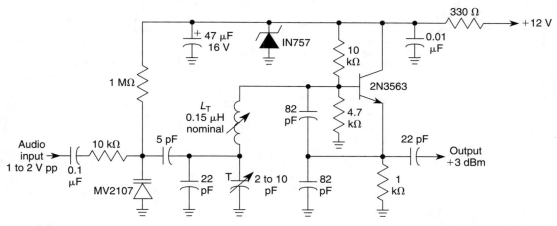

WILLIAM SHEETS

Fig. 63-1

This circuit can be used for FM wireless audio, microphone, and part-15 applications where a stable frequency modulated oscillator is needed. LT can be varied to cover 75 to 150 MHz, as needed.

JFET VARIABLE-FREQUENCY OSCILLATOR

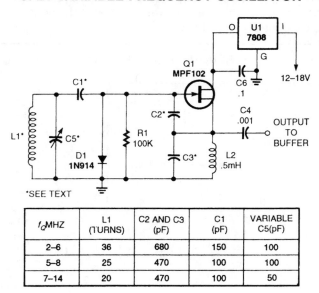

f_o MHZ	L1 (TURNS)	C2 AND C3 (pF)	C1 (pF)	VARIABLE C5(pF)
2–6	36	680	150	100
5–8	25	470	100	100
7–14	20	470	100	50

POPULAR ELECTRONICS

Fig. 63-2

This simple JFET-based variable-frequency oscillator can be used in receiver and transmitter circuits.

AM OSCILLATOR FOR WIRELESS MICROPHONES

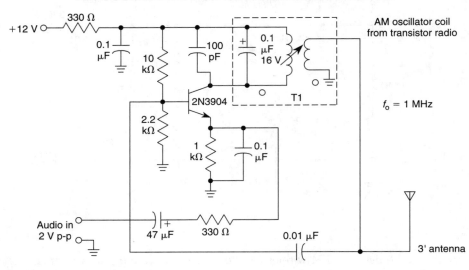

Fig. 63-3

This circuit will generate an AM-modulated signal in the AM broadcast band that can be picked up on a receiver. About 2 V of audio input will produce about 30% modulation of the oscillator signal. An old AM broadcast oscillator coil or other two-winding coil with about a 10:1 turn ratio and about 50 to 150 µH inductance can be used for T1.

REINARTZ OSCILLATOR

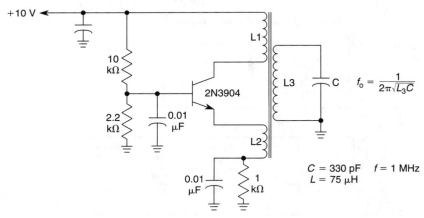

Fig. 63-4

This oscillator uses inductivity coupled emitter and collector windings to its main tank circuit. Take care so that L1 and L2 are not coupled to each other, otherwise this circuit is susceptible to parasitic oscillation at other frequencies. Typically, L1 has 5 to 10 times the number of turns that L2 has. L1, L2, L3 are wound on same coil form. This oscillator is more suited to lower frequencies, ≤10 MHz.

REMOTE-OSCILLATOR HIGH-FREQUENCY VFO

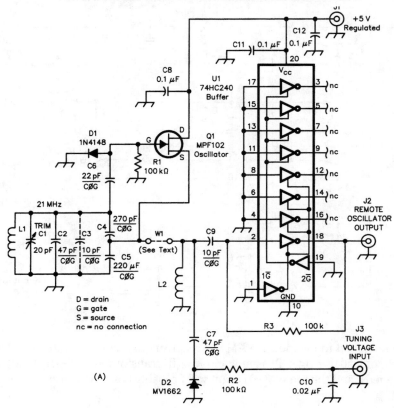

(A)

D = drain
G = gate
S = source
nc = no connection

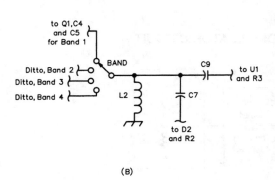

(B)

Band-Specific Oscillator Component Values

Band (m)	L1 (turns*)	Approximate Tuning Range† (kHz)
10	3	110
12	4	100
15	5	80
17	6	65
20	7	50
30	10	40

*Close-wound #20 enameled wire on a
³/₈-inch plastic rod; see text.

†With an MV1662 diode used at D1, and
C7 (Figure 2) equal to 47 pF. See text,
Note 2 and the Figure 2 parts list.

QST   *Fig. 63-5*

A remote VFO is sometimes used to control a transmitter or receiver. The circuit shown uses an MPF102 FET and is controlled by a dc voltage at J3. The table shows values for L_1 for various bands from 30 to 10 meters. U1 serves as a buffer amplifier.

BEAT-FREQUENCY OSCILLATOR FOR AM/SW RADIOS

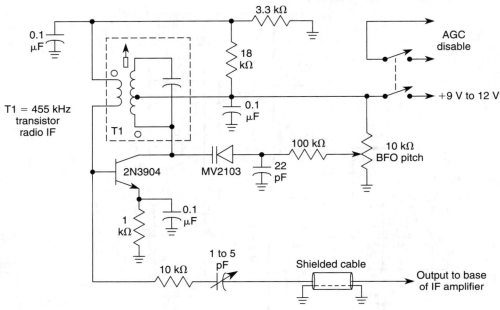

WILLIAM SHEETS

Fig. 63-6

This BFO can be added to inexpensive AM/SW receivers to enable reception of CW signals. Output couples to base of last IF stage. T1 is any 455-kHz IF transformer. The BFO switch should be a DPDT type (as needed), and the radio AGC circuit will probably have to be disabled for CW reception.

BUTLER OSCILLATOR CIRCUIT

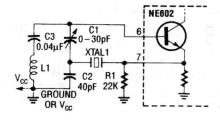

RADIO-ELECTRONICS

Fig. 63-7

This circuit uses an overtone crystal in a Butler oscillator. L1 is approximately 1300 μH, and the crystal frequency should be from 20 to 50 MHz.

455-kHz OSCILLATOR

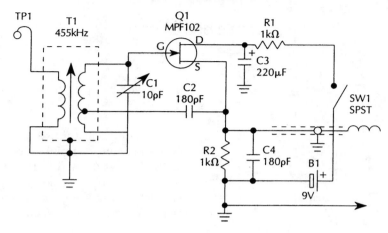

McGRAW-HILL

Fig. 63-8

The 455-kHz oscillator circuit uses a field-effect transistor (FET) for Q1. The output signal is taken from the source circuit of Q1. T1 is a 455-kHz IF transformer.

MODIFIED HARTLEY OSCILLATOR

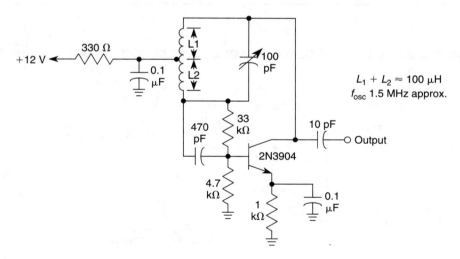

$L_1 + L_2 \approx 100\ \mu H$

f_{osc} 1.5 MHz approx.

WILLIAM SHEETS

Fig. 63-9

This oscillator uses a tapped coil in the collector circuit, with the tap grounded for the signal. L1 and L2 are coupled inductively and typically have a 3:1 turn ratio, and generally are sections of one entire winding.

VLF LC OSCILLATOR

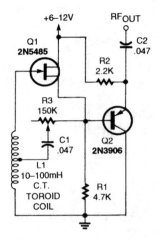

The VLF oscillator uses a large toroid coil as the frequency-determining component and a 2N5485 FET as the active device. R3 is used as a feedback control and also by running the circuit with slightly less feedback than needed for oscillation, can serve as a regenerative amplifier or detector.

POPULAR ELECTRONICS

Fig. 63-10

GROUNDED-BASE TUNED COLLECTOR OSCILLATOR FOR AM BROADCAST BAND

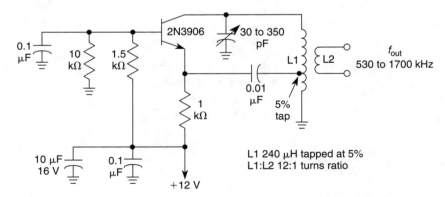

WILLIAM SHEETS

Fig. 63-11

HF VFO CIRCUIT

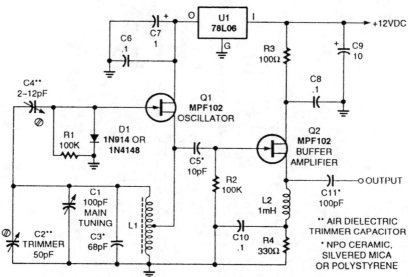

POPULAR ELECTRONICS

Fig. 63-12

This typical HF VFO circuit has several stability-enhancing features, including well-chosen capacitor types. The frequency of the VFO is approximately $2\pi (C_1 + C_2 + C_3) L_1$. L1 should be an air-core type coil, rigidly mounted, with high (>200 value) value of Q.

DARLINGTON TRANSISTOR OSCILLATOR

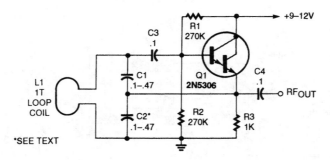

POPULAR ELECTRONICS

Fig. 63-13

This oscillator uses a very large capacitance-to-inductance ratio. L1 is a one-turn coil consisting of a loop of #12 wire 12" in diameter. This circuit is useful for metal detectors, etc., where a loop antenna is used.

FM HF OSCILLATOR WITH NO VARACTOR

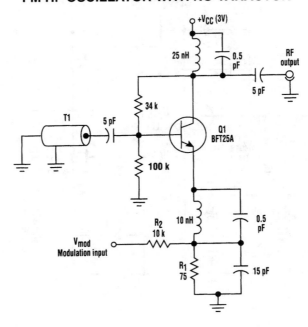

ELECTRONIC DESIGN

Fig. 63-14

Instead of using a varactor to frequency-modulate a high-frequency oscillator, this circuit uses base-charging capacitance modulation. Consequently, the large voltage change required by a varactor, which can be a major problem in battery-powered systems with limited supply voltages, is eliminated. T1 is a ceramic coaxial quarter-wave resonator.

TUNABLE UHF OSCILLATOR

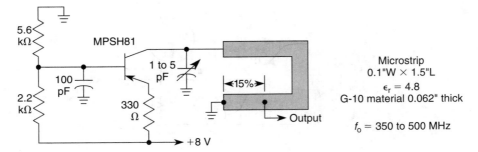

WILLIAM SHEETS

Fig. 63-15

This oscillator is typical for 350- to 500-MHz operation. The microstrip inductor is a PC board trace. The tap is typically 15% from the bottom end. The output power is 55 to 100 mW into 50 Ω, with the frequency stability typically 0.1% over 0 to 50°C.

"UNIVERSAL" VFO

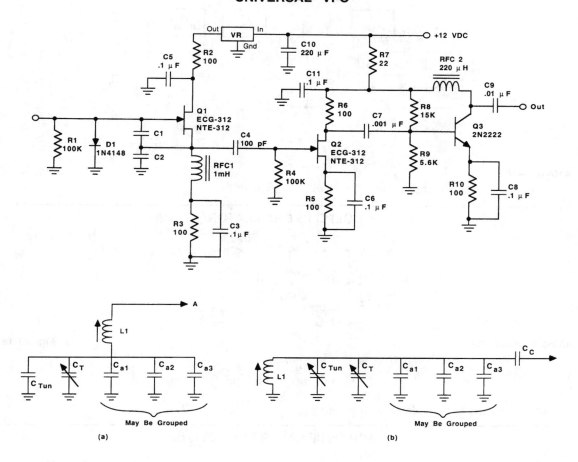

73 AMATEUR RADIO TODAY

Fig. 63-16

Figure 63-16A shows the basic circuit for the VFO, except for the tuning circuits (which are shown in Fig. 63-16B). Transistor Q1 is a junction field-effect transistor (JFET) oscillator stage. The device to use at Q1 includes MPF-102, 2N4416, and the replacement devices from the popular lines of "service" parts e.g., ECG and NTE).

Two different oscillator configurations can be accommodated by this design (i.e., both Clapp and Colpitts oscillators can be built). Both oscillators are the same from point A in Fig. 63-16C forward, and both depend on a capacitor voltage-divider feedback network. The Clapp oscillator (Fig. 63-16A) is series-tuned and the Colpitts oscillator is parallel-tuned (Fig. 63-16B).

The dc voltage supplied to the oscillator transistor (Q1) is voltage-regulated. The voltage regulator can be any 78Lxx series from 78LO5 to 78LO9.

OSCILLATOR CIRCUITS

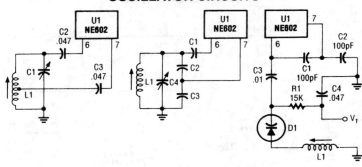

RADIO CRAFT

Fig. 63-17

These are methods of using an NE602 with a tunable VFO.

COLPITTS OSCILLATOR

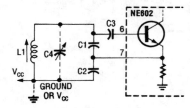

RADIO-ELECTRONICS

Fig. 63-18

$$L_1 \approx 7\ \mu\text{H}/f(\text{in MHz})$$
$$C_1 \approx C_2 \approx C_3 \approx 2400\ \text{pF}/f$$

In this circuit, the oscillator is free-running.

CLAPP OSCILLATOR FOR 100 kHz

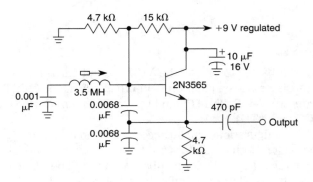

WILLIAM SHEETS

Fig. 63-19

This Colpitts oscillator is very stable and usable where good stability is needed, but crystal control is not desirable. It is capable of 1 part in 104 to 105 with good-quality components.

458

TUNED COLLECTOR OSCILLATOR

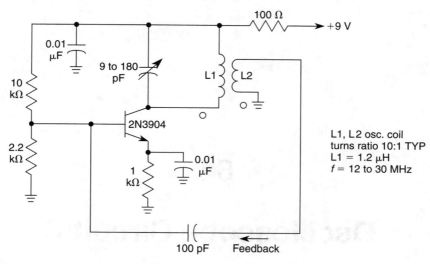

L1, L2 osc. coil
turns ratio 10:1 TYP
L1 = 1.2 μH
f = 12 to 30 MHz

WILLIAM SHEETS

Fig. 63-20

HARTLEY OSCILLATOR

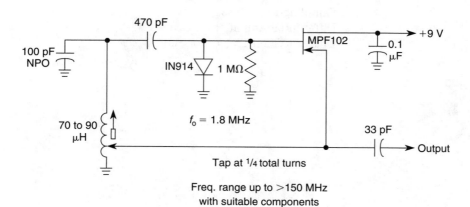

$f_o = 1.8$ MHz

Tap at 1/4 total turns

Freq. range up to >150 MHz
with suitable components

WILLIAM SHEETS

Fig. 63-21

This circuit uses a tapped inductor in a Hartley oscillator circuit. The tap is generally at 25 to 35% total turns in most instances.

64

Oscilloscope Circuits

The sources of the following circuits are contained in the Sources section, which begins on page 707. The figure number in the box of each circuit correlates to the entry in the Sources section.

SCOPE VOLTAGE CURSOR ADAPTER

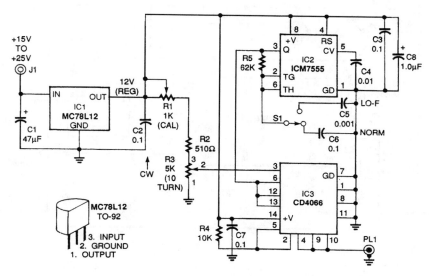

Fig. 64-1

The voltage cursor adapter superimposes horizontal cursor lines on the top and bottom of the waveform—a kind of electronic calipers—to permit direct readout of the voltage value. The cursor lines extend across the entire screen. The MC78L12 voltage regulator (IC1) supplies regulated 12-Vdc to the rest of the circuit. The ICM7555 timer (IC2) drives the CD4066B, a CMOS bilateral switch (IC3). This drive frequency can either be a normal frequency (NORM) of 100 Hz or a low-frequency (LO-*F*) of 10 kHz, depending on the setting of switch S1. Set S1 to LO-*F* for inputs below 500 Hz.

The dc reference voltage supplied to pin 3 of IC3 is set by R3, a 10-turn, 5000-Ω precision potentiometer. The voltage can be read directly from a turns counter dial coupled directly to the potentiometer's wiper. The accuracy of this reading can be 1% or better. Trimmer potentiometer R1 permits the voltage to R3 to be calibrated to precisely 10 V.

The circuit is calibrated by setting the digital reading on the turns counter of R3 to the full clockwise position and adjusting R1 for a reading of 10 V at the wiper of R3 with a digital voltmeter.

Bilateral switch IC3 converts the dc reference to a square wave with exactly the same wiper amplitude. The square-wave output appears on common pins 4, 9, and 10 of IC3 and coaxial plug PL1.

SAMPLING-RATE PHASE LOCK

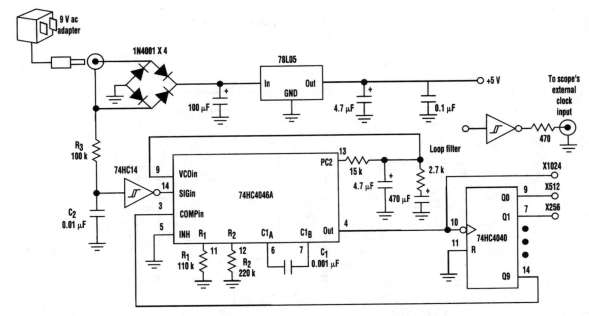

ELECTRONIC DESIGN

Fig. 64-2

Most digital scopes have record lengths that are power of 2 (e.g., 1024 points) and sampling rates constrained to a 1-2-5 sequence. This can lead to measurement errors on power-line waveforms because an integral number of line cycles can't be captured. Digital scopes that calculate measurements, such as the rms level, across the entire record will be in error.

One solution to this problem is to phase-lock the scope's sampling rate to the line frequency by exploiting the external clock input found on some digital scopes. Phase-locking the sampling to line frequency also tracks variations in the power-line frequency.

A 9- or 12-Vac wall transformer provides the circuit's power and the frequency reference. The negative output of the diode bridge refines the circuit ground. The 78L05 regulator provides the +5-V supply for the three ICs. R3 and C2 create a low-pass filter on the half-cycles from one of the floating transformer outputs. R3 also limits the current into the internal diode clamps of the inverter gate. The inverter output becomes the power-line frequency reference and is one input (SIG in to the phase comparator) of the Signetics 74HC4046A phase-locked loop (PLL). The 74HC4040 divides the PLL output frequency by 1024 and feeds the divided clock back to the other PLL phase-comparator input (COMP in). The phase-comparator output (PC2) is filtered and drives the PLL's control voltage (VCO in) so that the output frequency is 1024 times the reference frequency.

With the loop filter shown, the output frequency locks to the line frequency in about 10 s. The oscillator is locked to both 50- and 60-Hz inputs using a 74HC4046A and the values shown for resistors R1 and R2 and capacitor C1.

The output signal is buffered and sent to the scope's external clock input, which is typically a TTL-compatible input. A different tap from the 74HC4040 can be selected to control the number of cycles captured in one scope record.

DIFFERENTIAL AMPLIFIER FOR SCOPES

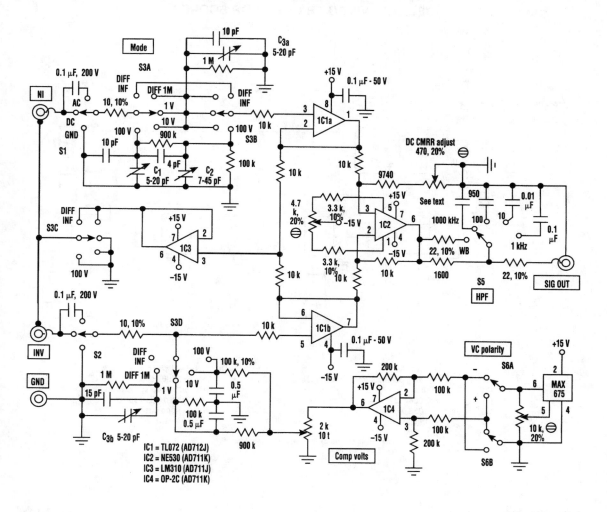

ELECTRONIC DESIGN

Fig. 64-3

Differential inputs and comparator modes can be added to any general-purpose oscilloscope using this circuit setup. Calibration doesn't change because the circuit operates in unity gain in most modes. Amplifier noise level is low enough not to degrade low-level signals, and its dynamic range can handle signals up to ±12 V peak. Notice that all of the resistors are 1%, unless specified otherwise.

DELAYED VIDEO TRIGGER FOR SCOPES

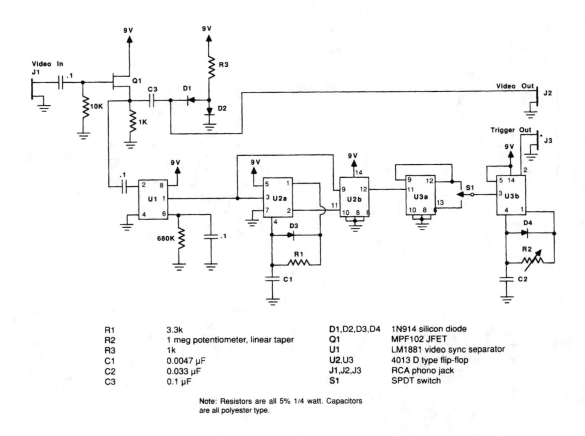

R1	3.3k	D1,D2,D3,D4	1N914 silicon diode
R2	1 meg potentiometer, linear taper	Q1	MPF102 JFET
R3	1k	U1	LM1881 video sync separator
C1	0.0047 µF	U2,U3	4013 D type flip-flop
C2	0.033 µF	J1,J2,J3	RCA phono jack
C3	0.1 µF	S1	SPDT switch

Note: Resistors are all 5% 1/4 watt. Capacitors are all polyester type.

73 AMATEUR RADIO TODAY

Fig. 64-4

This circuit will extract vertical sync from a video signal, produce a vertical sync pulse, and add an adjustable delay. This permits a delayed sweep effect to enable a scope to look at any particular horizontal line. It is useful for older scopes.

65

Photography-Related Circuits

The sources of the following circuits are contained in the Sources section, which begins on page 707. The figure number in the box of each circuit correlates to the entry in the Sources section.

Charger for Photoflash Capacitor
Slide Stepper
Photo Super Strobe

CHARGER FOR PHOTOFLASH CAPACITOR

C1—0.2 µF ±20%, 100 V
C (Load Capacitor)—480 µF, 500 V
D1, D2—MR814 (Fast-Recovery Rectifier)
Q1—MPS6520 (Selected)
Q2—MPS6563 (Selected)
Q3—MPS6562 (Selected)
Q4—MP3613 (Selected)
VR—Neon Lamp (Selected 5 AG)
R1—39K
R2—100Ω
R3—1.0K
R4—120Ω
R5—150Ω
R6—270Ω ±5%
R7—7.5Ω ±5%
R8—1.0 MΩ
R9—2.0 MΩ Pot
R10—390K ±5%
Note: All resistors ±10%, ¼ W, Unless
Otherwise Specified

L1: Timing Inductor

Core: Ferroxcube 266T125-3E2A
Winding: 145 Turns, No. 36 Wire

L2: Drive-Oscillator Transformer

Core: Ferroxcube No. 18/11PL00-3B7
Bobbin: 1811F2D
Air Gap: 0.005 in
Windings: W1: 40 Turns, No. 28 Wire
          W2: 20 Turns, No. 30 Wire
          W3: 140 Turns, No. 36 Wire

L3: Output Transformer

Core: Ferroxcube No. 26/16P-L00-3B7
Bobbin: Ferroxcube No. 26/16F2D
Windings: N1: 11 Turns, No. 18 Wire
          N2: 1100 Turns, No. 38 Wire
Air Gap: 0.030 in

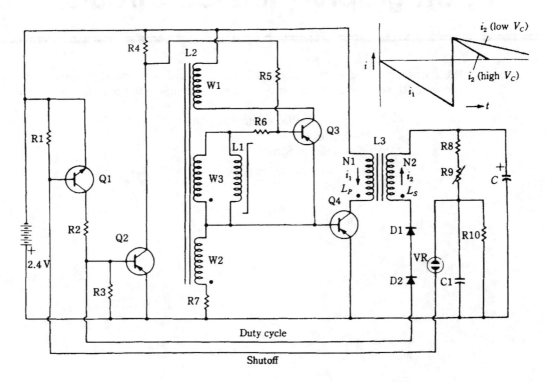

McGRAW-HILL

Fig. 65-1

This circuit charges photoflash capacitor C (480 µF, 500 V) for photoflash usage.

SLIDE STEPPER

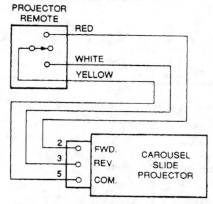

Fig. 1—ORIGINAL CONFIGURATION of the slide projector's remote control.

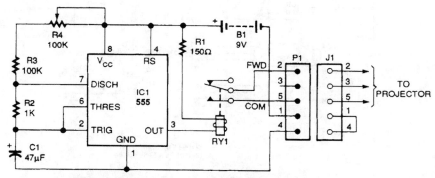

Fig. 2—SCHEMATIC DIAGRAM. The stepper circuit replaces the remote and will automatically advance the slides with a variable time delay.

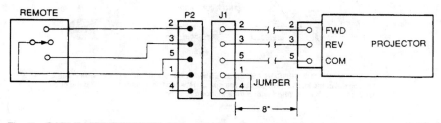

Fig. 3—CABLE MODIFICATION. This will allow the remote and the stepper circuit to be swapped easily.

ELECTRONICS NOW

Fig. 65-2

This stepper circuit replaces remote controls and will automatically advance slides in a projector. The time delay is variable with R4. The cable connections are for a Kodak carousel slide projector.

PHOTO SUPER STROBE

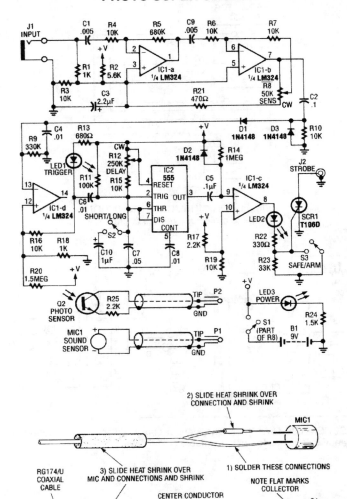

2) SLIDE HEAT SHRINK OVER
CONNECTION AND SHRINK

MIC1

RG174/U
COAXIAL
CABLE

3) SLIDE HEAT SHRINK OVER
MIC AND CONNECTIONS AND SHRINK

1) SOLDER THESE CONNECTIONS

CENTER CONDUCTOR

NOTE FLAT MARKS
COLLECTOR

Q1

SHIELD BRAID

R25

MAKE THE SENSOR ASSEMBLIES with heat-shrink tubing and small diameter coaxial cable such as RG-174/U. The space between the coaxial cable and the outer
heat-shrink tubing is filled with a little silicone rubber.

ELECTRONICS NOW!

Fig. 65-3

A change in audio or light level on the sensor connected to J1 is amplified by IC1-a and IC1-b
(rectified), and used to trigger IC2. R12 sets the delay between the trigger and the flash. IC1-c drives
indicator LED2 and triggers SCR1, which sets off the strobe connected to J2. A photo cell or a microphone can be used as a sensor.

66

Piezo Circuits

The sources of the following circuits are contained in the Sources section, which begins on page 707. The figure number in the box of each circuit correlates to the entry in the Sources section.

Piezoelectric Driver Circuit
Piezoelectric Buffer

PIEZOELECTRIC DRIVER CIRCUIT

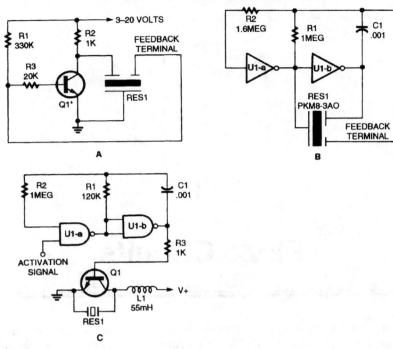

POPULAR ELECTRONICS

Fig. 66-1

Three-terminal piezoelectric elements are typically driven by transistor circuits (A), or logic gates (B). Two-terminal devices can be driven by two NAND gates. A booster coil is used to compensate for the sound-pressure attenuation caused by the case.

PIEZOELECTRIC BUFFER

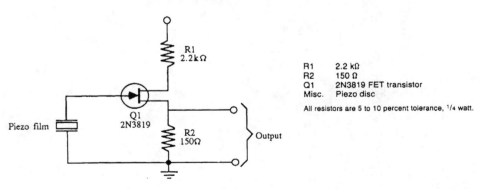

R1   2.2 kΩ
R2   150 Ω
Q1   2N3819 FET transistor
Misc. Piezo disc

All resistors are 5 to 10 percent tolerance, 1/4 watt.

McGRAW-HILL

Fig. 66-2

This circuit will serve as a buffer for experiments with Kynar film, a piezoelectric material, or with piezo devices.

67

Power Line Circuits

The sources of the following circuits are contained in the Sources section, which begins on page 707. The figure number in the box of each circuit correlates to the entry in the Sources section.

ac Power Controller
ac Power-Line Monitor
Power-Line Modem for Computer Control
Low-Voltage Power Controller

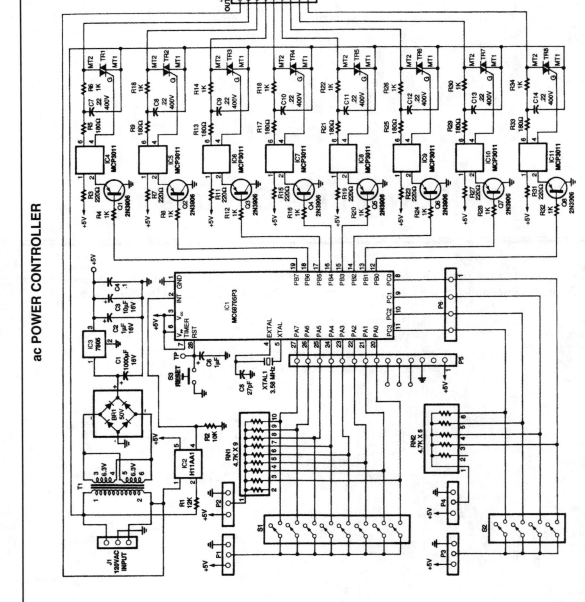

ac POWER CONTROLLER

Fig. 67-1

ELECTRONICS NOW

472

ac POWER CONTROLLER (*Cont.*)

This circuit is used to vary the power delivered to a 120-Vac load under software control. A 68705 micro controller can control eight discrete power triacs, each of which delivers power in 32 smoothly graduated steps, ranging from 0 to 97% of full power. The value delivered to one channel is independent of the value delivered to any other channel. Loads can include light displays, universal motors, heaters, and other appliances.

The power level is set by software, not a potentiometer. The software includes a basic set of routines for processing interrupts and setting the power level. The software also includes five test and demonstration routines for putting the circuit through its paces. Moreover, there's plenty of room to add your own routines to the 68705's built-in EPROM.

The basic circuit is simple, yet versatile enough to accept inputs from on-board DIP switches; alternatively, the inputs can be driven from a microcomputer bus or parallel port, or a stand-alone device with TTL-compatible outputs. There are 12 input bits to set modes and specify values.

ac POWER-LINE MONITOR

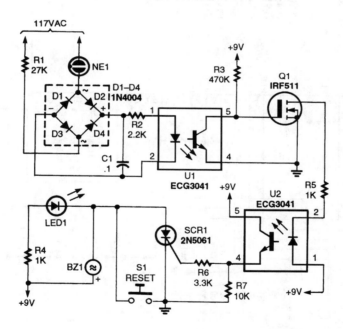

Fig. 67-2

When the power-line voltage source fails, Q1 turns on, activates optoisolator U2, and triggers SCR1. For small SCRs, U1 might directly trigger SCR1.

POWER-LINE MODEM FOR COMPUTER CONTROL

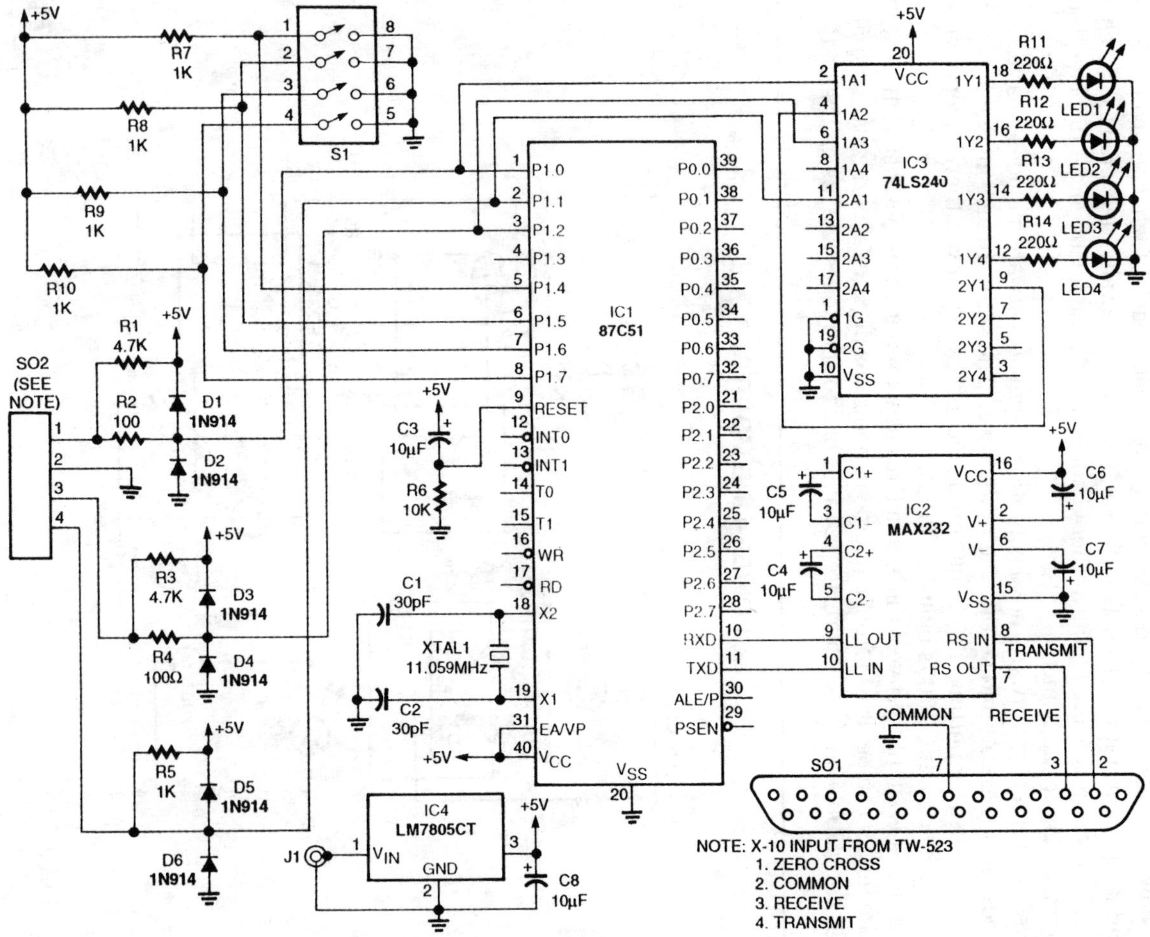

Fig. 67-3

NOTE: X-10 INPUT FROM TW-523
1. ZERO CROSS
2. COMMON
3. RECEIVE
4. TRANSMIT

ELECTRONICS NOW

This circuit uses an 87C57 microcontroller and a few peripherals to condition X-10 power-line carrier-code formats from a personal computer to use an X-10 power-line interface in a home-control system. Software details are available in the reference.

LOW-VOLTAGE POWER CONTROLLER

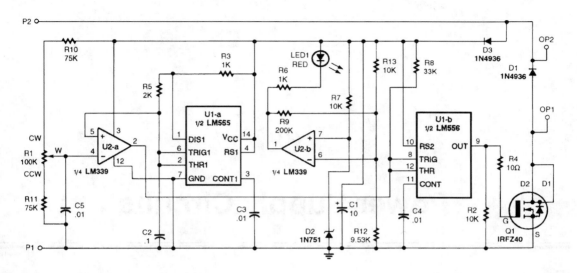

Fig. 67-4

The circuit has a duty-cycle generator that will produce an output varying from fully off to fully on and pulses of any duty cycle in between the two extremes.

This method of operation is called *PWM (pulse width modulation)*. The circuit can be fed from any dc supply source of between 10 to 15 V. Half of an LM556 dual oscillator/timer and U2-a (¼ of an LM339 quad comparator) combine to form a voltage-to-pulse-width converter. The first half of the dual oscillator/timer (U1-a) is configured as an astable oscillator, generating a continuously oscillating ramp voltage. Op amp U2-a compares the voltage at its noninverting input (pin 5)—which is connected to pins 2 and 6 of U1-a—to the voltage at its inverting input (pin 4). The op amp will produce a low output if R1's wiper voltage is higher than the instantaneous voltage that is present at pins 2 and 6 of U1-a. The output of U2-a at pin 2 will have an on/off ratio that is proportional to the voltage at R1's wiper.

The output of U2-a is fed to U1-b, which is used to buffer the signal. The low-impedance, pulsed output of U1-b at pin 9 is fed to the gate of MOSFET Q1, driving it on or off. The circuit also has a power-input detector, built around U2-band and LED1. If the input power is OK, LED1 will shut off.

Diode D1 is used to suppress the reverse voltage spikes that are generated by inductive loads during turn off; without that diode, the MOSFET might be destroyed. If the circuit will not be used to drive inductive loads (motors), D1 can be eliminated.

68

Power Supply Circuits

The sources of the following circuits are contained in the Sources section, which begins on page 707. The figure number in the box of each circuit correlates to the entry in the Sources section.

FOLDBACK CURRENT LIMITER

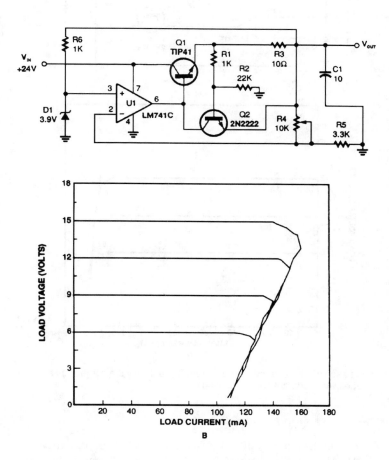

POPULAR ELECTRONICS

Fig. 68-1

This regulator uses the drop across R3 to sense current draw, turning on Q2, removing drive from Q1, and lowering the output voltage. Limiting occurs when Q2 has 0.65 V across the base-emitter junction. This circuit has foldback characteristics as seen from the figure.

CURRENT-LIMITING REGULATOR CIRCUIT

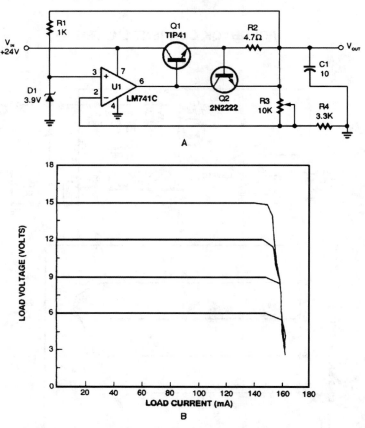

As shown in B, maximum load current is practically the same at all supply voltages with constant current limiting.

POPULAR ELECTRONICS

Fig. 68-2

This regulator uses the drop across R2 to turn on Q2, removing base drive from Q1 and reducing the current through R1. About 0.65 V must be dropped across R2 before limiting occurs. To set limit current,

$$R_2 \approx \frac{0.65}{I_{\text{LIMIT (amps)}}}$$

$$Output\ voltage = V_{\text{OUT}} = (3.9)\ \frac{(R_3 + R_4)}{R_4}$$

SWITCHING POWER SUPPLY

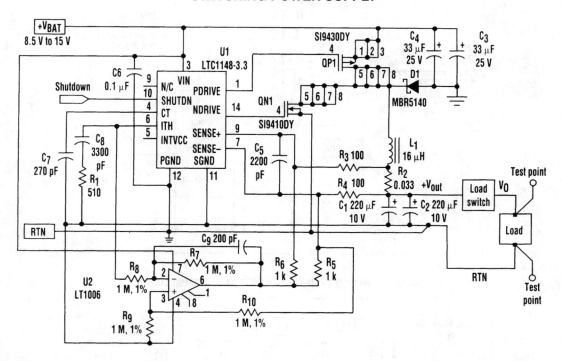

ELECTRONIC DESIGN   *Fig. 68-3*

In many switching-regulator applications for portable computers, the microprocessor is located some distance from the power supply. With the latest processors, total load currents range into several amperes. Thus, regulation at the load can become a problem.

The I_{TH} pin (pin 6) of the LTC1148 is approximately proportional to the load current. It scales nearly linearly from 0 V at no load to 2.0 V at current limit. U2, acting as a unity-gain differential amplifier, inverts the U1 pin 6 voltage (referenced to SENSE—, pin 7) and causes a current proportional to load current to flow in resistors R5 and R6. A small voltage drop appears across current-sense filter resistors R3 and R4. This makes the voltage measured by the internal feedback divider appear low. The duty factor is adjusted to bring this back to the correct voltage. As a result, the output is increased slightly as a function of load. Capacitor C9 rolls off the high-frequency gain of the correction amplifier.

$$R_{comp} = \frac{(V_{pin6} \times R_{filt})}{V_{corr}}$$

where: $R_{comp} = R_5 = R_6$
   $R_{filt} = R_3$ and R_4
   $V_{corr} = Measured\ drop.$

TRANSFORMERLESS dc POWER SUPPLY I

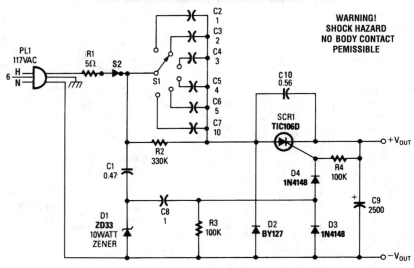

WARNING!
SHOCK HAZARD
NO BODY CONTACT
PEMISSIBLE

TABLE 1—OUTPUT CURRENT/VOLTAGE AT SPECIFIED LOADS

Capacitance (μF)	100 Ohms		200 Ohms		1000 Ohms	
	V_{OUT} (V)	Current (mA)	V_{OUT} (V)	Current (mA)	V_{OUT} (V)	Current (mA)
1	3.2	31	6.0	29	25	24
2	6.4	61	11.2	54	41	41
3	9.0	87	16.1	78	52	52
4	11.8	113	20.7	100	61	61
5	15.5	147	24.7	120	67	67
6	17.8	169	28.8	140	68	68
7	18.5	176	31.9	155	69.4	68
8	20.3	195	36.8	173	70	71
9	22.8	220	41.0	193	70	71
10	24.9	238	42.0	204	71	71
11	27.1	259	44.9	219	—	—
14	33.0	317	52.7	257	—	—
20	43.5	422	65.8	322	—	—

POPULAR ELECTRONICS

Fig. 68-4

An SCR fires on the positive half cycles of the ac line voltage. Switched capacitors are used to select the output voltage. These must all be ac-rated, nonpolarized types.

+5-V AT 1.5- TO 3-A SUPPLY, +6- TO +15-V INPUT

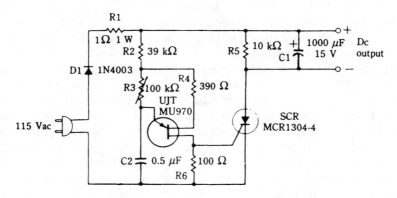

MAXIM   **Fig. 68-5**

Operating efficiencies of 80 to 90% are possible using the MAX741D and this circuit.

TRANSFORMERLESS dc POWER SUPPLY II

McGRAW-HILL   **Fig. 68-6**

Although it is simple, this supply can provide 10 to 15 V at 100 mA directly from the ac lines. This circuit has no isolation from the ac line; therefore, there is a shock hazard and it should only be used where no possibility of contacting external devices, circuits, or personnel exists.

FAST 3.3-V REGULATOR

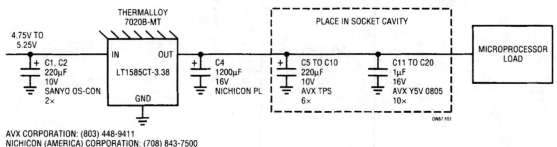

AVX CORPORATION: (803) 448-9411
NICHICON (AMERICA) CORPORATION: (708) 843-7500
SANYO VIDEO COMPONENTS (USA) CORPORATION: (619) 661-6322
THERMALLOY INCORPORATED: (214) 243-4321
FOR CORRECT OPERATION OF MICROPROCESSOR, DO NOT SUBSTITUTE COMPONENTS

LINEAR TECHNOLOGY   *Fig. 68-7*

New high-performance microprocessors require a fresh look at power-supply transient response. The LT1585 linear regulator features 1% initial accuracy, excellent temperature drift and load regulation, and virtually perfect line regulation. Complementing superb dc characteristics, the LT1585 exhibits extremely fast response to transients. Transient response is affected by more than the regulator itself. Stray inductances in the layout and bypass capacitors, as well as capacitor ESR dominate the response during the first 400 ns of transient.

The figure shows a bypassing scheme developed to meet all the requirements for the Intel P44C-VR microprocessor. Input capacitors C1 and C2 function primarily to decouple load transients from the 5-V logic supply. The values used here are optimized for a typical 5-V desktop computer "silver box" power-supply input. C5 to C10 provide bulk capacitance at low ESR and ESL, and C11 to C20 keep the capacitance at low ESR and ESL low at high (>100 kHz) frequencies. C4 is a damper and it minimizes ringing during setting. Trace C is the load current step, which is essentially flat at 4 A with a 20-ns rise time.

Trace A is the output settling response at 20 mV per division. Cursor trace B marks –46 mV relative to the initial output voltage. At the onset of load current, the microprocessor socket voltage dips to –38 mV as a result of inductive effects in the board and capacitors, and the ESR of the capacitors. The inductive effects persist for approximately 400 ns. For the next 3 μs, the output droops as the load current drains the bypass capacitors. The trend then reverses as the LT1585 catches up with the load demand, and the output settles after approximately 50 μs. Running 4 A with a 1.7-V drop, the regulator dissipates 6.8 W.

POWER SUPPLY FOR HIGH-POWER AUTOSOUND AMP

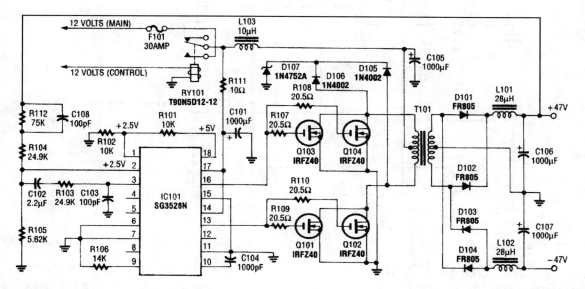

RADIO-ELECTRONICS

Fig. 68-8

A switching mode inverter is used with a pulse-width modulation voltage regulator (SG35260). Four IRF240 power MOSFETs are used as switches. The output is ±47 V at about 5 A peak. Transformer T101 is a four-turn center tapped primary, and 16-turn center tapped secondary on a Ferrox cube ETD-34 core.

IC REGULATOR PROTECTION

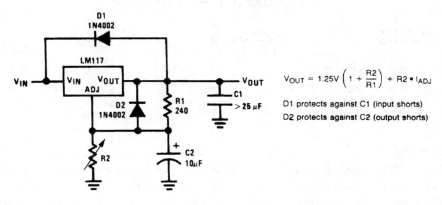

$$V_{OUT} = 1.25V \left(1 + \frac{R2}{R1} \right) + R2 \cdot I_{ADJ}$$

D1 protects against C1 (input shorts)
D2 protects against C2 (output shorts)

NATIONAL SEMICONDUCTOR

Fig. 68-9

This circuit protects an IC regulator against various fault conditions.

3.3-V SWITCHING REGULATOR

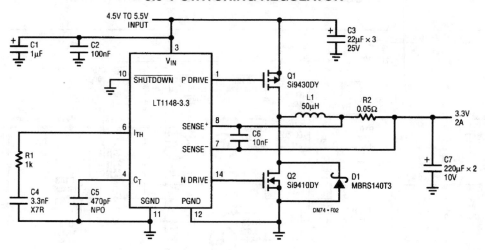

LINEAR TECHNOLOGY

Fig. 68-10

For the LT1129-3.3, dissipation amounts to a little under 1.5 W at full output current. The 5-lead surface-mount DD package handles this without the aid of a heatsink, provided that the device is mounted over at least 2500 mm^2 of ground or power-supply plane. Efficiency is around 62%; dissipation in linear regulators becomes prohibitive at higher current levels, where they are supplanted by high-efficiency switching regulators. The synchronous buck converter is implemented with an LTC1148-3.3 converter. The LTC1148 uses both Burst Mode™ operation and continuous, constant off-time control to regulate the output voltage, and maintain high efficiency across a wide range of output loading conditions.

NE602 POWER-SUPPLY OPTIONS

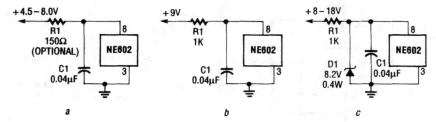

a   *b*   *c*

RADIO-ELECTRONICS

Fig. 68-11

Here, Figs. 68-11A through 68-11C show an RC-filter used as both current limiter (R1) and integrator (C1), as well as for isolation. In Fig. 68-11A, +4.5 to 8.0 Vdc is the normal operating range of the NE602. In Fig. 68-11B, R1 drops voltage, and is used because a +9-V battery can go higher, and a +9-V wall supply can produce up to 11 V. In Fig. 68-11C, a +8- to 18-Vdc supply is regulated using a 8.2-V Zener for D1.

SIMPLE 9-V SUPPLY

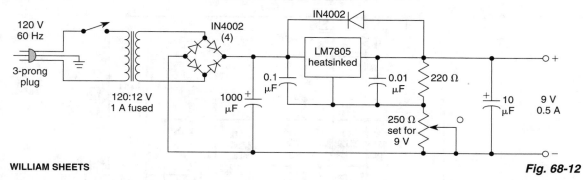

WILLIAM SHEETS

Fig. 68-12

This circuit uses an LM7805 with a resistive voltage divider in the common leg of the regulator. The regulator can be "fooled" into producing an apparent higher output voltage in this manner. This supply is useful for running radios, tape recorders, or other 9-V devices.

TRACKING POWER SUPPLY

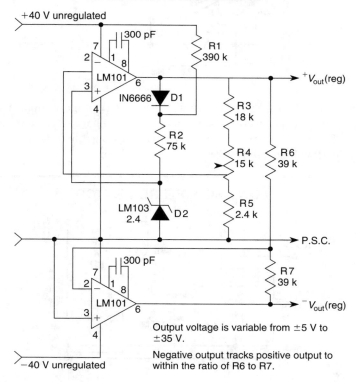

Output voltage is variable from ±5 V to ±35 V.

Negative output tracks positive output to within the ratio of R6 to R7.

NATIONAL SEMICONDUCTOR

Fig. 68-13

Two op amps are used in this basic op-amp regulator circuit. The outputs can be fed to current amplifier stages or emitter followers, if needed.

POWER EFFICIENT VOLTAGE REGULATOR

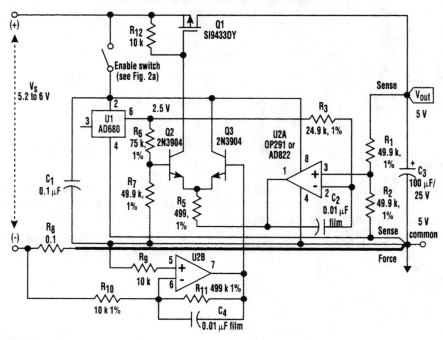

ELECTRONIC DESIGN   **Fig. 68-14**

Included in the many features of this power-efficient, voltage-regulator circuit is shutdown power control with a current output up to several hundred milliamperes (expandable to amperes, if desired). Current limiting can be preset to a fixed level for controlled dissipation in Q1 and the circuit requires no auxiliary voltage supply for the pass transistor.

LOW DROP-OUT REGULATOR

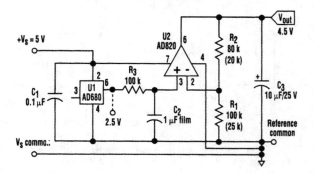

ELECTRONIC DESIGN   **Fig. 68-15**

This low-dropout reference produces a 4.5-V output from a supply just a few hundred millivolts greater. With 1-mA dc loading, it maintains a stable 4.5-V output for inputs down to 4.7 V.

SCR SWITCHING SUPPLY FOR COLOR TV RECEIVERS

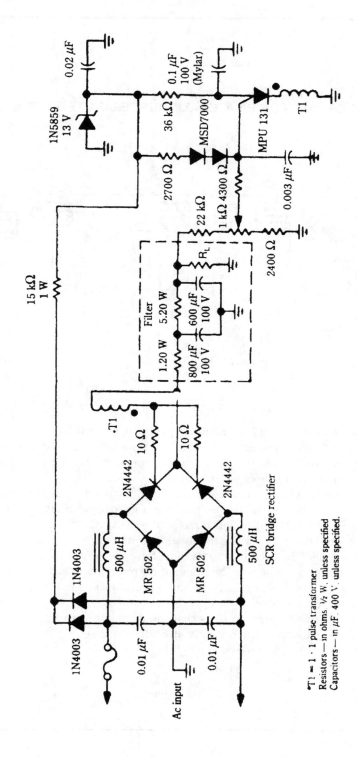

Fig. 68-16

McGRAW-HILL

An output +80 V at up to 1.5 A is available from this supply. A minimum load of 200 mA is required because of the SCR holding current. Notice that no ac line isolation is provided and a shock hazard exists.

REGULATOR CIRCUIT FOR BILATERAL SOURCE/LOAD POWER SYSTEM

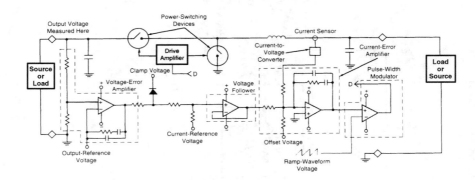

Fig. 68-17

The figure shows a circuit that regulates an output voltage, regardless of the direction of flow of output current. More specifically, it regulates the voltage at the left source or load, which can either supply power to or receive power from the right load or source, depending on the voltages and the direction of flow of current at the load/source terminals.

The overall system can be characterized as a voltage-controlled current source with bilateral current capability. The current flowing between the two source/loads, averaged over a power-switching cycle, is made to depend on the pulse-width modulation that governs the operation of the two power-switching devices, and this pulse-width modulation is, in turn, a function of amplified current-error and voltage-error signals. The voltage error is the difference between the actual output voltage and the output-reference voltage, which is the nominal output voltage at zero current. The pulse-width modulation is varied to increase or decrease the current, as needed, to limit the excursion of output voltage from the reference value.

An additional feature of this control circuit is that the maximum current in either direction can be limited by limiting the excursion of the output voltage from the zero-current value. Thus, external current-limiting circuitry is not necessary.

FAST 3.3-V ADJUSTABLE REGULATOR

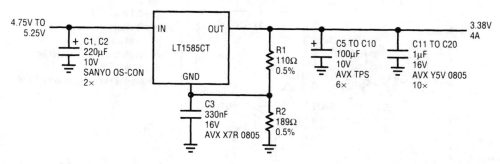

LINEAR TECHNOLOGY

Fig. 68-18

The adjustable version of the LT1585 makes it relatively easy to accommodate multiple microprocessor power-supply voltage specifications. To retain the tight tolerance of the LT1585 internal reference, a 0.5% resistor adjustment is recommended. R1 is sized to carry approximately 10 mA idling current ($\leq 124\ \Omega$), and R2 is calculated from:

$$R_2 = \frac{V_o - V_{\text{ref}}}{\dfrac{V_{\text{ref}}}{R_1}} + 1_{\text{ADJ}}$$

where:

$$1_{\text{ADM}} = 60\ \mu \text{ and } V_{\text{ref}} = 1.250 \text{ V}.$$

TWO-TERMINAL 100-mA CURRENT REGULATOR

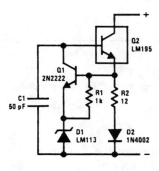

NATIONAL SEMICONDUCTOR   **Fig. 68-19**

The circuit has a low temperature coefficient and operates down to 3 V. The reverse base current of the LM195 biases the circuit.

A 2N2222 is used to control the voltage across current-sensing resistor, R2 and diode D1, and therefore the current through it. The voltage across the sense network is the V_{BE} of the 2N2222 plus 1.2 V from the LM113. In the sense network, R2 sets the current and D1 compensates for the V_{BE} of the transistor. Resistor R1 sets the current through the LM113 to 0.6 mA.

TWO-PHASE RECTIFIER

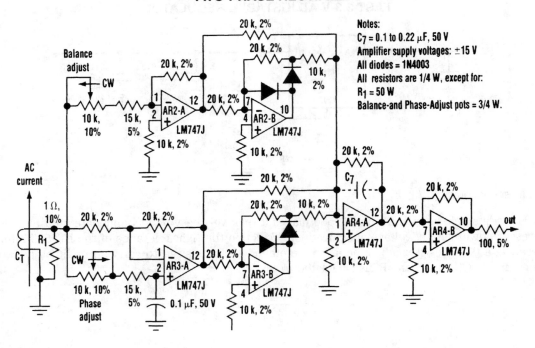

Notes:
C_7 = 0.1 to 0.22 μF, 50 V
Amplifier supply voltages: ±15 V
All diodes = 1N4003
All resistors are 1/4 W, except for:
R_1 = 50 W
Balance-and Phase-Adjust pots = 3/4 W.

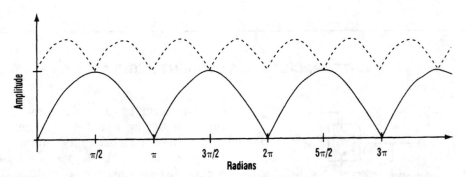

A single-phase AC signal can be converted to two phase with this circuit. It rectifies and sums the signal to a dc voltage level.

ELECTRONIC DESIGN

Fig. 68-20

The waveform generated by the two-phase rectifier illustrates that the ripple is less than half that of a conventional single-phase circuit's waveform. Also, the ripple frequency is double that of the conventional circuit. The circuit will follow amplitude changes in the ac input signal very rapidly, and it works equally well with current or voltage inputs.

LOW-NOISE 5-V SUPPLY

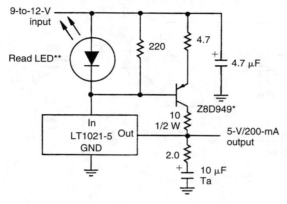

*Zetex Inc.
**Glows in current limit. Do not omit.

ELECTRONIC DESIGN

Fig. 68-21

Standard three-terminal regulator ICs can be noisy. The key is the noise over the 10-Hz to 10-kHz band; measurements revealed a 40-dB improvement over standard three-terminal regulators.

The regulator is built around a 5-V buried-Zener reference. It's the buried Zener's inherently low noise that makes the finished supply so quiet. Measured over a 10-Hz to 10-kHz band, the 5-V output contains just 7 μV rms of noise at full load. The 10-Hz to 10-kHz noise can be further reduced to 2.5 μV rms by adding a 100-μH, 1000-μF output filter. The noise characteristics of the reference are tested and guaranteed to a maximum of 11 μV over the band of interest.

An external boost transistor, the ZBD949, provides gain to meet a 200-mA output current requirement. Current limiting is achieved by ballasting the pass transistor and clamping the base drive. Although the oscillator only requires 200 mA, it's possible to extend the output current to at least 1 A.

POSITIVE REGULATOR WITH 0- TO 70-V OUTPUT

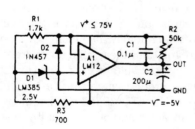

NATIONAL SEMICONDUCTOR

Fig. 68-22

The op amp has one input at ground and a reference current drawn from its summing junction. With this arrangement, the output voltage is proportional to setting resistor R2. A negative supply is used to operate the op amp within its common-mode range, providing zero output with sink current and power a low-voltage bandgap reference, D1. The current drawn from this supply is under 150 mA, except when sinking a load current. The output load capacitor, C2, is part of the op-amp frequency compensation.

SIMPLE 12-V POWER SUPPLY

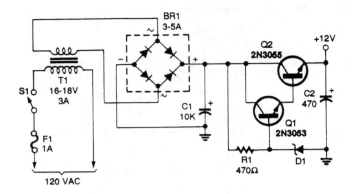

POPULAR ELECTRONICS

Fig. 68-23

This 12-V power supply is easy-to-build, and it produces a smooth output. D1 is a 14-V, ½-W Zener diode. The voltage can be varied by a few volts up or down to change the output voltage.

3.3 V FROM 5-V LOGIC SUPPLIES

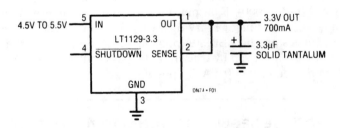

LINEAR TECHNOLOGY

Fig. 68-24

Microprocessor chip sets and logic families that operate from 3.3-V supplies are gaining acceptance in both desktop and portable computers. Computing rates, and in most cases, the energy consumed by these circuits, show a strong improvement over 5-V technology. The main power supply in most systems is still 5 V, necessitating a local 5-V to 3.3-V regulator. Linear regulators are viable solutions at lower ($I_o \leq 1$ A) currents, but they must have a low dropout voltage in order to maintain regulation with a worst-case input of only 4.5 V. The figure shows a circuit that converts a 4.5-V minimum input to 3.3 V with an output tolerance of only 3% (100 mV). The LT1129-3.3 can handle up to 700 mA in surface-mount configurations, including both 16-µA shutdown and 50-µA standby currents for system sleep modes. Unlike other linear regulators, the LT1129-3.3 combines both low-dropout and low-voltage operation. Small input and output capacitors facilitate compact, surface-mount designs.

12-V SUPPLY

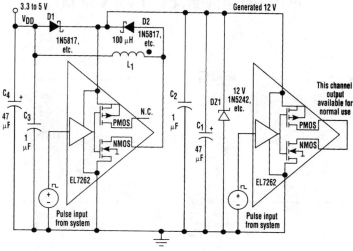

ELECTRONIC DESIGN

Fig. 68-25

When driving a power MOSFET from a 5-V or 3.3-V system, a significant number of components usually are needed to generate an extra +12 V.

It's possible, however, to apply the second channel in a typical dual MOSFET driver to derive a +12-V power supply. By using a driver with the drains brought to separated pins, you can connect an inductor between the n-channel drain and the logic supply without connecting the p-channel device.

The driver operates as a standard flyback-style switched-mode circuit (see the figure). When the output n-channel device is on, current starts flowing in the inductor, which stores energy. When the n-channel device is turned off, current must continue flowing. Therefore, it flows through diode D2 to charge up C1 and C2. As the cycle repeats, the C1 and C2 voltage rises until the Zener diode prevents further voltage rise. This is needed to prevent the driver's derived supply from exceeding the part's maximum voltage rating.

4- TO 70-V REGULATOR

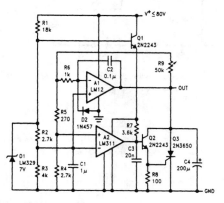

This regulator operates from a single supply. If the op amp is not able to control an overvoltage condition, the SCR will crowbar the output.

NATIONAL SEMICONDUCTOR

Fig. 68-26

SWITCHED POWER-CONTROL CIRCUITS

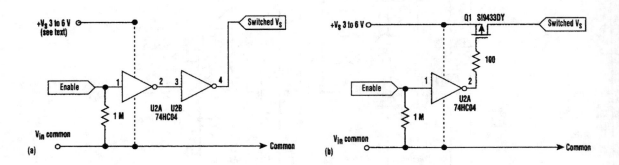

ELECTRONICS DESIGN

Fig. 68-27

Load currents of a few milliamperes to amperes can be turned on and off by these switched power-control circuits. The U2 CMOS inverter stage works as a simple power switch for load currents less than 5 mA (Fig. 68-27A), allowing easy reference shutdown. If appreciably higher switched output currents are called upon, an alternate CMOS inverter driving a low-threshold PMOS device can be used to switch currents of up to 1 A or more (Fig. 68-27B).

MULTIPLE ON-CARD REGULATOR ADJUSTER

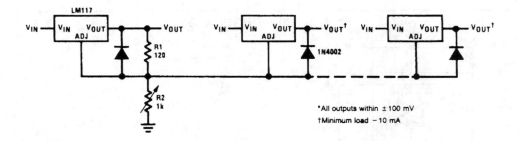

NATIONAL SEMICONDUCTOR

Fig. 68-28

This circuit allows one pot to control several on-card regulators for adjustment within ±100 mV of each other.

SIMPLE 9-V POWER SUPPLY

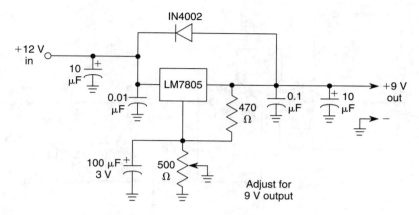

Fig. 68-29

This supply will provide 9-V transistor radios or cassettes from a 12-V auto electrical system.

+5-V AT 1-A SUPPLY WITH +3- TO +5-V INPUT

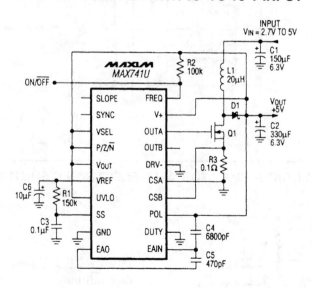

Fig. 68-30

A MAX741U switching-mode power-supply controller and a switching FET Q1 are used to provide +5 V at 1 A.

5-Vdc REGULATED SUPPLY

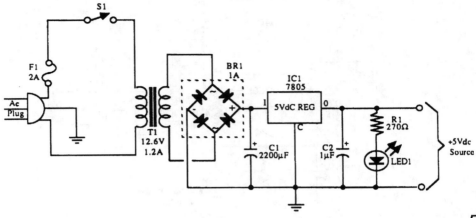

Fig. 68-31

BUFFERED REFERENCE SUPPLY

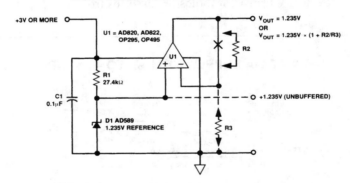

ANALOG DEVICES   *Fig. 68-32*

This buffered reference (for 1.23 V or more) uses a supply voltage of greater than 3 V.

5-V LOGIC REGULATOR WITH ELECTRONIC SHUTDOWN

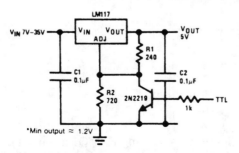

The circuit will shut down to 1.2 V under fault conditions.

NATIONAL SEMICONDUCTOR   *Fig. 68-33*

496

12-Vdc REGULATED SUPPLY

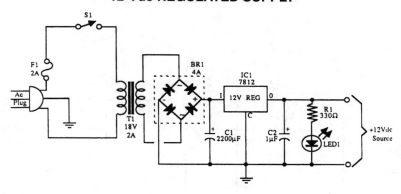

McGRAW-HILL

Fig. 68-34

JUNKED TRANSISTOR REGULATORS

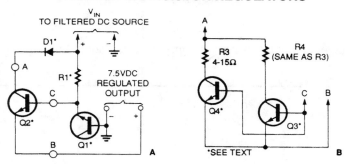

POPULAR ELECTRONICS

Fig. 68-35

Old transistors can make excellent regulators. Simply use one as a Zener to control the base current to another transistor (Fig. 68-35A). If the pass transistor cannot supply enough current, you can use two pass transistors in its place (Fig. 68-35B).

TELEPRINTER LOOP SUPPLY

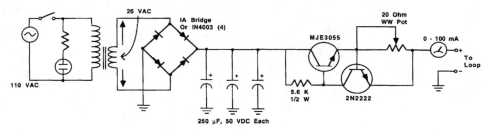

73 AMATEUR RADIO TODAY

Fig. 68-36

A circuit to power a teleprinter, using transistors as current-controlling devices. The power supply used provides a constant current in a loop, normally 60 mA or 20 mA, depending on the machine.

5-A CONSTANT-VOLTAGE SUPPLY

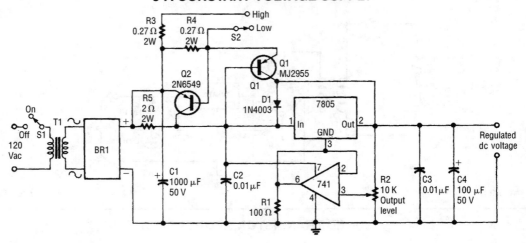

ELECTRONICS NOW

Fig. 68-37

This constant-voltage supply has a variable output. It can supply more than 5 A, and has two switchable current limits.

69

Power Supply Circuits (High Voltage)

The sources of the following circuits are contained in the Sources section, which begins on page 707. The figure number in the box of each circuit correlates to the entry in the Sources section.

Fluorescent Lamp 12-V Supply
High-Voltage Regulator
Night-Vision Scope Power Supply
High-Voltage Power-Supply Control Circuit
−100-Vdc Supply
ac-Operated He-Ne Power Supply
HV Regulator with Foldback Current Limit
Kirlian Device Supply
High-Voltage Tripler
200-V Regulator
Pulse-Width Modulated Laser Supply

FLUORESCENT LAMP 12-V SUPPLY

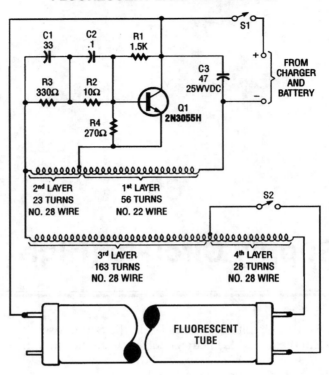

POPULAR ELECTRONICS

Fig. 69-1

This high-voltage power supply can operate fluorescent tubes from a 12-V source, even if the tube has a defective filament. It essentially is an oscillator that excites a home-made autotransformer. T1 is wound on a ferrite rod 5/16" diameter by 1 7/8" long, in layers. S2 is an optional lamp filament switch.

HIGH-VOLTAGE REGULATOR

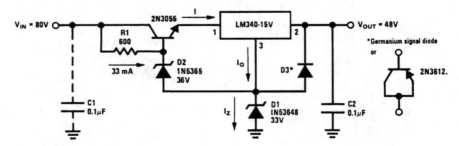

NATIONAL SEMICONDUCTOR

Fig. 69-2

This circuit produces 48 V from an 80-V input.

NIGHT-VISION SCOPE POWER SUPPLY

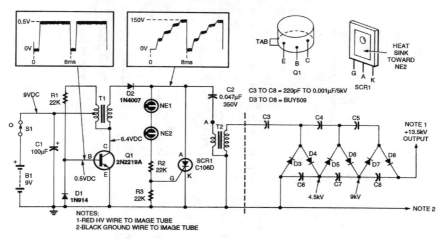

ELECTRONICS NOW

Fig. 69-3

This high-voltage power supply has an inverter around Q1 that supplies 150-V pulses to the converter of SCR1 and C2. The output of T2 is a 4.5-kV pulse that is multiplied by the voltage-tripler network (right) to produce 13.5 kV.

T1 is a 3-kΩ to 500-Ω CT transistor audio transformer, T2 is a flash tube trigger transformer with a 6-kV secondary.

HIGH-VOLTAGE POWER-SUPPLY CONTROL CIRCUIT

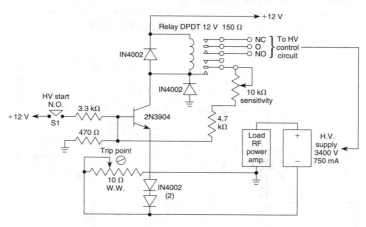

WILLIAM SHEETS

Fig. 69-4

To start the HV supply, S1 is pressed, latching the relay. The 10-kΩ pot is set so that the relay just latches. When HV current becomes excessive (arc-over, etc.), an excessive voltage is developed across the 10-Ω WW pot, cutting off the 2N3904, causing the relay to unlatch. This circuit was used in a 1000-W linear RF power amplifier.

−100-Vdc SUPPLY

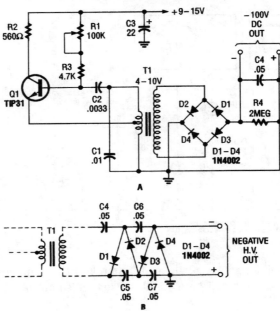

POPULAR ELECTRONICS

Fig. 69-5

The combination Hartley oscillator/step-up transformer shown in A can generate significant negative high-voltage—especially if the voltage output of the transformer is multiplied by the circuit in Fig. 69-5B. T1 is a small low-voltage filament transformer of around 4- to 10-Vac output, 120-V primary.

ac-OPERATED HE-NE POWER SUPPLY

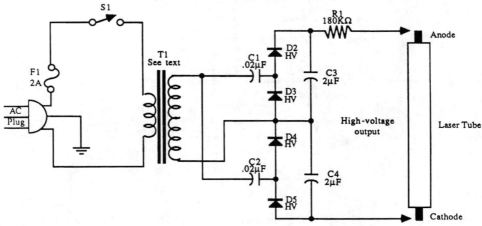

McGRAW-HILL

Fig. 69-6

T1 is a 120-V to 1000-V step-up 60-Hz transformer. C1, C2, C3, C4 and D2 through D5 form a voltage quadrupler. The initial voltage is 4 to 5 kV, which drops when the laser tube fires.

HV REGULATOR WITH FOLDBACK CURRENT LIMIT

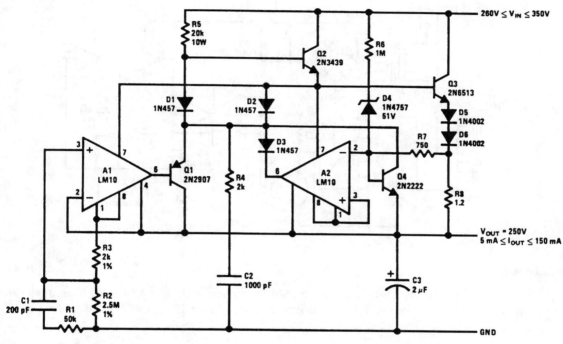

Fig. 69-7

The output current is sensed across R8. This is delivered to the current-limit amplifier through R7, across which the foldback potential is developed by R6 with a threshold determined by D4. The values given limit the peak power below 20 W and shut off the pass transistors when the voltage across them exceeds 310 V. With unregulated input voltages above this value, start-up is initiated solely by the current through R5. Q4 is added to provide some control on current before A2 has time to react.

The circuit is stable with an output capacitor greater than about 2 µF. Spurious oscillations in current limit are suppressed by C2 and R4, while a strange, latch-mode oscillation coming out of current limit is killed with C1 and R1.

KIRLIAN DEVICE SUPPLY

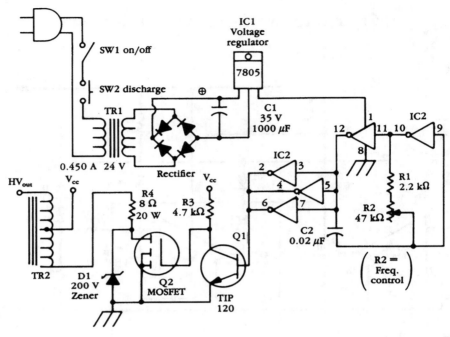

Fig. 69-8

This device is essentially a high-voltage variable-frequency ac supply. A CD4049 IC multivibrator circuit drives a Darlington connected transistor pair, which drives TR2, an HV transformer.

HIGH-VOLTAGE TRIPLER

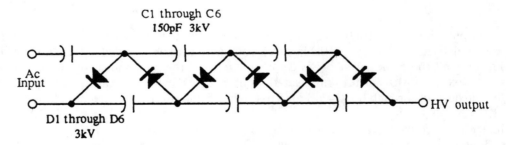

Fig. 69-9

This tripler is useful for low-current and high-voltage applications. The capacitors can be 0.001-µF, 3- to 6-kV discs, and the diode's 3-kV units, or three each IN4007 in series.

200-V REGULATOR

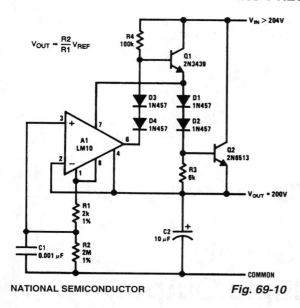

$$V_{OUT} = \frac{R2}{R1} V_{REF}$$

With high-voltage regulators, powering on the IC through the drive resistor for the pass transistors can become quite inefficient. This is avoided with the circuit shown. The supply current for the IC is derived from Q1. This allows R4 to be increased by an order of magnitude without affecting the dropout voltage.

Selection of the output transistors will depend on voltage requirements. For output voltages above 200 V, it might be more economical to cascade lower-voltage transistors.

NATIONAL SEMICONDUCTOR   *Fig. 69-10*

PULSE-WIDTH MODULATED LASER SUPPLY

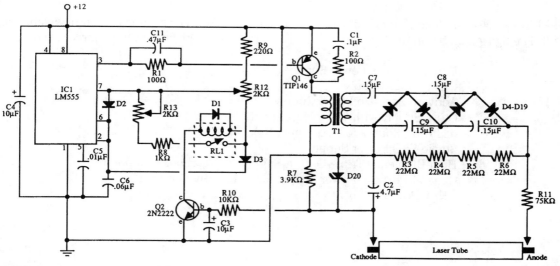

McGRAW-HILL   *Fig. 69-11*

IC1 initially provides drive for Q1 and HV transformer T1, and it rectifies D4 through D19. When the laser tube ignites, Q2 is triggered; this activates relay RL1, reducing the duty cycle. R13 controls the duty cycle of the pulses through the laser tube.

70

Power Supply
Circuits (Multiple Output)

The sources of the following circuits are contained in the Sources section, which begins on page 706. The figure number in the box of each circuit correlates to the entry in the Sources section.

Experimenter's Power Supply
Quad Power Supply
Activate Back-Up Power Supply
CCFL Supply with Variable Contrast
dc Power Source for Experiments
Stable VFO Power Supply
High-Efficiency Triple-Output Supply for Notebook Computers
General-Purpose Power Supply for Automotive Projects
±15-V Power Supply

EXPERIMENTER'S POWER SUPPLY

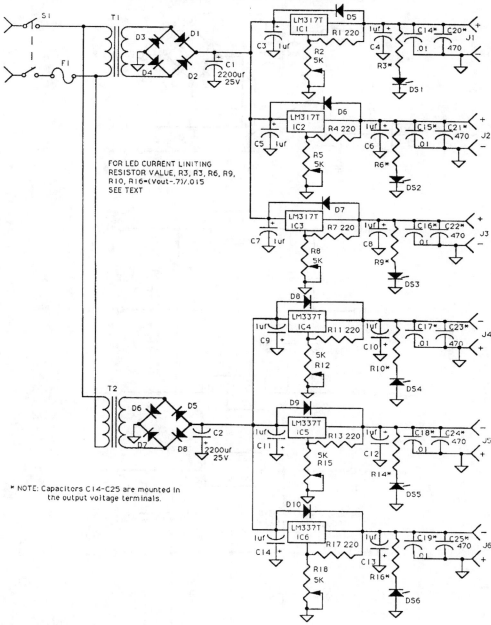

FOR LED CURRENT LIMITING
RESISTOR VALUE, R3, R3, R6, R9,
R10, R16=(Vout-.7)/.015
SEE TEXT

* NOTE: Capacitors C14–C25 are mounted in
the output voltage terminals.

Fig. 70-1

Passive linear IC regulators are used to make up a supply delivering +12, +9, +5, –5, –9, and –12 Vdc. T1 and T2 are 12-V, 3-A transformers.

QUAD POWER SUPPLY

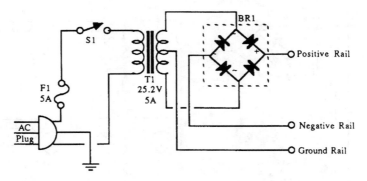

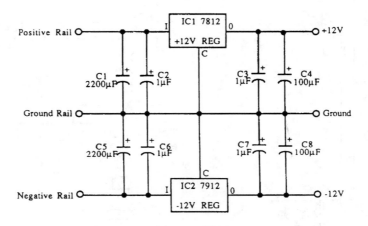

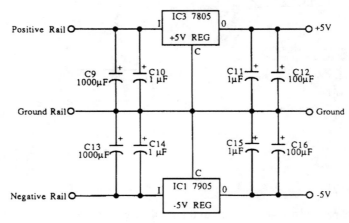

Fig. 70-2

ACTIVATE BACK-UP POWER SUPPLY

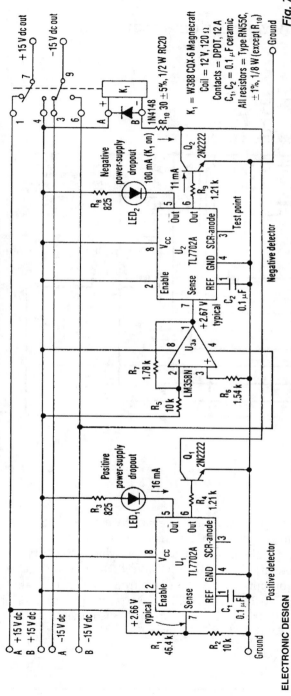

Fig. 70-3

ELECTRONIC DESIGN

A circuit, which can be built around two TI TL7702 chips, monitors a 15-V power supply and activates a relay to turn on a back-up supply if the voltage drops below ±14.1 V. With the back-up in place, the malfunctioning supply can be replaced without any down time. The TL7702 power-supply voltage supervisor chips are intended for use as reset controllers in microprocessor systems, but they work well in their modified form for this application.

One chip detects the positive supply (U1), and the other is used for the negative (U2). A pair of output-relay drive transistors, Q1 and Q2, form a wired OR circuit so that relay K1 is activated by the positive or negative voltage detector that switches U1 or U2 on.

The supervisor chips have a direct connection to the input comparator so that the trigger level is set by a resistor-divider network (R1 and R2) at the sense-input pin. These chips also have an internal, stable, reference-voltage source set at +2.53 V, typical. The positive-sensed voltage drops enough to activate the comparator, its output goes low, switches the internal gate, and triggers the silicon-controlled rectifier (SCR). The output comparator then forces the two output transistors to switch, one high and one low. The output transistor (pin 5) turns on the light-emitting diode (LED), and the output transistor (pin 6) turns on the relay driver, Q1.

The negative detector is preceded by half of the dual op amp LM358N (U3a)—an inverting amplifier with a gain of −0.178. R5 connects to the −15 Vdc being sensed. The output of U3a is usually set at 2.67 V, higher than the +2.53-V reference voltage. Therefore, no switching occurs. If the −15-Vdc voltage decreases, U2 switches Q2 and activates K1 in the same manner as described for the U1-Q1 positive detector. R3, R4, and R8 through R10 serve as current-limiting resistors.

509

CCFL SUPPLY WITH VARIABLE CONTRAST

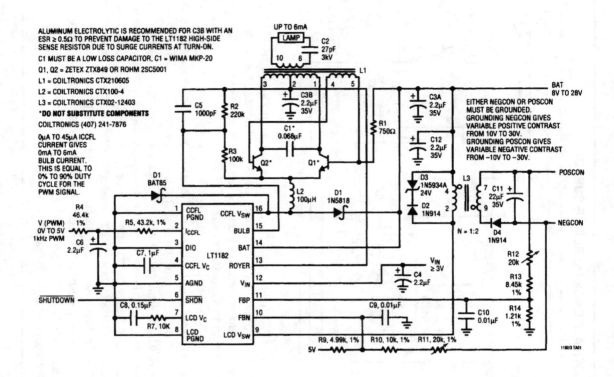

ALUMINUM ELECTROLYTIC IS RECOMMENDED FOR C3B WITH AN
ESR ≥ 0.5Ω TO PREVENT DAMAGE TO THE LT1182 HIGH-SIDE
SENSE RESISTOR DUE TO SURGE CURRENTS AT TURN-ON.

C1 MUST BE A LOW LOSS CAPACITOR, C1 = WIMA MKP-20

Q1, Q2 = ZETEX ZTX849 OR ROHM 2SC5001

L1 = COILTRONICS CTX210605

L2 = COILTRONICS CTX100-4

L3 = COILTRONICS CTX02-12403

DO NOT SUBSTITUTE COMPONENTS

COILTRONICS (407) 241-7876

0μA TO 45μA ICCFL
CURRENT GIVES
0mA TO 6mA
BULB CURRENT.
THIS IS EQUAL TO
0% TO 90% DUTY
CYCLE FOR THE
PWM SIGNAL.

LINEAR TECHNOLOGY

Fig. 70-4

The figure is a complete floating CCFL circuit with variable negative/variable positive-contrast voltage capability, based on the LT1182. Lamp current is programmable from 0 mA to 6 mA using a 0- to 5-V 1-kHz PWM signal at 0% to 90% duty cycle. LCD contrast output voltage polarity is determined by which side of the transformer secondary (either POSCON or NEGCON) the output connector grounds. In either case, LCD contrast output voltage is variable from an absolute value of 10 V to 30 V. The input supply voltage range is 8 V to 28 V. The CCFL converter is optimized for photometric output per watt of input power. CCFL electrical efficiency up to 90% is possible and requires strict attention to detail. LCD contrast efficiency is 82% at full power.

dc POWER SOURCE FOR EXPERIMENTS

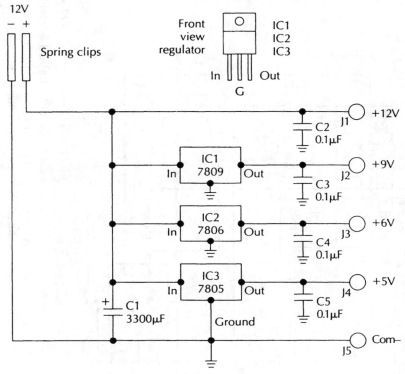

McGRAW-HILL

Fig. 70-5

This supply uses IC regulators to supply +5, +6, +9, and +12 volts regulated from a nominal 12-V supply.

STABLE VFO POWER SUPPLY

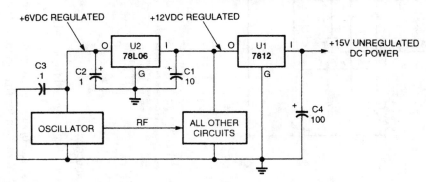

POPULAR ELECTRONICS

Fig. 70-6

A dc power-distribution system for a stable oscillator should use a separate voltage regulator just for the oscillator circuit.

HIGH-EFFICIENCY TRIPLE-OUTPUT SUPPLY FOR NOTEBOOK COMPUTERS

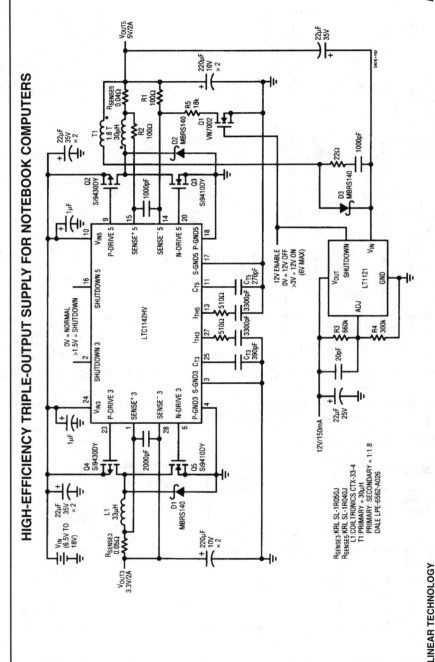

Fig. 70-7

LINEAR TECHNOLOGY

The circuit is configured to provide output voltages of 3.3 V, 5 V, and 12 V. The current capability of both the 3.3-V and 5-V outputs is 2 A (2.5 A peak). The logic controlled 12-V output can provide 150 mA (200 mA peak), which is ideal for flash memory applications. The operating efficiency shown in the figure exceeds 90% for both the 3.3-V and 5-V sections.

The 3.3-V section for the circuit in the figure is comprised of the main switch Q4, synchronous switch Q5, inductor L1, and current shunt R_{SENSE}. Current-sense resistor R_{SENSE} monitors the inductor current and is used to set the output current according to the formula $OUT = 100 \text{ mV}/R_{SENSE}$. Advantages of current control include excellent line and load transient rejection, inherent short-circuit protection, and controlled start-up currents. Peak inductor currents for L1 and T1 for the circuit in the figure are limited to 150 mV/R_{SENSE} or 3.0 A and 3.75 A, respectively.

512

GENERAL-PURPOSE POWER SUPPLY FOR AUTOMOTIVE PROJECTS

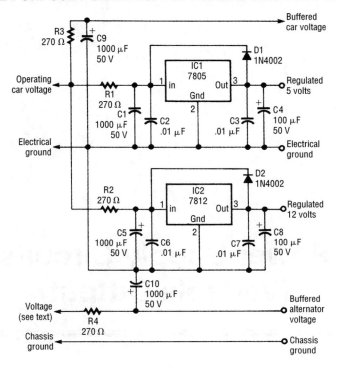

ELECTRONICS NOW

Fig. 70-8

This supply produces 12 V and 5 V for a variety of automotive projects. F4 is connected directly to the alternator field winding (usable only if your car has a separate regulator).

±15-V POWER SUPPLY

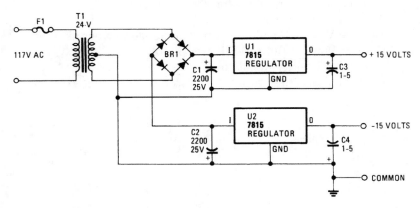

RADIO-ELECTRONICS

Fig. 70-9

A simple bridge rectifier feeds two IC regulators. This circuit should be useful for op-amp circuitry.

71

Power Supply Circuits
(Variable Output)

The sources of the following circuits are contained in the Sources section, which begins on page 707. The figure number in the box of each circuit correlates to the entry in the Sources section.

VARIABLE VOLTAGE REGULATOR WITH CURRENT CROWBAR LIMITING

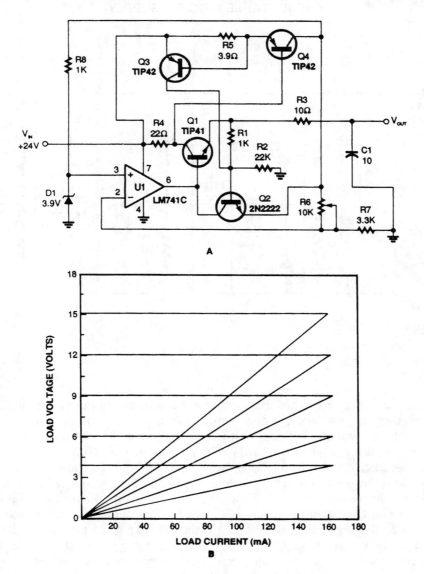

A

B

POPULAR ELECTRONICS

Fig. 71-1

The variable voltage regulator with current-crowbar limiting, shown in Fig. 71-1A, overcomes the disadvantages of constant and foldback limiting. As you can see in the graph (Fig. 71-1B), the current crowbar quickly shuts down the supplied power when a preset current is exceeded. It also has excellent load regulation over its operating range.

ADJUSTABLE 0- TO 5-V SUPPLY

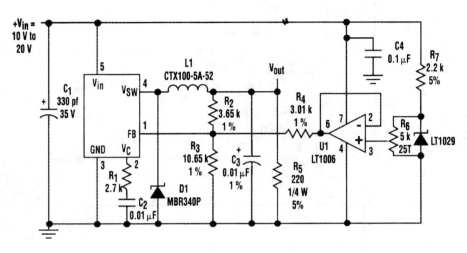

ELECTRONIC DESIGN

Fig. 71-2

Although linear-regulator ICs are frequently used in variable power-supply applications, they might not always be the best choice. At low output voltages, power losses in these regulators can cause headaches for designers. For example, if an output current of 1.25 A is required at 1.25 V from an input of 8 V. a regulator (such as the LT317) dissipates more than 10 W.

The figure depicts a dc-dc converter that functionally replaces a linear regulator in the just described application. The converter not only eliminates the problem of power loss, but it can be adjusted for output voltages (as low as 25 mV) while delivering an output current of 1.5 A.

The circuit uses a basic positive-buck topology with one exception. A control voltage is applied through R4 to the feedback summing node at pin 1 of the LT1076 regulator IC, making it possible to adjust the output from 0 V to approximately 6 V. This range encompasses the 3.3-V and 5-V logic supply voltages for portable and desktop equipment, as well as battery-pack combinations of one to four cells.

As R4 is driven from 0 to 5 V by the buffer (U1), more or less current is required from R2 to satisfy the loop's desire to hold the feedback summing point at 2.37 V. This forces the converter's output to swing over the range of 0 to 6 V.

The LT1076 is capable of 1.75-A guaranteed output current in this application, and 2 A is typical. If more current is required, the LT1074 can be substituted for the LT1076.

TRANSCEIVER POWER SUPPLY FOR VARIABLE LAB SOURCE

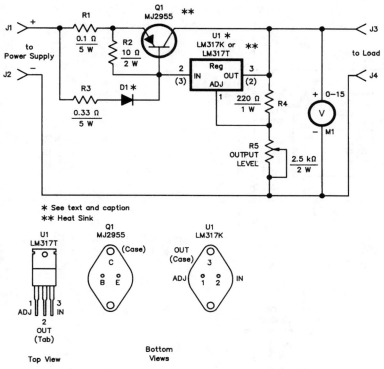

QST

Fig. 71-3

A variable voltage regulator provides 1 to 11 Vdc for lab bench work, using an existing 13.8-V transceiver supply.

ADJUSTABLE POWER SUPPLY

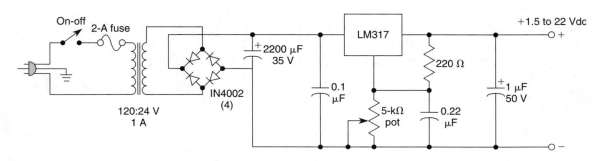

WILLIAM SHEETS

Fig. 71-4

Using an LM317, this supply delivers 1.25 to 22 Vdc for various purposes. The LM317 should be heatsinked. This supply will deliver 600-mA output current.

VARIABLE-VOLTAGE REGULATOR WITH WIDE-RANGE CURRENT LIMITING

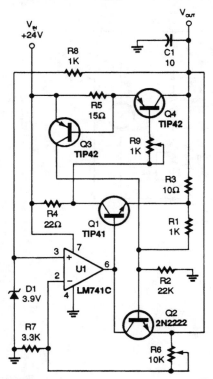

In this circuit, R9 acts as a control to set current limiting. If $R_9 = 0$, limiting occurs at 47 mA. Input is 24 V, output is

$$\frac{(R_6 + R_7)}{R_7} \times \qquad (3.9)$$

depending on the setting of R6.

POPULAR ELECTRONICS

Fig. 71-5

GENERAL-PURPOSE 0- TO 30-V POWER SUPPLY

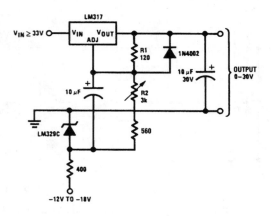

NATIONAL SEMICONDUCTOR

Fig. 71-6

ADJUSTABLE POSITIVE REGULATOR

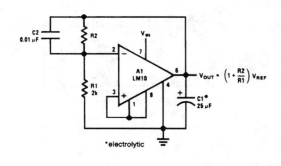

NATIONAL SEMICONDUCTOR

Fig. 71-7

ADJUSTABLE BIAS REGULATOR

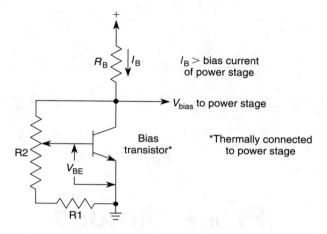

WILLIAM SHEETS

Fig. 71-8

If the wiper of R2 is set at $N\%$ rotation, the bias voltage will be:

$$V_{BIAS} = V_{BE}\left(\frac{R_1 + NR_2}{R_1 + R_2}\right)$$

$$V_{BIAS\ MIN} = V_{BE} \qquad V_{BIAS\ MAX} = \left(\frac{R_1}{R_1 + R_2}\right)\ V_{BE}$$

This method derives a bias voltage that tracks V_{BE} of this bias transistor. If the bias transistor is thermally linked to the power stage, tracking over a wide temperature range will result.

72

Probe Circuits

The sources of the following circuits are contained in the Sources section, which begins on page 707. The figure number in the box of each circuit correlates to the entry in the Sources section.

CURRENT PROBE AMPLIFIER

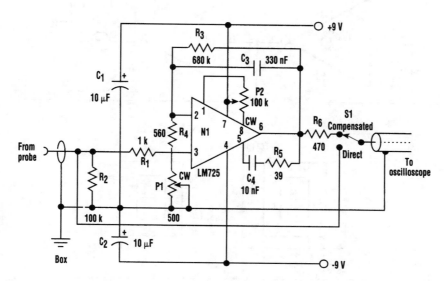

ELECTRONIC DESIGN

Fig. 72-1

A clamp-on current probe, such as the Tektronix P6021, is a useful means of displaying current waveforms on an oscilloscope. A less-expensive and simple alternative is shown in the figure.

The more sensitive range on the P6021 is 2 mA/mV, but it has a roll-off of 6-dB per octave below 450 Hz. The purpose of the compensator is to counteract the low-frequency attenuation, which is achieved by means of C3 and R4 + P1 in the feedback around op amp N1. It's important that the latter is a low-noise type, such as the LM725 shown in the figure. On top of that, it's necessary at some point to limit the increasing gain with decreasing frequency; otherwise, amplifier noise and drive will overcome the signal. The values shown for C_3 and R_3 give a lower limit of less than 1 Hz.

A test square wave of ±1 mA is fed to the current probe so that P1 can be adjusted for minimum droop or overshoot in the output waveform. It's vital that the sliding core on the probe is fully closed. At high frequencies, the response begins to fall off at 100 kHz. Therefore, for most waveforms, switch S1 is moved to "direct," above a fundamental frequency of, for example, 10 kHz.

This circuit's current consumption is quite low, and it can be battery powered. If a mains power supply is built-in, it must be well screened to prevent hum problems.

SIMPLE LOGIC PROBE WITH ALPHANUMERIC DISPLAY

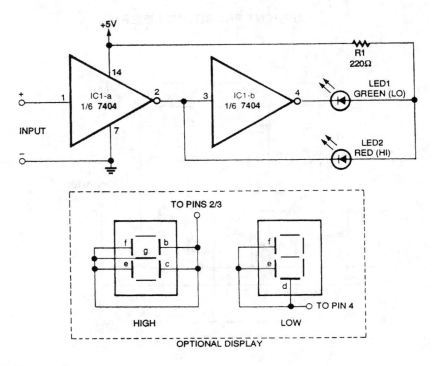

ELECTRONICS NOW

Fig. 72-2

A logic probe also includes BCD decoder module. The red LED lights to indicate a logic high, and the green LED lights to indicate a logic low. This probe circuit will light a green (low) or red (high), and if desired, an alphanumeric display can be obtained with two 7-segment LED displays.

SIMPLE RF PROBE

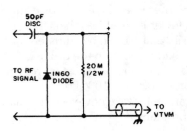

73 AMATEUR RADIO TODAY

Fig. 72-3

Your VTVM can measure peak voltage up to 200 MHz by using this probe. The maximum RF that can be measured is determined by the diode; with a 1N60, the probe is limited to 30 V. To increase the capacity, substitute a higher-voltage small-signal detector diode. House the circuit in a metal enclosure and use shielded wire.

125-MHz LOGIC PROBE

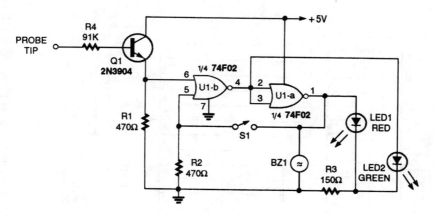

POPULAR ELECTRONICS

Fig. 72-4

This logic probe features either high-low (LED) indication or latching operation. When S1 is closed, the indication of a pulse is latched and the red LED1 stays on. Piezoelectric buzzer BZ1 is used as a beeper to sound that a logic high is preset.

pH PROBE AMPLIFIER

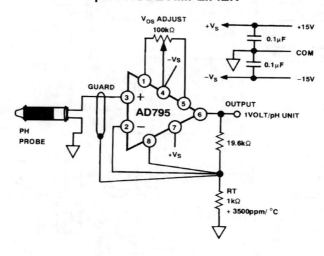

ANALOG DEVICES

Fig. 72-5

The low-noise precision FET op-amp AD795 has around 1014-Ω common-mode impedance, low-offset voltage ($250\ \mu V_{max}$) and L13 μV_C drift make this device ideal for low-voltage measurements from high-impedance sources.

8-DIGIT 100-MHz FREQUENCY PROBE

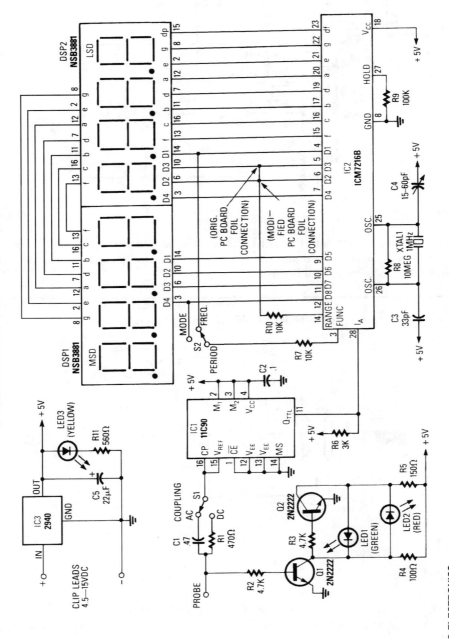

Fig. 72-6

RADIO-ELECTRONICS

Small enough to mount in a probe, this frequency counter circuit is good to 100 MHz. It operates from +5 to +15 Vdc. An 11C90 prescaler drives a 10-MHz counter chip (ICM7216B). Note the dotted line connecting R10 with pins 5 and 6 of IC2; that variable connection controls the decimal point and total count appearing on DSP1 and DSP2. The relative intensities and durations of ON/OFF time for LED1 (green) and LED2 (red) give a rough indication of logic level and duty cycle.

73

Protection Circuits

The sources of the following circuits are contained in the Sources section, which begins on page 707. The figure number in the box of each circuit correlates to the entry in the Sources section.

Short-Circuit Protection Circuit
Polarity Protector

SHORT-CIRCUIT PROTECTION CIRCUIT

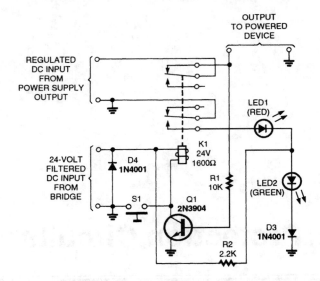

Fig. 73-1

When S1 is pressed, the coil of K1 is energized, closing its normally open contacts. If the regulated dc input is between 1 and 24 V, that voltage feeds the base of Q1 through R1, turning on the transistor, and latching the relay. When that occurs, LED2 glows indicating that all is okay.

If there is a short to ground at the circuit's output (i.e., in the device being powered), the voltage that feeds the base of Q1 goes to zero, turning off the transistor. Then, LED1 glows because K1 is de-energized to indicate the short circuit.

POLARITY PROTECTOR

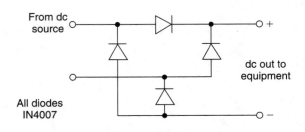

The use of a four-diode bridge guarantees correct polarity irrespective of input dc polarity. Remember that two diode drops (about 1.2 to 1.5 V) are lost from the input voltage using this circuit.

Fig. 73-2

74

Radar Detector Circuits

The sources of the following circuits are contained in the Sources section, which begins on page 707. The figure number in the box of each circuit correlates to the entry in the Sources section.

Display Board for Radar Gun
2.6-GHz Oscillator for Radar Speed Gun

DISPLAY BOARD FOR RADAR GUN

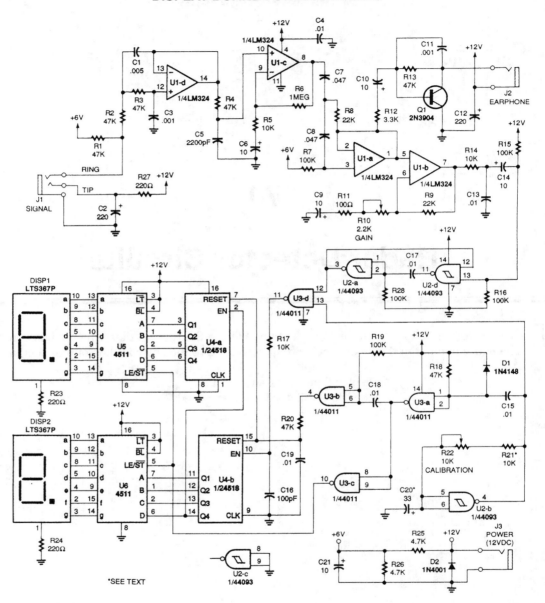

POPULAR ELECTRONICS

Fig. 74-1

This circuit takes signal (doppler) from a radar gun, amplifies and limits it, and feeds the frequency into a counter (U4) and display circuit (DISP1, DISP2, U5, U6). Counter calibration is set by clock circuit U2B. Calibration is obtained via R21 and R22. R21 can be changed if kilometers/hour readout is desired.

2.6-GHz OSCILLATOR FOR RADAR SPEED GUN

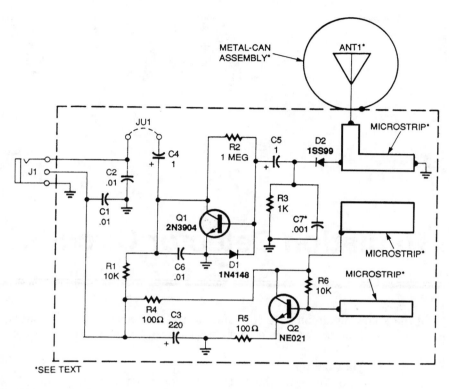

POPULAR ELECTRONICS

Fig. 74-2

This circuit consists of 2.6-GHz oscillator Q2, a coupling microstripline to ANT1, a 1.1" ¼-wave probe, detector D2, and audio amp Q1. The oscillator feeds power to the antenna, which radiates the signal. The reflected signal from a moving target mixes with the oscillator signal in D2. The resultant beat note (doppler shift) is amplified by Q1 and fed to jack J1, which is used to feed the circuit 12 Vdc.

75

Radiation Detector Circuits

The sources of the following circuits are contained in the Sources section, which begins on page 707. The figure number in the box of each circuit correlates to the entry in the Sources section.

Geiger Counter
Voltage Tripler for Radon Detector Ionization Chamber
Flyback Power Supply for Radon Monitor
Radon Monitor Amplifier and Head
Ion Detector

GEIGER COUNTER

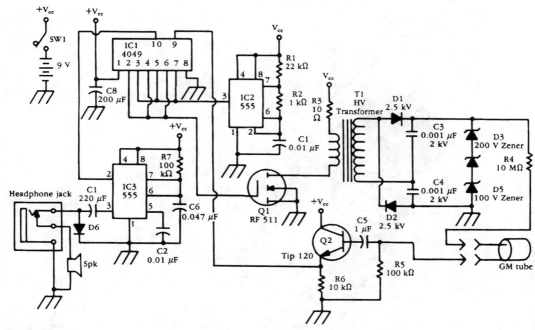

McGRAW-HILL

Fig. 75-1

An HV generator (IC1, IC2, Q1, T1, and associated components) power a G-M tube. A pulse from the GM tube is interfaced through Q2 and IC1 to pulse generator IC3, which drives a speaker.

VOLTAGE TRIPLER FOR RADON DETECTOR IONIZATION CHAMBER

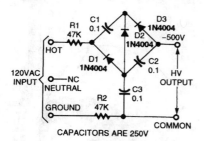

ELECTRONICS NOW

Fig. 75-2

The voltage tripler charges the ionization chamber capacitor. It is powered from the 120-Vac line. Warning: Shock hazard exists.

FLYBACK POWER SUPPLY FOR RADON MONITOR

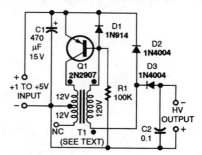

ELECTRONICS NOW

Fig. 75-3

This blocking-oscillator flyback circuit is an alternative for charging the ionization chamber capacitor.

RADON MONITOR AMPLIFIER AND HEAD

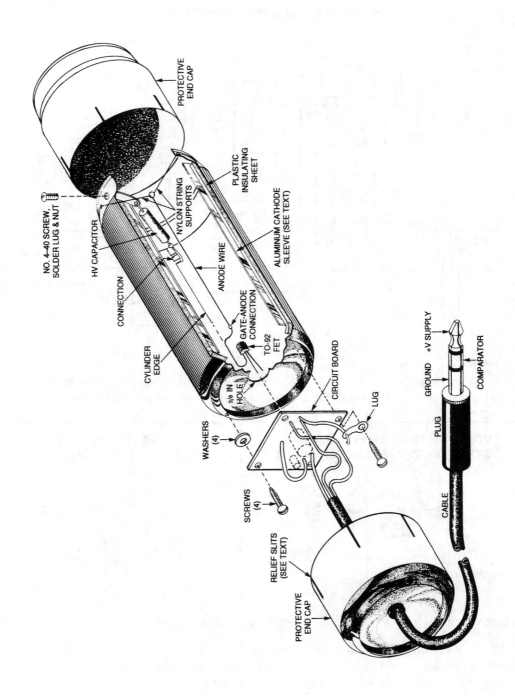

PROTECTIVE END CAP

NO. 4-40 SCREW, SOLDER LUG & NUT

HV CAPACITOR

CONNECTION

CYLINDER EDGE

WASHERS (4)

SCREWS (4)

RELIEF SLITS (SEE TEXT)

PROTECTIVE END CAP

NYLON STRING SUPPORTS

PLASTIC INSULATING SHEET

ANODE WIRE

ALUMINUM CATHODE SLEEVE (SEE TEXT)

GATE-ANODE CONNECTION

TO-92 FET

3/8 IN HOLE

CIRCUIT BOARD

LUG

GROUND

+V SUPPLY

PLUG

COMPARATOR

CABLE

RADON MONITOR AMPLIFIER AND HEAD (Cont.)

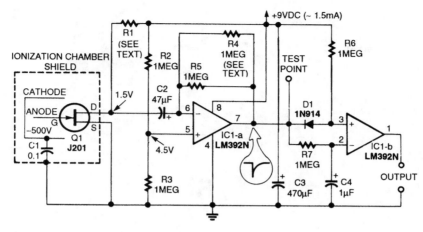

ELECTRONICS NOW

Fig. 75-4

A positively charged anode wire attracts electrons and a negatively charged cathode attracts positively charged ions. The recombination of electrons and ions causes a current that produces a voltage pulse. The cathode is maintained at –500 V by a charge on the 0.1-μF capacitor.

A beverage can forms the chamber, an aluminum can forms the cathode, and half cans form protective end covers. The amplifier circuit board is shown to the left of center.

ION DETECTOR

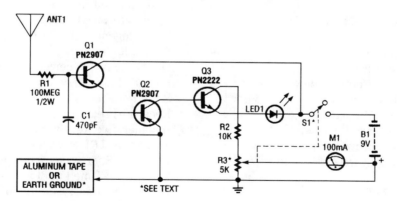

POPULAR ELECTRONICS

Fig. 75-5

ANT1 is a short whip antenna from a junked radio or other device. R_3 is adjusted to bring the meter on scale. This device should be grounded to operate properly. A length of aluminum or copper foil tape attached to the instrument case makes contact with the hand, and the body serves as a ground via hand contact with this tape.

76

Receiving Circuits

The sources of the following circuits are contained in the Sources section, which begins on page 707. The figure number in the box of each circuit correlates to the entry in the Sources section.

AM RADIO

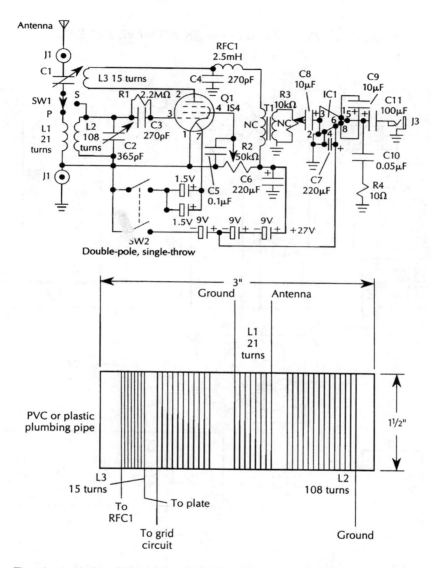

The primary winding of L1 has 21 turns of #24 or #26 enameled wire; L2 has 108 turns; and L3 has 15 turns of wire. All are wound on 1½-inch PVC pipe form.

Fig. 76-1

A 1S4 regenerative detector feeds an LM386 audio IC (IC1). 1.5-V D cells and three 9-V batteries are used for a power supply.

535

ac/dc VACUUM-TUBE AM AND SHORTWAVE RECEIVER

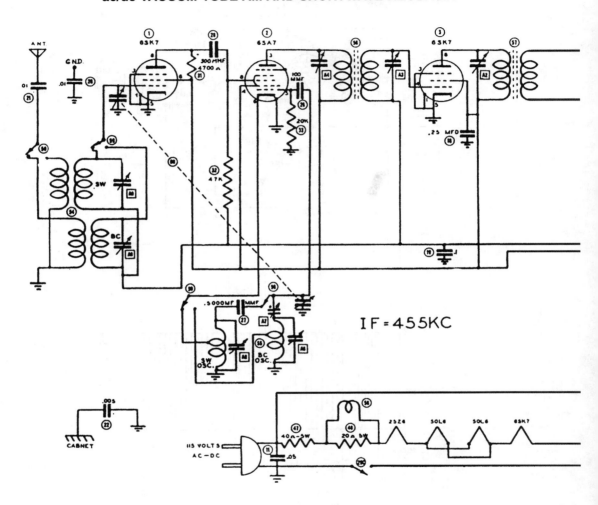

IF = 455KC

This circuit was used in a World War II vintage AM/SW (6 to 18 MHz) receiver and shows typical circuits used in receivers at that time.

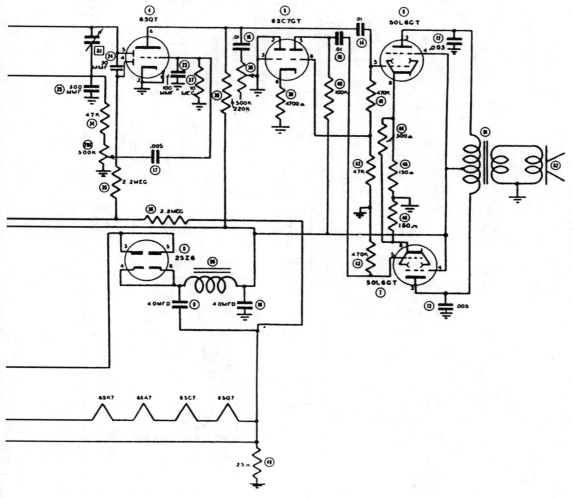

Fig. 76-2

WWV RECEIVER

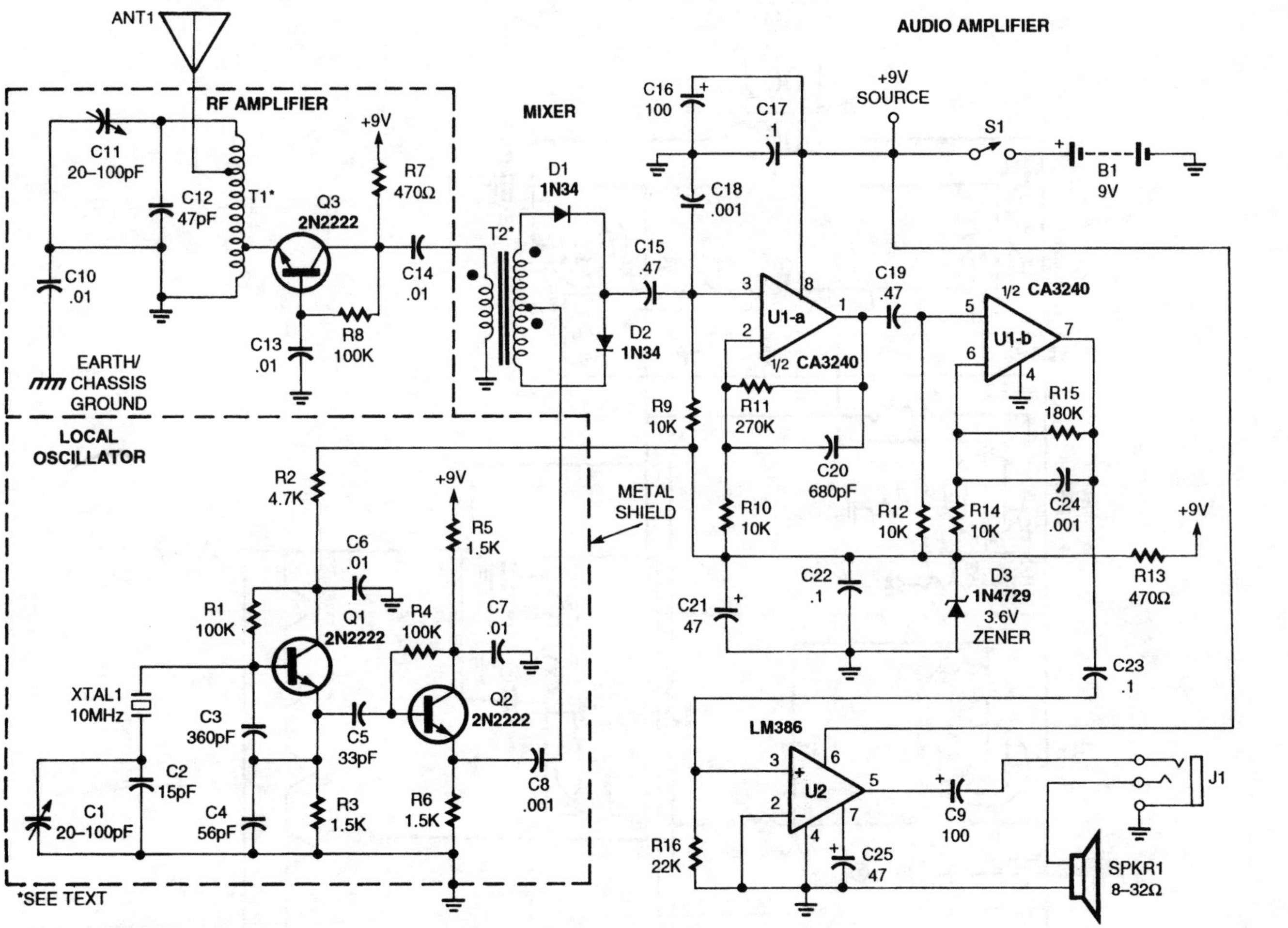

AUDIO AMPLIFIER

Fig. 76-3

POPULAR ELECTRONICS

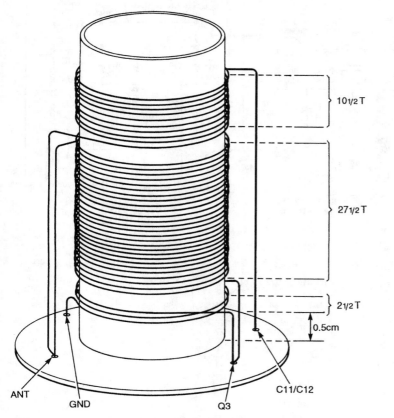

10½ T

27½ T

2½ T

0.5cm

ANT

GND

Q3

C11/C12

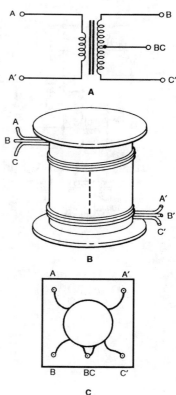

A

B

C

Transformer T1 is a home-made unit comprised of 40 closely wound turns of #26 AWG enameled wire on a ¼-inch diameter air-core form, with taps at 2½ and 10½ turns from each end; after each tap, the winding continues in the same direction.

Construction details for T2 (the mixer transformer) are shown here. The diagram in A is a schematic representation of the unit once completed; B illustrates how the three lengths of wire are wound as a set on the bobbin; and C shows how the bobbin is connected to the pinned base.

RF amplifier Q3 feeds diode mixer D1–D2 and Q1–Q2 provide 10-MHz L.O. injection to D1 through T1 and T2. U1A, U1B and U2 are audio amplifiers. Details of T1 and T2 are shown.

SHORTWAVE RECEIVER

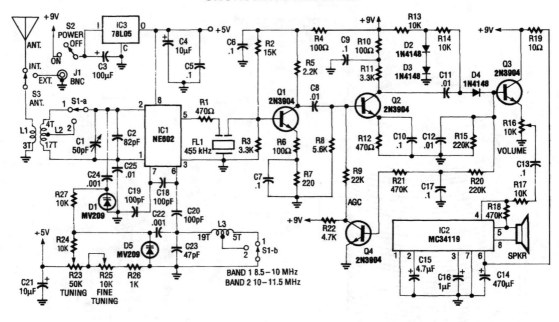

TABLE 2

Frequency (MHz)	C2 (pF)	C18, C19 (pF)	C23 (pF)	L1, L2 (Ant) (# of turns on T-37-2 core)	L3 (Osc)
5	100	120	68	5, 41	45
6	100	120	68	4, 30	34
7	82	100	47	4, 26	29
8	82	100	47	3, 22	24
10	82	100	47	3, 17	19
12	82	100	47	2, 15	17
14	68	82	33	2, 14	15
15	68	82	33	2, 13	14

TABLE 3

	NE602	MC34119
Pin 1	1.27 V	0V
Pin 2	1.27 V	4.15 V
Pin 3	0V	4.11 V
Pin 4	3.64 V	3.97 V
Pin 5	3.59 V	4.14 V
Pin 6	4.99 V	9.09 V
Pin 7	4.33 V	0V
Pin 8	5.05 V	4.20 V

	Q1	Q2	Q3	Q4
Emitter	0.95 V	0.80 V	0.27 V	0V
Base	1.61 V	1.45 V	0.82 V	0.58 V
Collector	2.56 V	3.30 V	9.17 V	7.41 V

RADIO-ELECTRONICS

Fig. 76-4

This receiver covers 8.5 to 11.5 MHz in two bands and has a sensitivity of under 1 μV. An NE602 mixer feeds a 455-kHz IF amplifier (Q1 and Q2), detector D4, and audio amplifier IC2. Q4 serves as an AGC amplifier coil data is given in the table. The LO is varactor tuned.

AM/FM RECEIVER CIRCUIT

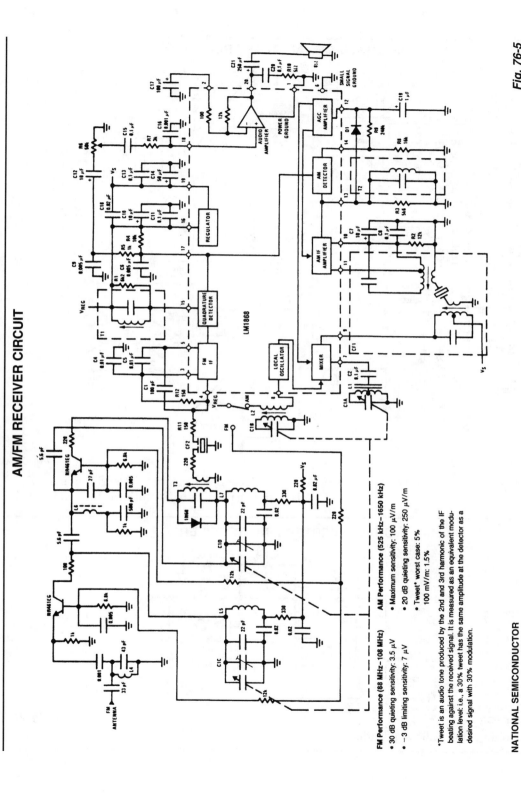

FM Performance (88 MHz–108 MHz)

- 30 dB quieting sensitivity: 3.5 µV
- −3 dB limiting sensitivity: 7 µV

AM Performance (525 kHz–1650 kHz)

- Maximum sensitivity: 100 µV/m
- 20 dB quieting sensitivity: 250 µV/m
- Tweet* worst case: 5%
  100 mV/m: 1.5%

*Tweet is an audio tone produced by the 2nd and 3rd harmonic of the IF beating against the received signal. It is measured as an equivalent modulation level: i.e., a 30% tweet has the same amplitude at the detector as a desired signal with 30% modulation.

NATIONAL SEMICONDUCTOR

Fig. 76-5

This circuit shows the LM1868 as a complete AM radio and FM IF section. An external FM front end is used for the 88- to 108-MHz band. Audio output is 0.5 W and either 9-V battery or line operated supply can be used.

541

118- TO 136-MHz AIRCRAFT RECEIVER

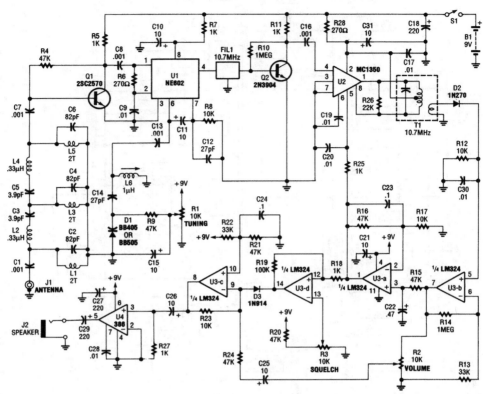

RADIO CRAFT

Fig. 76-6

This receiver covers the 118- to 136-MHz AM aviation band. It has a 10.7-MHz IF amplifier. L1, L3, and L5 are 1½ turns of #24 wire. F1L1 is a 10.7-MHz ceramic filter. IF bandwidth will be about 250 kHz.

DUAL-INVERTER LINE RECEIVER

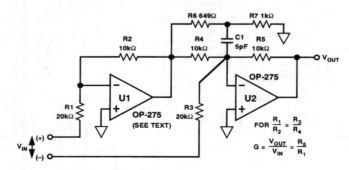

ANALOG DEVICES

Fig. 76-7

This circuit is for audio applications.

TOROIDAL CORE TRF SHORTWAVE RECEIVER

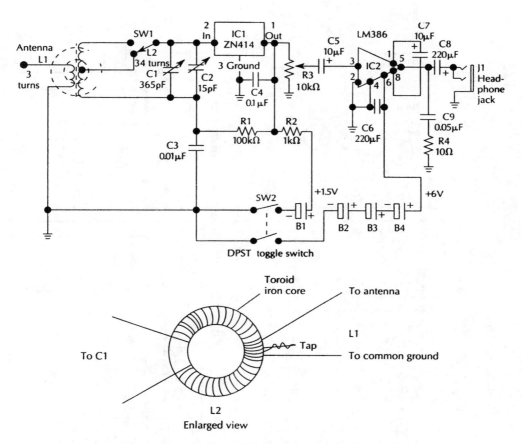

L1—3 turns of #24 enameled magnet wire wound over center of L2.
L2—34 turns of #24 enameled wire—coils wound on toroid iron core
form T-50-2—tapped at the 17th turn. Form has only ½-inch diameter.

Wind 34 turns, with a tap at the 17th turn, for L2 on the small ½-inch diameter iron core form.

McGRAW-HILL   *Fig. 76-8*

A ZN414 IC feeds an LM386 audio amplifier in this TRF circuit. SW1 is a band-switch. Coverage is up to 18 MHz.

NINE-BAND SHORTWAVE RECEIVER

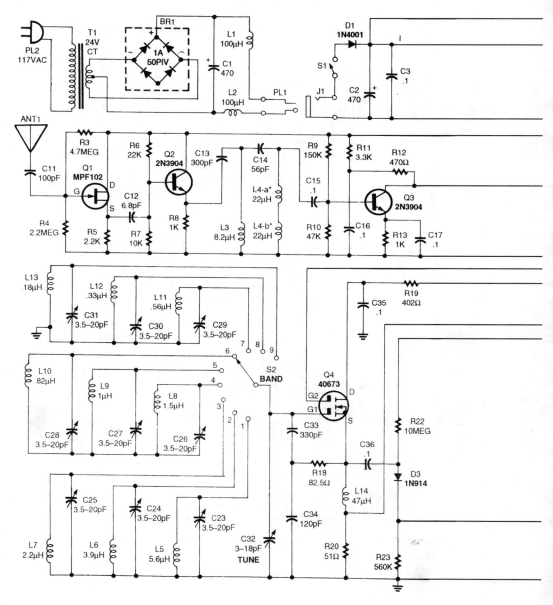

POPULAR ELECTRONICS

Dual-gate MOSFET Q4 is used as a regenerative amplifier in this circuit. An active antenna feeds the signal to Q4, and a short whip antenna is adequate. Detector Q5 feeds volume control R24, and audio amplifier U5, an LM386. The frequency range is 49 to 11 meters in nine bands (6 to 27 MHz).

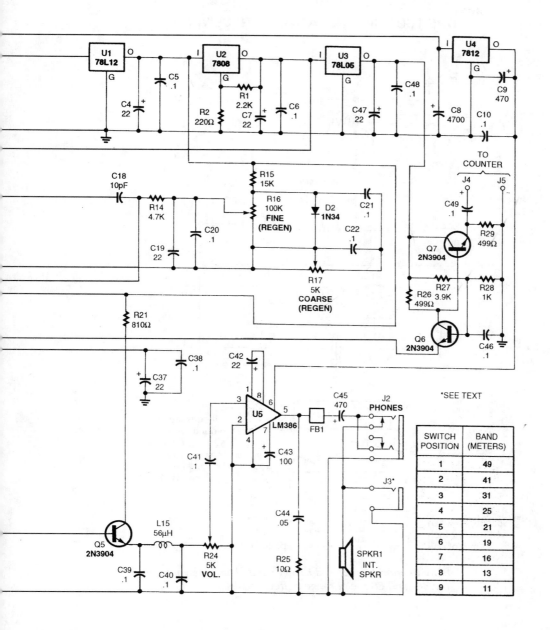

SWITCH POSITION	BAND (METERS)
1	49
2	41
3	31
4	25
5	21
6	19
7	16
8	13
9	11

ONE-TUBE REGENERATIVE SW RECEIVER

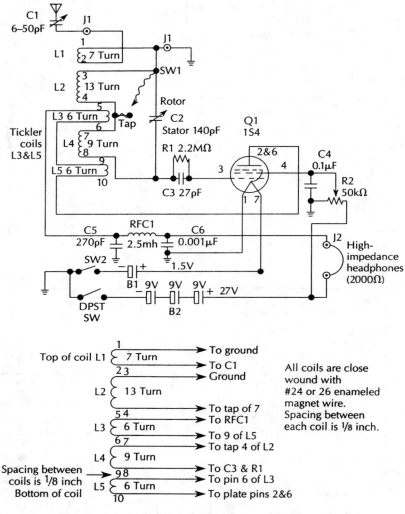

How to wind L1, L2, and L3 with taps. This receiver tunes in the 40- and 80-meter range of frequencies.

McGRAW-HILL

Fig. 76-10

A 154 tube is used in a regenerative detector circuit. Details for coils are shown and frequency range can be shifted within 1.5 to 20 MHz by proportionally adjusting the number of turns on coils.

ONE-TUBE REGENERATIVE AM RECEIVER

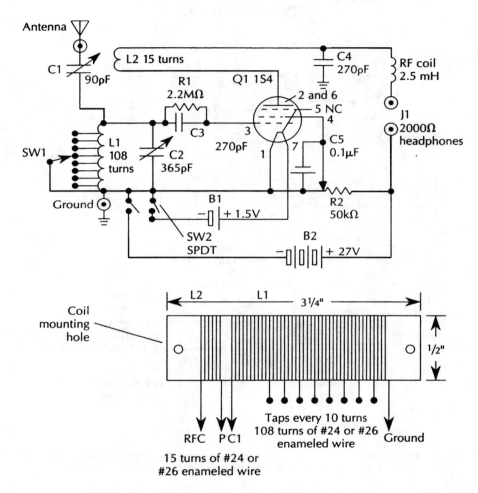

Wind both coils on PVC pipe using either #24 or #26 enameled wire. For coil L1, wind 108 turns on the pipe, and tap every 10 turns.

McGRAW-HILL

Fig. 76-11

Suitable for AM reception and as a simple radio project, this circuit uses a single tube as a re-generative detector.

TWO-BAND RADIO

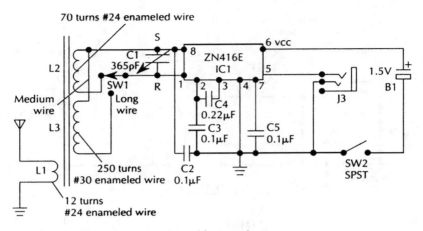

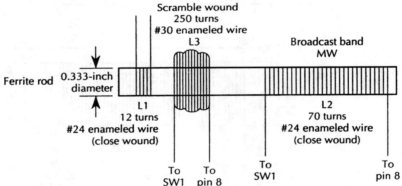

All three coil windings are wound on one long ferrite form. Two different coils are switched into the circuit, covering the longwave (lw) and medium-wave (mw) broadcast band.

McGRAW-HILL

Fig. 76-12

This TRF receiver covers the AM broadcast band and longwave bands (used in Europe and Asia for broadcasting). A loop antenna is used for reception and an external antenna can be connected. Frequency coverage is 150 to 1600 kHz.

SIMPLE CRYSTAL RADIO

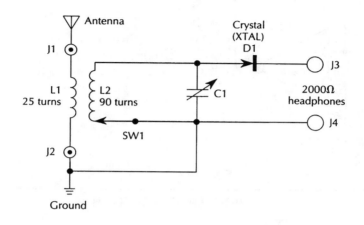

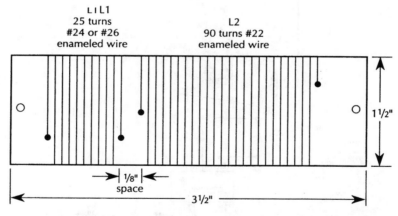

Wind 90 turns or 2½ inches of #22 enameled wire for L2, and 25 turns of #24 or #26 enameled wire for L1 on a 1½-inch PVC plastic pipe form.

Fig. 76-13

An IN34A (D1) is used as a detector in this crystal radio. A good outdoor antenna should be used.

VIDEO LINE RECEIVER

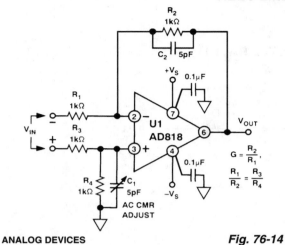

This circuit can achieve 46-dB common-mode rejection if R1, R2, R3, and R4 are matched to 1%. C1 is adjusted for best CMR above 1 MHz.

$$G = \frac{R_2}{R_1},$$

$$\frac{R_1}{R_2} = \frac{R_3}{R_4}$$

ANALOG DEVICES

Fig. 76-14

TWO-CHIP AM RECEIVER

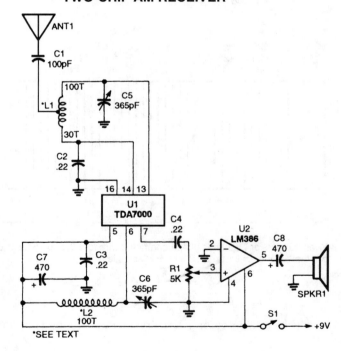

POPULAR ELECTRONICS

Fig. 76-15

This receiver is comprised of a TDA7000 single-chip FM receiver (U1), an LM386 low-voltage audio-power amplifier (U2), a pair of hand-wound coils (L1 and L2), and a few additional components. L1 and L2 are 100 turns of #28 wire on toroidal cores (about 240 µH each). L1 is tapped at 30 turns.

RELAY INTERFACE TO RC RECEIVERS

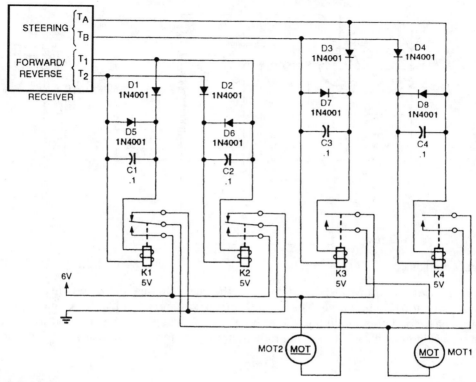

POPULAR ELECTRONICS

Fig. 76-16

You can add relays to some inexpensive RC receivers to operate your own chassis.

BASIC REGENERATIVE RECEIVER

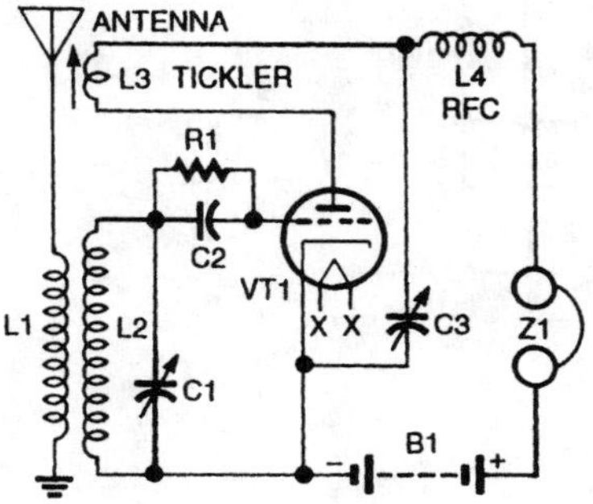

ELECTRONICS NOW

Fig. 76-17

SIMPLE RADIO RECEIVER

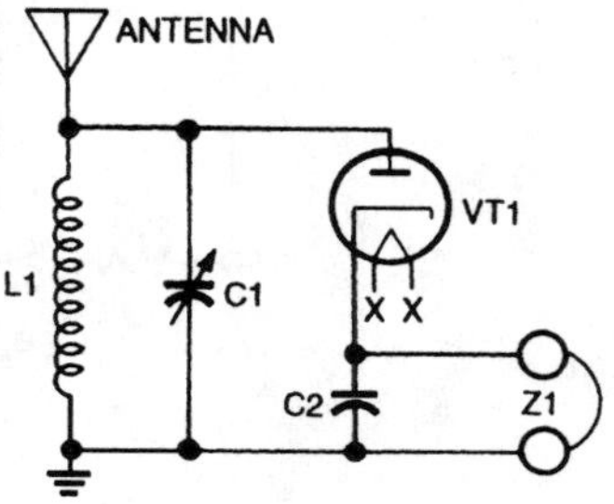

ELECTRONICS NOW

Fig. 76-18

Vacuum-tube detector receiver.

551

ONE-TUBE AM RECEIVER

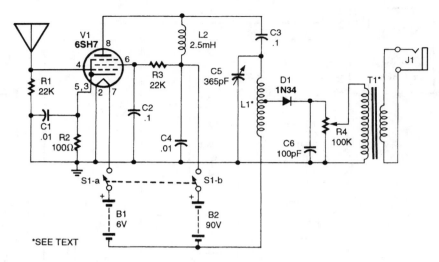

POPULAR ELECTRONICS

Fig. 76-19

This radio uses an untuned RF stage to boost the signal voltage up to the linear portion of the crystal diode's characteristic curve. The circuit's distortion and wide bandpass and a good-quality transformer make for a great-sounding AM radio. L1 is a winding of #22 enamelled wire 2" long on a 2" diameter plastic pipe. T1 is a tube-type radio output transformer, rated at 2000 Ω to the speaker voice coil.

BALANCED LINE RECEIVER

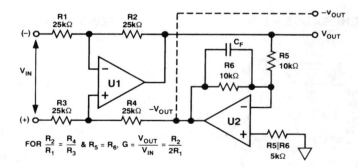

$$\text{FOR } \frac{R_2}{R_1} = \frac{R_4}{R_3} \text{ \& } R_5 = R_6, \ G = \frac{V_{OUT}}{V_{IN}} = \frac{R_2}{2R_1}$$

ANALOG DEVICES

Fig. 76-20

Unity-gain inverter U2 drives R4 (usually grounded at $-V_{out}$, equalizing currents in ±input legs, and provides a choice of balanced p-p output with a gain of R_2/R_1).

SUPERHET FRONT END

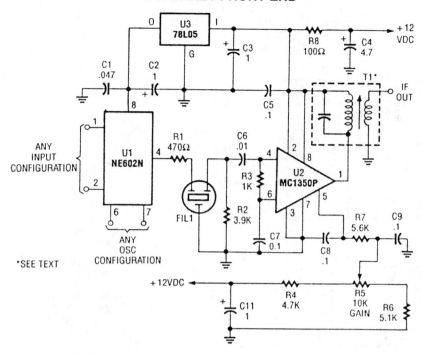

RADIO CRAFT

Fig. 76-21

This superhet receiver front end is simple and uses an NE602 followed by an MC1350 IF amplifier.

RECEIVER PREAMP

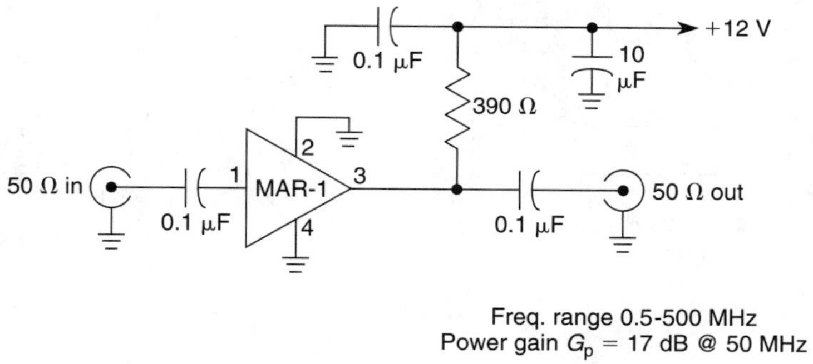

Freq. range 0.5-500 MHz
Power gain G_p = 17 dB @ 50 MHz

WILLIAM SHEETS

Fig. 76-22

Suitable for HF and VHF receivers, this preamplifier can be mounted on the back of the receiver for a boost in gain. Useful gain is about 17 dB at 50 MHz.

REGENERATIVE RECEIVER FOR 6 TO 17 MHz

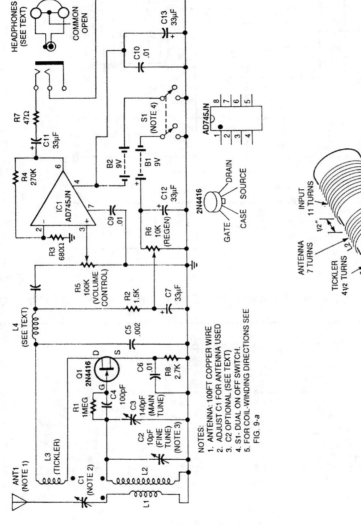

NOTES:
1. ANTENNA: 100FT COPPER WIRE
2. ADJUST C1 FOR ANTENNA USED
3. C2 OPTIONAL (SEE TEXT)
4. S1- DUAL ON OFF SWITCH.
5. FOR COIL-WINDING DIRECTIONS SEE
   FIG. 9-a

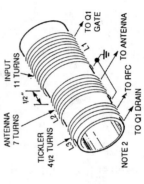

NOTES:
1. ALL WINDINGS ARE NO.22
   INSULATED STRANDED COPPER
   HOOKUP WIRE
2. 1½-IN. OD PVC PIPE

ELECTRONICS NOW!

Fig. 76-23

The headphones are 32-Ω stereo types. The common lead is left floating so that the two sides are in series, giving 64 Ω.

TWO-STAGE TRF REGENERATIVE RECEIVER

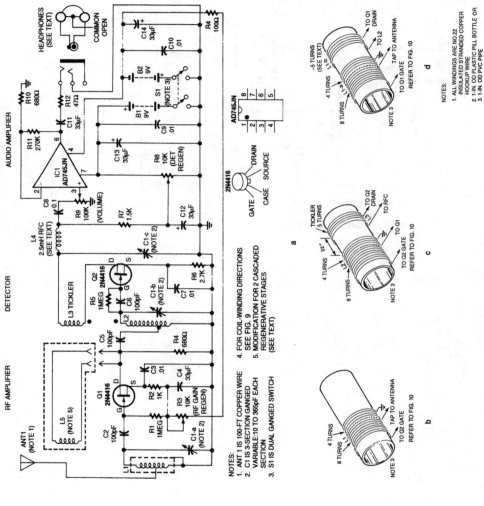

NOTES:
1. ANT 1 IS 100-FT COPPER WIRE
2. C1 IS 3-SECTION GANGED VARIABLE:10 TO 365pF EACH SECTION
3. S1 IS DUAL GANGED SWITCH
4. FOR COIL-WINDING DIRECTIONS SEE FIG. 9
5. MODIFICATION FOR 2 CASCADED REGENERATIVE STAGES (SEE TEXT)

NOTES:
1. ALL WINDINGS ARE NO.22 INSULATED STRANDED COPPER HOOKUP WIRE
2. 1-IN. OD PLASTIC PILL BOTTLE OR
3. 1-IN. OD PVC PIPE

This regenerative receiver uses a tuned RF stage to improve performance. The coil in Fig. 76-24D is for the purpose of adding a second regenerative stage (RF amp). This coil is L5 in the schematic.

ECONOMY SHORTWAVE RECEIVER

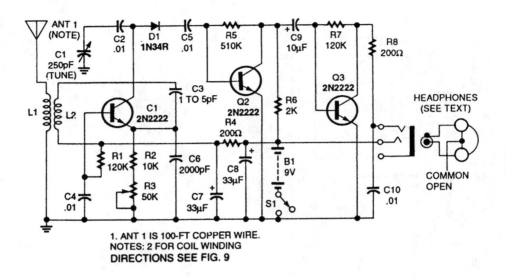

NOTES:
1. ANT 1 IS 100-FT COPPER WIRE.
2. FOR COIL WINDING DIRECTIONS SEE FIG. 9

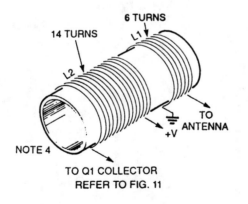

NOTES:

1. ALL WINDINGS ARE NO.22 INSULATED STRANDED COPPER HOOKUP WIRE
2. 1-IN. OD PLASTIC PILL BOTTLE
3. 1-IN. OD PVC PIPE OR NOTE 3

ELECTRONICS NOW

Fig. 76-25

Using three transistors, this receiver covers the range of 6 to 17 MHz. Coils can be altered to change the range to a lower or higher frequency.

VARIOMETER-TUNED RADIO

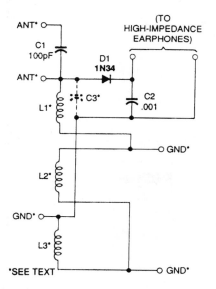

The two fixed coils of the variometer, L1 and L3, are wound on an 8½-inch-long piece of 1-inch-diameter plastic pipe (its outer diameter is about 1¼ inches). Each coil is 2¼ inches long. The number of turns is not critical, but 86 tightly wound turns of #22 enameled wire were used. When winding the coils, make sure you start at a point that will allow them to be placed 2 inches apart on the pipe. Drill holes in the pipe and run the leads of the coils out the end of the pipe that is closest to each.

The movable coil, L2, is wound on a piece of 1½-inch plastic pipe (its outer diameter is about 1⅞ inches). The winding is 2 inches long. Like L1 and L3, the actual number of windings of this coil are not critical, as long as the winding is approximately the right length.

POPULAR ELECTRONICS   *Fig. 76-26*

OLD-FASHIONED CRYSTAL RADIO

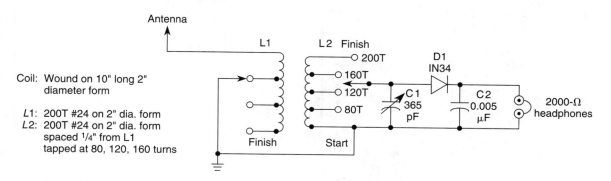

WILLIAM SHEETS   *Fig. 76-27*

L1 and L2 are wound on 4" diameter 10" form and are 200 turns of #24 wire. PVC pipe can be used.

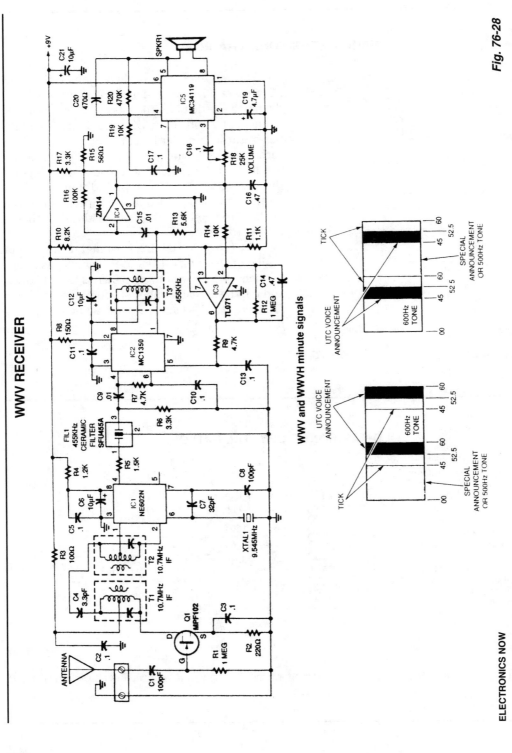

WWV RECEIVER

WWV and WWVH minute signals

Fig. 76-28

ELECTRONICS NOW

This receiver for 10-MHz WWV signals uses a 10.7-MHz FM receiver IF transformers as front-end components. It is a super-het with a 455-kHz IF frequency. By changing the front-end components 5- or 15-MHz reception could be obtained. A 3- to 6-foot antenna is usually adequate.

77

Reference Circuits

The sources of the following circuits are contained in the Sources section, which begins on page 707. The figure number in the box of each circuit correlates to the entry in the Sources section.

Low-Voltage Reference
Positive Voltage Reference
Negative Voltage Reference

LOW-VOLTAGE REFERENCE

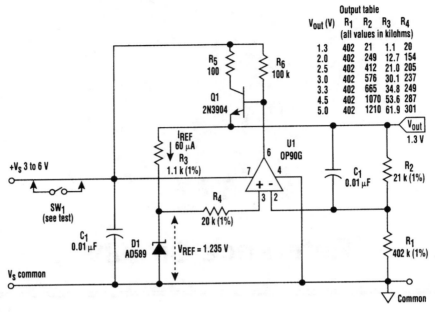

Output table

V_{out} (V)	R_1	R_2	R_3	R_4
	(all values in kilohms)			
1.3	402	21	1.1	20
2.0	402	249	12.7	154
2.5	402	412	21.0	205
3.0	402	576	30.1	237
3.3	402	665	34.8	249
4.5	402	1070	53.6	287
5.0	402	1210	61.9	301

ELECTRONIC DESIGN

Fig. 77-1

This circuit illustrates a number of techniques that are useful for low-voltage, series-mode, power-efficient references. Intended for output currents of up to 10 mA, this design has an enabled standby current of about 100 µA; it can be easily programmed over a wide range of output voltages.

POSITIVE VOLTAGE REFERENCE

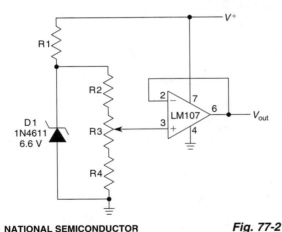

NATIONAL SEMICONDUCTOR

Fig. 77-2

D1 is used as a reference. R2, R3, and R4 provide desired output voltage to the op-amp voltage follower.

NEGATIVE VOLTAGE REFERENCE

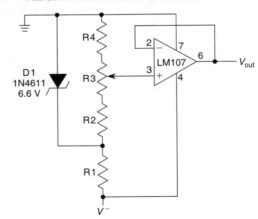

NATIONAL SEMICONDUCTOR

Fig. 77-3

D1 is used as a reference. R2, R3, and R4 are voltage dividers to obtain desired output voltage to the op-amp voltage follower.

78

Regulator Circuits

The sources of the following circuits are contained in the Sources section, which begins on page 707. The figure number in the box of each circuit correlates to the entry in the Sources section.

3.3-V 1-A Surface-Mount Regulator
Logic Control of 78XX Regulator
Low-Cost Step-Down Regulator
Dual-Output Regulator
Low-Noise Regulator (5 to 3.3 V)
Reducing Ripple in a Switching Voltage Regulator
Low-Dropout Three-Terminal Regulators for New Microprocessor Applications
Low-Dropout Regulator
Positive Regulator Sinks Current
5- to 3.3-V Surface-Mount Switching Regulator

3.3-V 1-A SURFACE-MOUNT REGULATOR

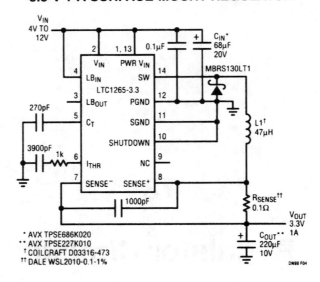

LINEAR TECHNOLOGY

Fig. 78-1

This figure shows a typical LTC1265 surface-mount application. It provides 3.3 V at 1 A from an input voltage range of 4 V to 12 V. The peak efficiency approaches 93% at mid-current levels.

LOGIC CONTROL OF 78XX REGULATOR

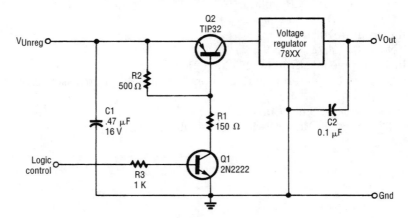

ELECTRONICS NOW

Fig. 78-2

Transistors can be used to control any 78xx series regulator with logic signals. Both transistors are controlled by the logic level present at the base of Q1.

LOW-COST STEP-DOWN REGULATOR

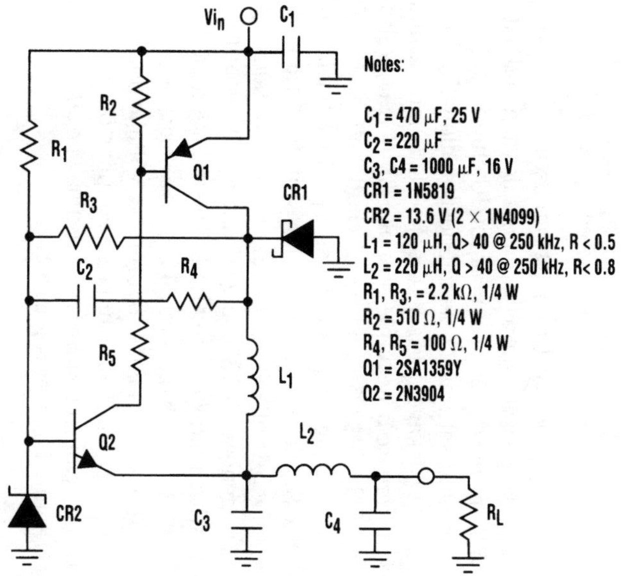

Vin
C1
R2
R1
Q1
R3
CR1
C2
R4
R5
L1
L2
Q2
CR2
C3
C4
RL
Notes:
C1 = 470 μF, 25 V
C2 = 220 μF
C3, C4 = 1000 μF, 16 V
CR1 = 1N5819
CR2 = 13.6 V (2 × 1N4099)
L1 = 120 μH, Q> 40 @ 250 kHz, R < 0.5
L2 = 220 μH, Q > 40 @ 250 kHz, R< 0.8
R1, R3, = 2.2 kΩ, 1/4 W
R2 = 510 Ω, 1/4 W
R4, R5 = 100 Ω, 1/4 W
Q1 = 2SA1359Y
Q2 = 2N3904

Input = 18.0 V, Output taken at C3

Output (V)	Load (Ohm)	Ripple (mVpp)	Ripple (kHz)	Eff. (%)
12.54	1k	50	4	67
12.52	90.9	25	112	86.5
12.49	47.6	40	58	88.9
12.45	24.4	70	31	91.8

Input = 18.0 V, Output taken at C4

Output (V)	Load (Ohm)	Ripple (mVpp)	Ripple (kHz)	Eff. (%)
12.53	1k	58	.08	67
12.46	90.9	1.5	—	86.5
12.37	47.6	1.5	—	88.4
12.20	24.4	1.5	—	90.4

Output taken at C3, Load = 24.4 Ohms

Input (V)	Output (V)	Ripple (mVPP)	Ripple (kHz)	Eff. (%)
15.0	12.35	73	17.8	93.4
18.0	12.45	70	31.0	91.8
21.0	12.53	75	43.3	90.8

ELECTRONIC DESIGN

Fig. 78-3

This inexpensive and efficient discrete step-down regulator is based on a complementary transistor arrangement that uses both positive and negative feedback and is referenced to a Zener diode. Inductor L1 is selected to maintain the switching frequency above the audible range for the intended operating load. The output filter L2 and C4 reduces ripple to less than 10 mV p-p over a large range of loads, with only a slight decrease in efficiency.

DUAL-OUTPUT REGULATOR

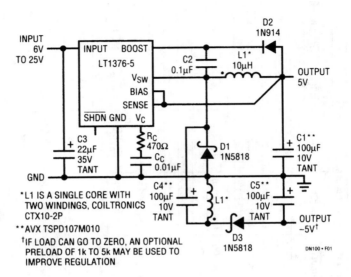

LINEAR TECHNOLOGY

Fig. 78-4

If load can go to zero, an optional preload of 1 to 5 kΩ can be used to improve regulation. Many modern circuit designs still need a dual polarity supply. Communication and data acquisition are typical areas where both 5 V and –5 V are needed for some of the IC chips.

The current mode architecture and saturating switch design allow the LT1376 to deliver up to 1.5-A load current from the 8-pin SO package. L1 is a 10-µH surface-mount inductor from Coiltronics. The second winding is used to create a negative-output SEPIC (Single-Ended Primary Inductance Converter) topology using D3, C4, C5, and the second half of F1. This converter takes advantage of the fact that the switching signal driving L1 as a positive buck converter is already the correct amplitude for driving a –5-V SEPIC converter. During switch-off time, the voltage across L is equal to the 5-V output plus the forward voltage of D1. An identical voltage is generated in the second winding, which is connected to generate –5 V using D3 and C5. Without C4, this would be a simple flyback winding connection with modest regulation. The addition of C4 creates the SEPIC topology. Note that the voltages swing at both ends of C4 is theoretically identical—even without the capacitor. The undotted end of both windings goes to a zero ac voltage node, so the equal windings will have equal voltages at the opposing ends. Unfortunately, coupling between windings is never perfect, and load regulation at the negative output suffers as a result. The addition of C4 forces the winding potentials to be equal and gives much better regulation.

LOW-NOISE REGULATOR (5 TO 3.3 V)

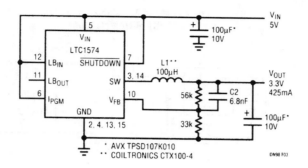

LINEAR TECHNOLOGY

Fig. 78-5

In some applications, it is important not to introduce any switching noise within the audio frequency range. To circumvent this problem, a feed-forward capacitor can be used to shift the noise spectrum up and out of the audio band with C2 being the feed-forward capacitor. The peak-to-peak output ripple is reduced to 30 mV over the entire load range. A toroidal surface mount inductor L1 is chosen for its excellent self-shielding properties.

REDUCING RIPPLE IN A SWITCHING VOLTAGE REGULATOR

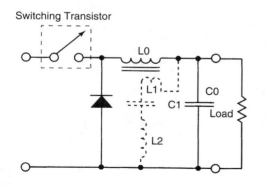

NASA TECH BRIEFS

Fig. 78-6

Simple additional circuitry that consists of relatively small components can reduce the output ripple by a factor of about 10. The additional components are indicated by the dashed lines.

A current opposing the ripple is injected into the filter capacitor. The essence of the present technique is to inject, into this capacitor, a current opposite to that which already flows into this capacitor. A small additional winding, L1, in inductor L0 provides transformer coupling to generate the current that opposes the original ripple current. The circuit from L1 through C0 is completed by a small additional external inductor L2 and coupling capacitor C1.

LOW-DROPOUT THREE-TERMINAL REGULATORS
FOR NEW MICROPROCESSOR APPLICATIONS

Recommended LT1584 Adjustable Circuit for
the Intel P54CT Microprocessor

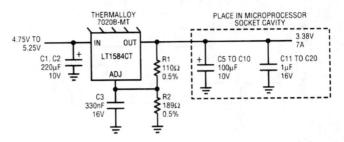

Fig. 78-7

The LT1584/LT1585/LTL1587 are high-performance, low-dropout regulators designed to meet the demands of the newest high speed, low voltage microprocessors. These devices are designed to regulate from 5-V supplies to output voltages between 1.25 V and 3.6 V. The LT1584 can provide up to 7 A of current, making it ideal for powerful Pentium processor or similar applications. The LT1585 can supply up to 4 A, while the LT1587 supplies up to 3 A. The excellent transient response capability allows them to maintain good regulation even with significant load steps. Fixed 3.3 V, 3.45 V, 3.6 V and adjustable output voltages are available.

LOW-DROPOUT REGULATOR

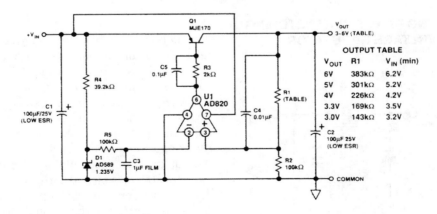

OUTPUT TABLE

V_{OUT}	R1	V_{IN} (min)
6V	383kΩ	6.2V
5V	301kΩ	5.2V
4V	226kΩ	4.2V
3.3V	169kΩ	3.5V
3.0V	143kΩ	3.2V

Fig. 78-8

This low-dropout reference produces a 4.5-V output from a supply just a few hundred mV greater. With 1-mA dc loading, it maintains a stable 4.5-V output for inputs down to 4.7 V.

POSITIVE REGULATOR SINKS CURRENT

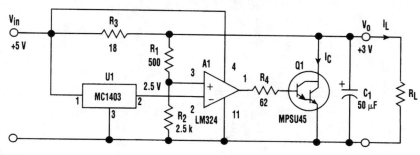

ELECTRONIC DESIGN

Fig. 78-9

Generally speaking, conventional positive voltage regulators can only source current; they can't sink it. However, the positive regulator shown breaks that rule because it can perform both functions. The idea is to have the control transistor Q1 in shunt so that the regulator can either source or sink current.

The circuit provides +3-V output from a +5-V supply. U1 is a bandgap reference that supplies a stable +2.5-V reference to the error amplifier (A1). The output voltage (V_O) is sampled by the resistor network (R1 and R2). If V_O were to increase, A1 will drive the base of Q1 harder, increasing the collector current (I_c).

This increases the drop across R3 and V_O decreases, thus regulating the output voltage. The output voltage is given by $V_O = 2.5(1 + R_1/R_2)$.

Under no load conditions, Q1 draws 110 mA [$(V_{in} - V_O)/R_3$]. With a load connected, and as the regulator begins to source load current (I_L), I_c decreases to keep the drop across R3 constant.

At $I_L = 100$ mA, Q1 carries 10 mA. If R_L is connected to the positive supply higher than V_O, then the regulator must sink current, and I_L becomes negative. At $I_L = -100$ mA, Q1 carries 210 mA while maintaining the output voltage at +3 V. The output voltage will remain constant at +3 V—even if the load current changes sign.

With the proper heatsink on Q1, the regulator can sink more than 300 mA. If a "sink only" option is desired, the dissipation in Q1 can be reduced by using a 180-Ω resistor for R3. R4 limits the base current drive for Q1 and prevents the output of A1 from being clamped at $2 V_{BE}$.

5- TO 3.3-V SURFACE-MOUNT SWITCHING REGULATOR

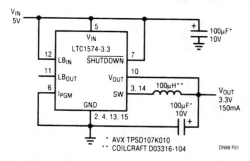

This converter provides 3.3 V at 150 mA from an input voltage of 5 V. Peak inductor current is limited to 340 mA by connecting pin 6 (I_{PGM}) to ground. For applications requiring higher output current, connect pin 6 to V_{in}. Under this condition, the maximum load current is increased to 425 mA.

LINEAR TECHNOLOGY

Fig. 78-10

79

Relay Circuits

The sources of the following circuits are contained in the Sources section, which begins on page 707. The figure number in the box of each circuit correlates to the entry in the Sources section.

LATCHING RELAY ALARM CIRCUIT

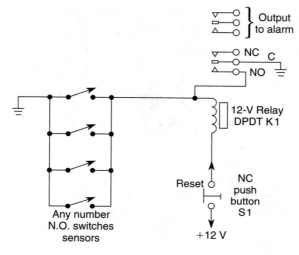

WILLIAM SHEETS

Fig. 79-1

Momentarily closing any sensors will cause K1 to latch. S1 must be depressed to reset circuit. If any sensor is still closed circuit will not reset.

MOMENTARY RELAY CIRCUIT

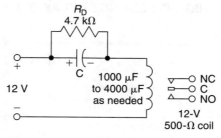

WILLIAM SHEETS

Fig. 79-2

The charging current of a capacitor can be used if a momentary relay-on circuit is needed. Depending on the relay characteristics, C will vary from 1000 to 4000 µF or so for a 1-s hold time if a 500-Ω relay is used. R_D discharges capacitor C to ready the circuit for the next operation. The value should be high enough so as not to maintain the relay closure at highest expected supply voltage.

LATCHING RELAY DRIVER FOR +12-V LOADS

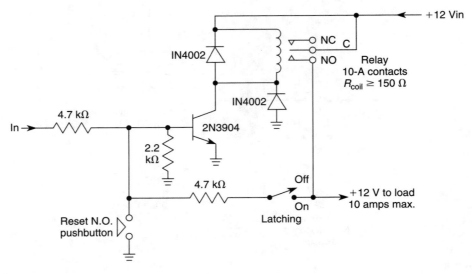

WILLIAM SHEETS

Fig. 79-3

A 4-V signal will cause the relay to pull in when Q1 turns on. Latching is obtained by feedback through a 4.7-kΩ resistor. A switch is used to select latching or nonlatching operation. A NO pushbutton releases the circuit.

HIGH-IMPEDANCE RELAY DRIVER

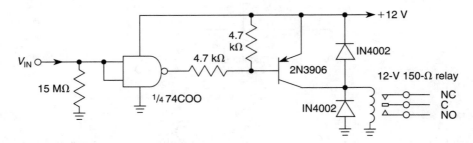

WILLIAM SHEETS

Fig. 79-4

A CMOS gate is used to drive a switching transistor and relay.

LATCHING RELAY DRIVER

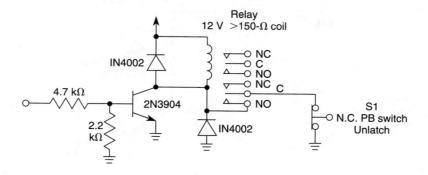

WILLIAM SHEETS

Fig. 79-5

An input of 4 V or greater will drive this circuit. When the relay pulls in, one pair of contacts is used to latch the relay closed. It will remain closed until S1 is pressed.

TRANSISTOR RELAY DRIVER

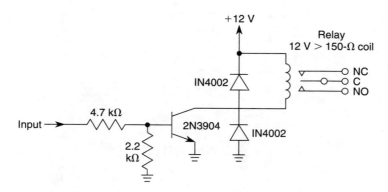

WILLIAM SHEETS

Fig. 79-6

An input of 4 V or greater will drive this relay circuit.

FAST TURN-ON/DELAYED-OFF RELAY CIRCUIT

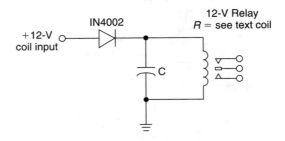

WILLIAM SHEETS

Fig. 79-7

C is a large capacitor that has a charge time of $R_{supply} C$, assuming $R_{supply} < R_{coil}$. The discharge time will be $R_{coil} C$ neglecting relay coil inductance. With $C = 10,000$ μF and $R_{coil} = 500$ Ω, a release time constant of 5 seconds might be obtained. Many relays will hold in until the coil current decays to 25% of the pull-in current so that the actual time constant depends on the relay holding current.

LOW-FREQUENCY RELAY OSCILLATOR

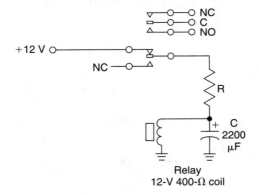

WILLIAM SHEETS

Fig. 79-8

Depending on the value of C and the resistance of the relay coil, and the difference in pull-in and drop-out voltage, this circuit will oscillate at a low frequency. R limits in rush current to capacitor C to a level that the relay contacts can handle. Typically, for a 400-Ω relay, R can be 20 to 440 ohms. Flash rate is approximately 1 cycle/second, depending on the relay.

80

Sample-and-Hold Circuits

The sources of the following circuits are contained in the Sources section, which begins on page 707. The figure number in the box of each circuit correlates to the entry in the Sources section.

Micropower 4-Channel Sample-and-Hold Circuit
Low-Drift Sample and Hold

MICROPOWER 4-CHANNEL SAMPLE-AND-HOLD CIRCUIT

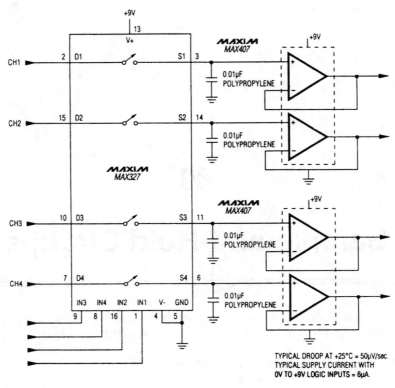

MAXIM

Fig. 80-1

Three Maxim ICs make up this sample-and-hold circuit. The supply current is only 6 μA.

LOW-DRIFT SAMPLE AND HOLD

Low Drift Sample and Hold*

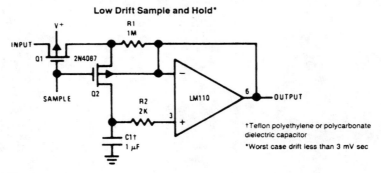

†Teflon polyethylene or polycarbonate dielectric capacitor

*Worst case drift less than 3 mV sec

NATIONAL SEMICONDUCTOR

Fig. 80-2

81

Sawtooth Generator Circuits

The sources of the following circuits are contained in the Sources section, which begins on page 707. The figure number in the box of each circuit correlates to the entry in the Sources section.

Op-Amp Linear Sawtooth Generator
Sawtooth Generator

OP-AMP LINEAR SAWTOOTH GENERATOR

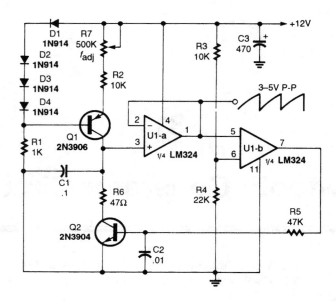

POPULAR ELECTRONICS

Fig. 81-1

Q1 is connected in a simple constant-current generator circuit. The value of Q1's emitter resistor sets the constant-current level flowing from the transistor's collector to the charging capacitor, C1.

One op amp of an LM324 quad op-amp IC, U1-a, is connected in a voltage-follower circuit. The input impedance on the voltage follower is very high and offers little or no load on the charging circuit. The follower's output is connected to the input of U1-b, which is configured as a voltage comparator. The comparator's other input is tied to a voltage-divider setting the input level to about 8 V.

The output of U1-b at pin 7 switches high when the voltage at its positive input, pin 5, goes above 8 V. That turns on Q2, discharging C1. The sawtooth cycle is repeated over and over as long as power is applied to the circuit.

The sawtooth's frequency is determined by the value of C_1 and the charging current supplied to that capacitor. As the charging current increases, the frequency also increases, and vice versa. To increase the generator's frequency range, decrease the value of C_1, and to lower the frequency, increase the value of C_1. The output is about 3 to 5 V.

SAWTOOTH GENERATOR

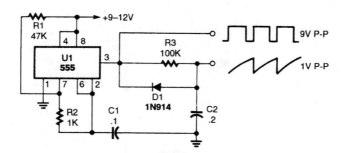

POPULAR ELECTRONICS

Fig. 81-2

A sawtooth waveform generator circuit using a 555 IC is shown. The IC is connected in an astable oscillator circuit with the majority of the output contained in the positive portion of the cycle. The negative output is a very brief pulse.

Capacitor C2 charges through R3 in a positive direction during the time that the IC's output (at pin 3) is high. When the output goes negative, C2 is rapidly discharged through D1 and the IC's output.

Peak-to-peak sawtooth output is about 1 V. The linearity of this circuit is best when R3 is as large as possible. The oscillator's frequency is about 200 Hz and can be increased by lowering either the value of R_1 or C_1; to decrease the frequency, increase the values of those components.

82

Scanner Circuits

The sources of the following circuits are contained in the Sources section, which begins on page 707. The figure number in the box of each circuit correlates to the entry in the Sources section.

FM Scanner Noise Squelch
Scanner Silencer
Shortwave Converters for Scanners

FM SCANNER NOISE SQUELCH

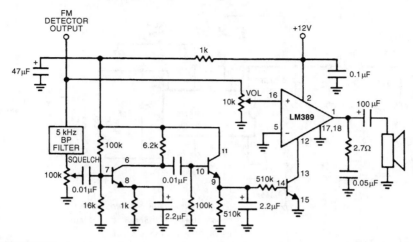

NATIONAL SEMICONDUCTOR

Fig. 82-1

The LM389 is operated in the cut-off mode with pin 12 grounded via one of the internal transistors. A sample of detected noise is taken through a 5-kHz filter. Upon reception of signal, the detector output quiets, and noise level drops. This increases impedance at pin 12 of the LM389, causing audio to be passed. The three transistors are part of the LM389.

SCANNER SILENCER

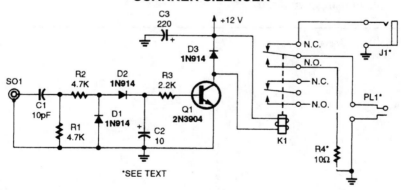

POPULAR ELECTRONICS

Fig. 82-2

When a scanner is used at amateur radio or CB stations, the scanner sometimes picks up transmitted signal and howls or squeals. When RF is detected, Q1 turns on, energizing K1 and disconnecting the scanner speakers. SO1 is connected to the transmit antenna lead via a tee fitting. C1 is optimum for 5- to 10-W 30-MHz use. For higher power or higher frequencies, reduce C1 to as low as needed. If C1 is so small as to be impractical, R1 can be shunted with a 10- or 22-pF capacitor, as needed.

SHORTWAVE CONVERTERS FOR SCANNERS

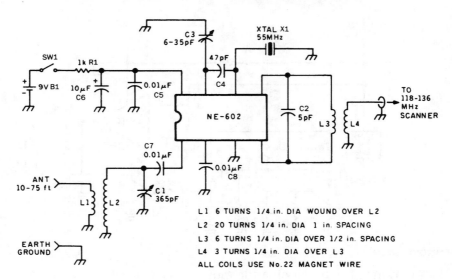

L1 6 TURNS 1/4 in. DIA WOUND OVER L2
L2 20 TURNS 1/4 in. DIA 1 in. SPACING
L3 6 TURNS 1/4 in. DIA OVER 1/2 in. SPACING
L4 3 TURNS 1/4 in. DIA OVER L3
ALL COILS USE No. 22 MAGNET WIRE

73 AMATEUR RADIO TODAY

Fig. 82-3

The AM aircraft band at 118 to 136 MHz is used in this converter design as an IF output. The second harmonic of the 55-MHz crystal (110 MHz) mixes with the shortwave input of 8 to 36 MHz. An NE602 IC is used for the mixer. Sensitivity is about 3 µV. If desired, a crystal tuning circuit for fine tuning can be obtained using a varactor.

83

Siren, Warbler, and Wailer Circuits

The sources of the following circuits are contained in the Sources section, which begins on page 707. The figure number in the box of each circuit correlates to the entry in the Sources section.

Fire Siren
Warble Oscillator
Electronic Siren
Wailing Sound Generator
Two-Tone Siren

FIRE SIREN

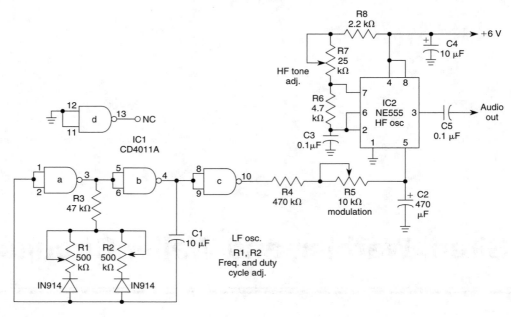

Fig. 83-1

IC1 is an LF oscillator that is variable in attack and decay time with R1 and R2. The LF output modulates HF oscillator IC2. R5 varies the modulation depth. By proper control adjustment, sirens of various types can be simulated.

WARBLE OSCILLATOR

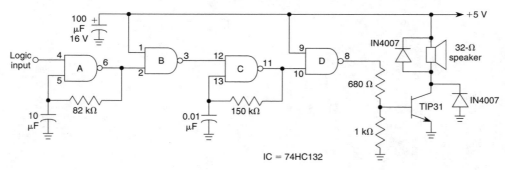

Fig. 83-2

Sections A & B form an oscillator running at 2 Hz, which gates sections C and D, a 1-kHz oscillator. This drives the TIP31 speaker driver.

ELECTRONIC SIREN

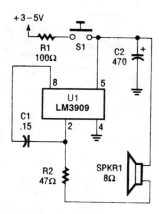

POPULAR ELECTRONICS

In this circuit, the LM3909 is used in a simple electronic siren. When S1 is closed, C2 begins to charge rapidly through R1. When the charge on C2 reaches about 1 V, the oscillator starts. As the voltage across C1 increases toward $+V$, the oscillator's output frequency also increases. Releasing (opening) S1 removes power from the circuit. The oscillator continues to operate, with a decline in output volume and frequency until C1 discharges to about the 1-V level.

Experiment with the siren circuit by selecting different R_1/C_2 combinations to obtain a desired rise and fall output. Change the value of C_1 to vary the oscillator's frequency. Keep the value of R_2 at or above 47 Ω to protect the IC from drawing too much current.

Fig. 83-3

WAILING SOUND GENERATOR

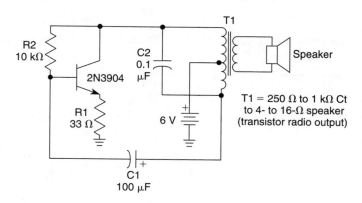

T1 = 250 Ω to 1 kΩ Ct
to 4- to 16-Ω speaker
(transistor radio output)

WILLIAM SHEETS

Fig. 83-4

In this circuit, C2 and T1 determine the tone generated and C_1/R_2 control the blocking rate. The signal produced is an interrupted tone, like a police whistle or toy ray gun, depending on C1 and C2.

TWO-TONE SIREN

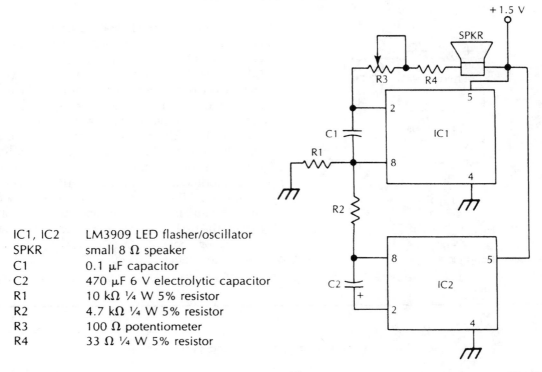

IC1, IC2	LM3909 LED flasher/oscillator
SPKR	small 8 Ω speaker
C1	0.1 μF capacitor
C2	470 μF 6 V electrolytic capacitor
R1	10 kΩ ¼ W 5% resistor
R2	4.7 kΩ ¼ W 5% resistor
R3	100 Ω potentiometer
R4	33 Ω ¼ W 5% resistor

McGRAW-HILL   *Fig. 83-5*

IC1 generates the main siren tone while IC2 generates a low-frequency square wave, switching IC1 between two different tones.

84

Sound-Effects Circuits

The sources of the following circuits are contained in the Sources section, which begins on page 707. The figure number in the box of each circuit correlates to the entry in the Sources section.

Complex Sound-Effect Generator
Dual-Tone Generator
Surf Man Sound Generator
Electronic Whistle
Bird-Chirp Sound-Effect Generator
Robotic Chatter Sound Generator
Electronic Wind Chime
Gunshot Sound-Effects Generator

COMPLEX SOUND-EFFECT GENERATOR

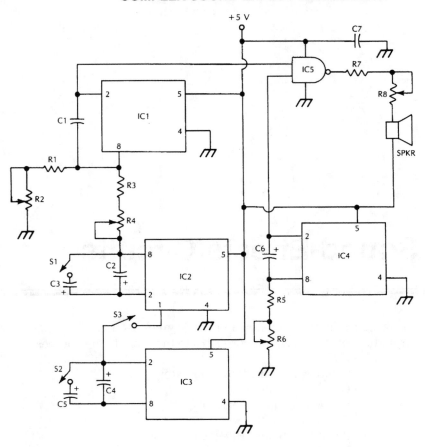

IC1–IC4	LM3909 LED flasher/oscillator	C4	47 μF 6 V electrolytic capacitor
IC5	7400 quad NAND gate	C5	33 μF 6 V electrolytic capacitor
SPKR	small loudspeaker	C6	4.7 μF 6 V electrolytic capacitor
S1, S2, S3	SPST switch	R1, R3, R5	2.2 kΩ ¼ W 5% resistor
C1	0.1 μF capacitor	R2, R4, R6	10 kΩ potentiometer
C2	22 μF 6 V electrolytic capacitor	R7	33 Ω ¼ W 5% resistor
C3	10 μF 6 V electrolytic capacitor	R8	100 Ω potentiometer

McGRAW-HILL

Fig. 84-1

This system uses four free running oscillators to produce a wide variety of complex sounds. LF oscillator IC3 modulates IC2, which modulates IC1. The audio from IC1 is combined with a variable frequency from IC4. Switches at various points allow oscillators IC3 to be switched in or out, IC1 and IC2 to be varied in frequency, and IC4 also can be varied in frequency. The circuit is not critical and different arrangements can be tried to produce various sound effects.

DUAL-TONE GENERATOR

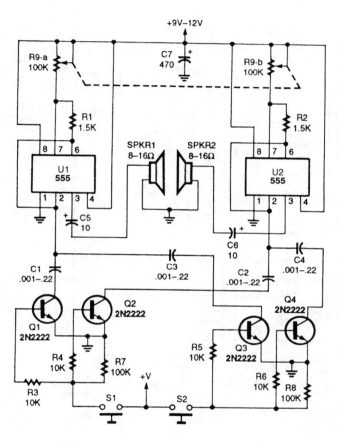

Fig. 84-2

Two 555 oscillator/timers are configured similarly as audio oscillators, with each oscillator feeding a separate speaker.

A dual 100-kΩ potentiometer is use to tune the two oscillators simultaneously. The oscillators' frequency range is controlled by a dual-transistor switch, which selects the timing capacitor for both oscillators. Although the circuit only shows two range-switching circuits, any number can be added by simply duplicating the two-transistor switching circuit.

SURF MAN SOUND GENERATOR

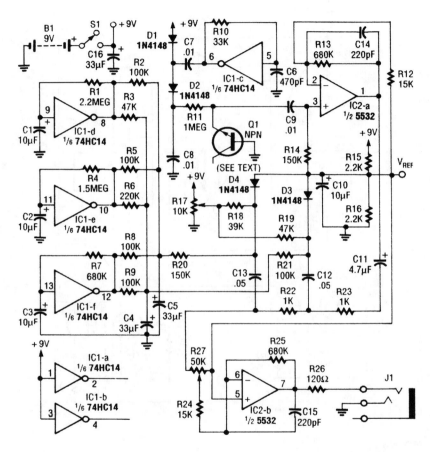

ELECTRONICS NOW

Fig. 84-3

Three low-frequency oscillators (IC1d, e, f) are used to simulate "wave action" of the surf. Q1 is an emitter-base junction used as a diode noise generator, biased by dc derived from oscillator IC1-c. The noise is fed into two voltage-controlled filters R22, R23, C12, C13, with D3 and D4 as "tuning" elements. The low-frequency oscillator signals randomly vary the filters, therefore, the spectrum of the noise signal fed through them. This simulates the sound of a surf.

ELECTRONIC WHISTLE

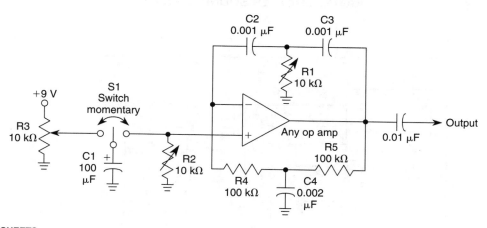

WILLIAM SHEETS

Fig. 84-4

The circuit shown is a twin-tee oscillator. R1 varies the pitch, R2 the duration, and R3 the format (bell, rise & fall time, etc.). Vary R4, R5, C4 and C2, C3 for large shifts in frequency.

BIRD-CHIRP SOUND-EFFECT GENERATOR

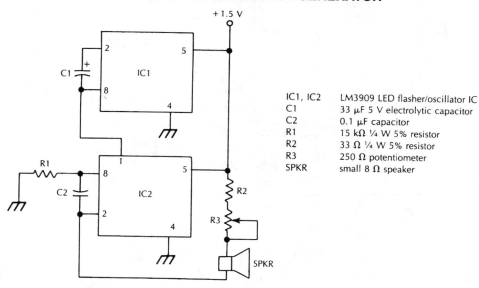

IC1, IC2	LM3909 LED flasher/oscillator IC
C1	33 µF 5 V electrolytic capacitor
C2	0.1 µF capacitor
R1	15 kΩ ¼ W 5% resistor
R2	33 Ω ¼ W 5% resistor
R3	250 Ω potentiometer
SPKR	small 8 Ω speaker

McGRAW-HILL

Fig. 84-5

A low-frequency oscillator modulates a higher frequency oscillator, which drives the speaker.

ROBOTIC CHATTER SOUND GENERATOR

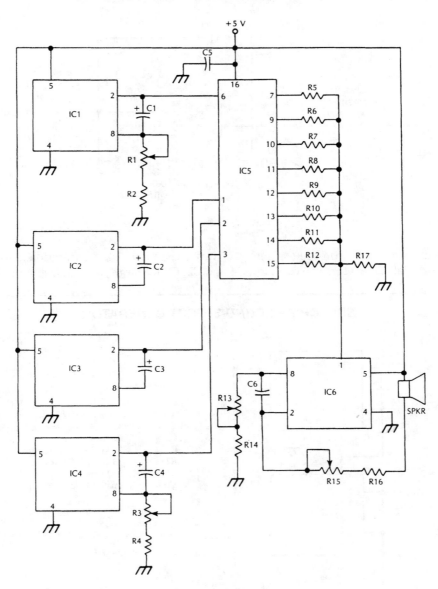

McGRAW-HILL

Fig. 84-6

This circuit simulates sound effects of a robot, for toy or novelty applications.

ROBOTIC CHATTER SOUND GENERATOR (*Cont.*)

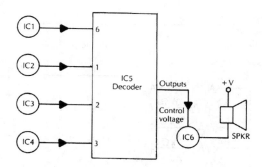

IC1–IC4, IC6	LM3909 LED flasher/oscillator	R5	6.8 kΩ ¼ W 5% resistor
IC5	74LS138 three-line to eight-line decoder	R6	10 kΩ potentiometer
SPKR	small loudspeaker	R7	2.2 kΩ ¼ W 5% resistor
C1	10 µF 10 V electrolytic capacitor	R8	33 kΩ ¼ W 5% resistor
C2	22 µF 10 V electrolytic capacitor	R9	3.9 kΩ ¼ W 5% resistor
C3	33 µF 10 V electrolytic capacitor	R10	4.7 kΩ ¼ W 5% resistor
C4	100 µF 10 V electrolytic capacitor	R11	100 kΩ ¼ W 5% resistor
C5	0.01 µF capacitor	R12	470 kΩ ¼ W 5% resistor
C6	0.1 µF capacitor	R15	100 Ω potentiometer
R1, R3, R14	10 kΩ potentiometer	R16	33 Ω ¼ W 5% resistor
R2, R4, R13	3.3 kΩ ¼ W 5% resistor	R17	1 MΩ ¼ W fixed resistor

ELECTRONIC WIND CHIME

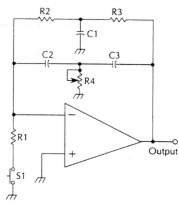

The value of R_4 controls the damping or decay time of the feedback circuit (a twin Tee oscillator). When S1 is closed, the circuit breaks into oscillation. When S1 is opened, the circuit stops oscillating generating a decaying tone like a bell. The frequency is approximately $\frac{1}{2}RC$. C1, C2, and C3 are typically in the 0.01-µF range.

McGRAW-HILL   *Fig. 84-7*

GUNSHOT SOUND-EFFECTS GENERATOR

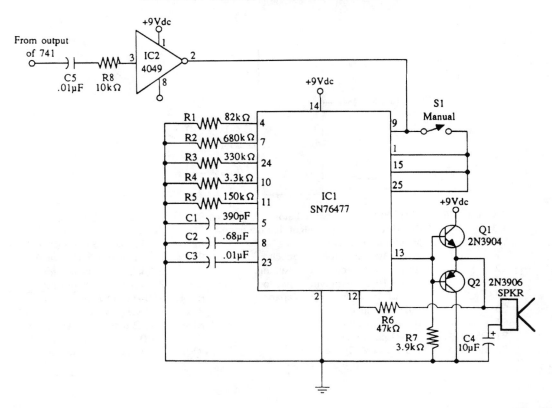

McGRAW-HILL

Fig. 84-8

Gunshot sound-effects generator built around a Texas Instruments SN76477 sound chip. An input pulse causes IC1 to generate a gunshot sound.

85

Square-Wave Generator Circuits

The sources of the following circuits are contained in the Sources section, which begins on page 707. The figure number in the box of each circuit correlates to the entry in the Sources section.

Square-Wave Generator
Sharp Square Waveforms from Multivibrator

SQUARE-WAVE GENERATOR

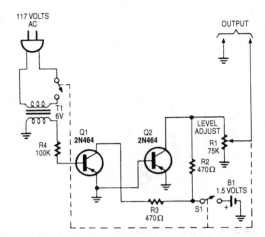

POPULAR ELECTRONICS

Fig. 85-1

A 60-Hz waveform from T1 drives an audio amplifier to clipping. Output is 60 Hz with about 0- to 1.4-V p-p amplitude.

SHARP SQUARE WAVEFORMS FROM MULTIVIBRATOR

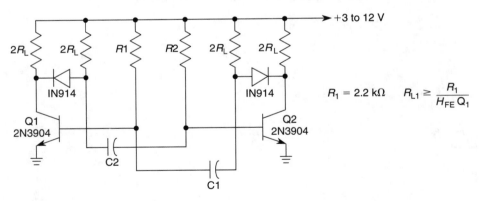

$$R_1 = 2.2 \text{ k}\Omega \qquad R_{L1} \geq \frac{R_1}{H_{FE}\,Q_1}$$

WILLIAM SHEETS

Fig. 85-2

By using diodes as shown, the loading effect on the collector of the transistors caused by the timing capacitors can be avoided. As the collector of the transistors rises toward V_{CC}, the diode disconnects the timing capacitors.

86

Staircase Generator Circuits

The sources of the following circuits are contained in the Sources section, which begins on page 707. The figure number in the box of each circuit correlates to the entry in the Sources section.

Stepped Triangle Waveform Generator
Video Staircase Generator
Free-Running Staircase Wave Generator
Up/Down Staircase Wave Generator

STEPPED TRIANGLE WAVEFORM GENERATOR

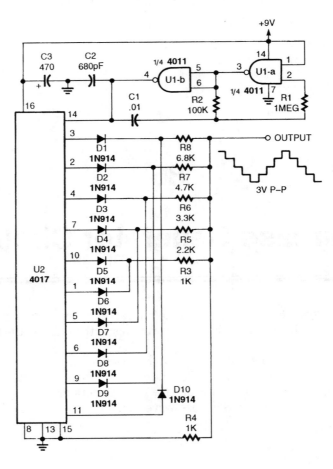

POPULAR ELECTRONICS

Fig. 86-1

Two gates of a 4011 quad two-input NAND gate (U1) are connected in a pulse generator circuit. The square output at pin 4 of U1-b, connects to the clock input, pin 14, of a 4017 decade counter IC (U2). For each input clock pulse, the 4017 takes a single step. Because the 4017 counter is set up to count ten and then repeat the count, the stepped output frequency will only be $\frac{1}{10}$ of the clock frequency. For a 100-Hz output, the clock generator must operate at 1 kHz.

The 4017's positive output pulses begin at pin 3 and progress to pin 11 in a serial manner. The first output pulse, at pin 3, passes through D1 and R8 and appears across R4 to produce the first step up the triangle. The second pulse is routed through D2 and R7 to produce the second step. The outputs at pins 10 and 1 form the top of the waveform and outputs at pins 5, 6, 9, and 11 produce the down steps.

VIDEO STAIRCASE GENERATOR

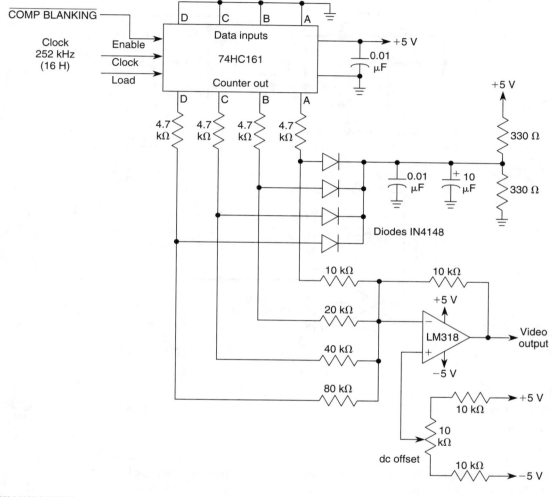

WILLIAM SHEETS

Fig. 86-2

Using a 74HC161 counter and a simple D-A converter using an op amp and resistor network, a very simple video staircase generator for gray-scale generation (12 bars) can be obtained. The output is clean and mostly free of "glitches."

FREE-RUNNING STAIRCASE WAVE GENERATOR

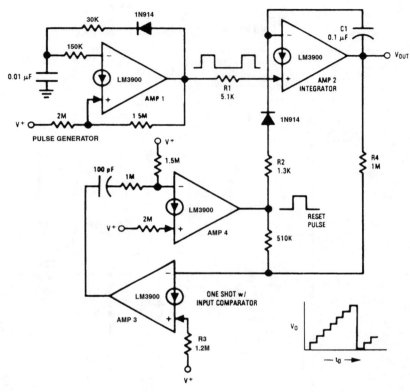

NATIONAL SEMICONDUCTOR

Fig. 86-3

This free-running staircase generator uses all four of the amplifiers, which are available in one LM3900 package. Amp 1 provides the input pulses that "pump up" the staircase via resistor R1. Amp 2 does the integrate and hold function and also supplies the output staircase waveform. Amps 3 and 4 provide both a compare and a one-shot multivibrator function. Resistor R4 is used to sample the staircase output voltage and to compare it with the power supply voltage (V+) via R3. When the output exceeds approximately 80% of V+ the connection of Amps 3 and 4 causes a 100-μs reset pulse to be generated. This is coupled to the integrator (AMP2) via R2 and causes the staircase output voltage to fall to approximately 0 V. The next pulse out of Amp 1 then starts a new stepping cycle.

UP/DOWN STAIRCASE WAVE GENERATOR

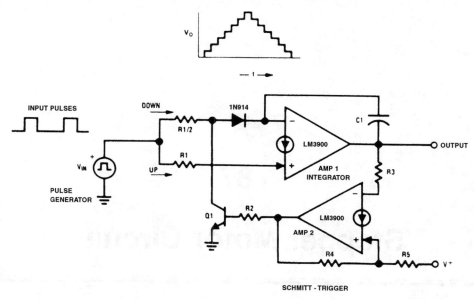

NATIONAL SEMICONDUCTOR

Fig. 86-4

This staircase waveform first steps up and then steps down by the circuit shown. An input pulse generator provides the pulses that cause the output to step up or down, depending on the conduction of the clamp transistor, Q1. When this is ON, the "down" current pulse is diverted to ground and the staircase then steps "up." When the upper voltage trip point of Amp 2 is reached, Q1 goes OFF and as a result of the smaller "down" input resistor (one-half the value of the "up" resistor, R1), the staircase steps "down" to the low-voltage trip point of Amp 2. The output voltage, therefore, steps up and down between the trip voltages of the Schmitt Trigger.

87

Stepper Motor Circuits

The sources of the following circuits are contained in the Sources section, which begins on page 707. The figure number in the box of each circuit correlates to the entry in the Sources section.

Stepper Motor Pulse Generator
Stepper Motor as Shaft Encoder
Stepper Motor Encoder Circuit

STEPPER MOTOR PULSE GENERATOR

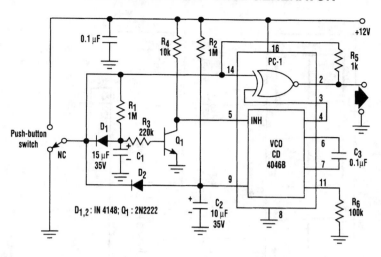

ELECTRONIC DESIGN

Fig. 87-1

When the switch is in its normally closed (NC) position, capacitors C1 and C2 are held discharged by diodes D1 and D2. Switching off transistor Q1 inhibits the voltage-controlled oscillator of the PLL. The two inputs and, hence, the output of the EX-OR gate (phase comparator 1) of the PLL remain at the logic 0 level.

When the pushbutton is pressed, C1 and C2 are allowed to charge via resistors R1 and R2. The VCO is enabled only after a time delay (≈ 0.5 second) set by R1, R3, and C1. During this delay period, the EX-OR gate output follows the logic level at the switch output. As a result, one-shot pulses can be generated by pressing the pushbutton, then releasing it within 0.5 second. R5 provides the switch-debouncing function.

If the pushbutton is pressed for more than 0.5 second, the VCO is enabled. The rising voltage at the control input (pin 9) causes a linear increase in VCO frequency and thus accelerates the stepper motor. Releasing the pushbutton discharges C1 and C2 and inhibits the VCO.

STEPPER MOTOR AS SHAFT ENCODER

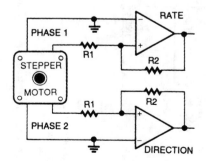

To use a stepper as a shaft encoder, the output signals must be converted to square waves with a pair of voltage comparators.

ELECTRONICS NOW

Fig. 87-2

STEPPER MOTOR ENCODER CIRCUIT

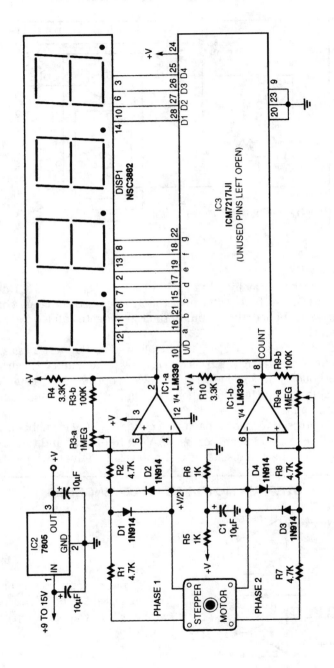

ELECTRONICS NOW

Fig. 87-3

This circuit translates shaft rotation and direction to a readout on an LED display. A stepper motor is used as an encoder.

88

Switching Circuits

The sources of the following circuits are contained in the Sources section, which begins on page 707. The figure number in the box of each circuit correlates to the entry in the Sources section.

ISOLATED SWITCH

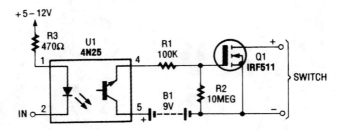

POPULAR ELECTRONICS

Fig. 88-1

This dc-controlled switch uses an optoisolator/coupler, U1, to electrically isolate the input signal from the output control device.

ANALOG SWITCHED INVERTER

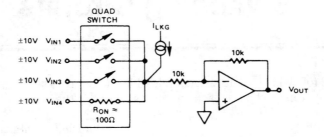

ANALOG DEVICES

Fig. 88-2

ANALOG SWITCH CIRCUIT

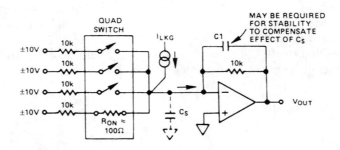

ANALOG DEVICES

Fig. 88-3

LOW OUTPUT IMPEDANCE MULTIPLEXER

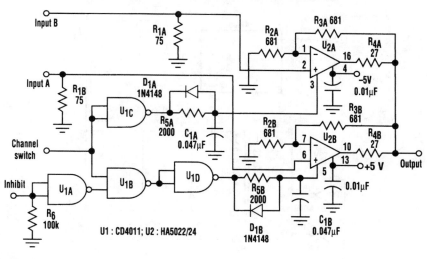

ELECTRONIC DESIGN

Fig. 88-4

Both inputs are terminated in their characteristic impedance; 75 Ω is typical for video applications. Because the output cables usually are terminated in their characteristic impedance, the gain is 0.5. Consequently, amplifiers U2A and U2B are configured in a gain of +2 to set the circuit gain at 1. R_2 and R_3 determine the amplifier gain; if a different gain is desired, R_2 should be changed according to the equation $G = (1 + R_3/R_2)$. R5, LCL1, and D1 make up an asymmetrical charge/discharge time circuit that configures U1 as a break-before-make switch to prevent both amplifiers from being active simultaneously. The multiplexer transition time is approximately 15 μs with the component values shown.

OP-AMP AND ANALOG SWITCH RON COMPENSATOR

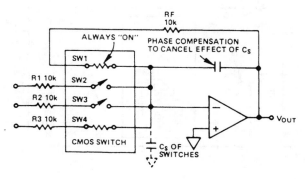

ANALOG DEVICES

Fig. 88-5

This switch is in series with feedback resistor to compensate gain.

OSCILLATOR TRIGGERED SWITCH

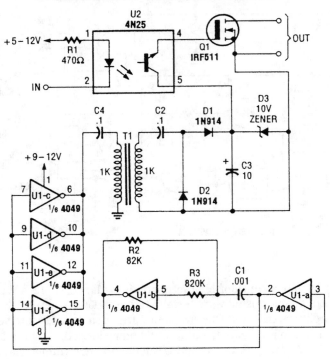

POPULAR ELECTRONICS

Fig. 88-6

In this circuit (the oscillator-triggered switch), the HEXFET's base bias is provided by a signal generated by an astable oscillator.

BASIC ZERO-CROSSING SWITCH CIRCUIT

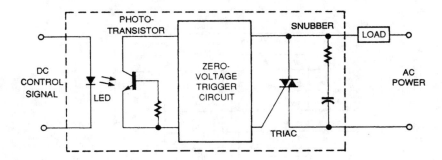

ELECTRONICS NOW

Fig. 88-7

Here is the schematic of a solid-state ac relay with zero-crossing. The triac permits the relay to switch to ac directly.

ANALOG SWITCH

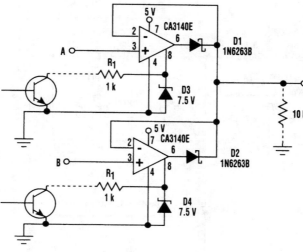

ELECTRONIC DESIGN

Fig. 88-8

This design takes advantage of the strobed output stage of a CA3140 amplifier. With the strobing capability, the circuit's output voltage can be set to either of the input voltages by grounding one of the control inputs, either A or B.

When the strobe input at pin 8 is taken below 1 V, that amplifier is disabled. The remaining amplifier then acts as a unity-gain high-impedance buffer.

The 10-kΩ output resistor enables the output voltage to swing down to 20 mV from ground. The Zener-diode clamps and associated resistors connected at the strobe inputs can be omitted for the lowest-cost applications. However, experience has shown that they allow the amplifiers to shrug off the effect of high transient voltages.

The circuit is particularly suited to 8-bit microcontroller applications, where the strobe inputs can be driven directly from two open-collector output ports under software control.

The use of Schottky diodes for D1 and D2 makes possible an output swing of 2.5 V when the circuit is powered from a 5-V supply.

SHUNT PIN-DIODE SWITCH

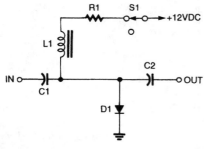

This PIN-diode switching circuit directs signals to ground when D1 is forward-biased. R1 is typically 470 Ω to 2.2 kΩ. $C_1 = C_2 = 0.1 \ \mu F$.

POPULAR ELECTRONICS

Fig. 88-9

RECEIVER BANDSWITCHING

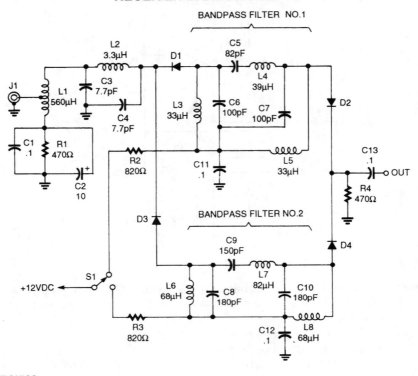

POPULAR ELECTRONICS

Fig. 88-10

Eight-band receiver front-end selection can be accomplished by using PIN diode switches.

RESISTOR PIN-DIODE SWITCH

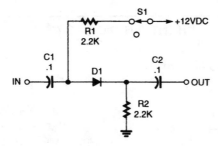

POPULAR ELECTRONICS

Fig. 88-11

This circuit uses resistors instead of RF chokes to keep costs low. The values of R_1 and R_2 should be no lower than about 1 kΩ to minimize loss.

DIGITALLY CONTROLLED ONE-OF-FOUR ANALOG SWITCH

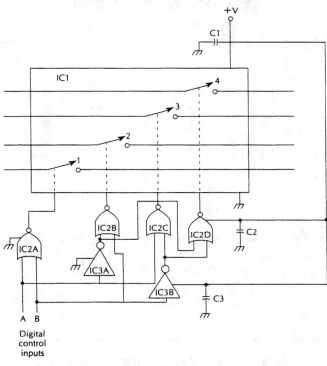

McGRAW-HILL

Fig. 88-12

IC1 = CD4066
IC2 = CD4001
IC3 = CD4049
All capacitors = 0.1 μF

PIN DIODE SWITCH

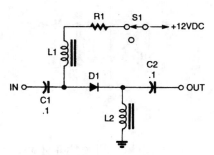

POPULAR ELECTRONICS

Fig. 88-13

This PIN diode switch uses RF chokes and a single diode. R1 is typically 470 Ω to 2.2 kΩ.

TRANSCEIVER T/R SWITCH

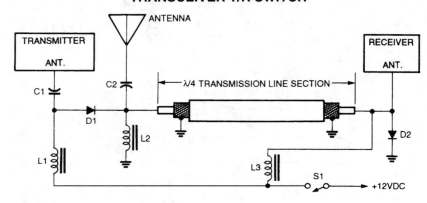

POPULAR ELECTRONICS

Fig. 88-14

This transceiver's transmit/receive switch uses PIN-diode instead of a relay. On receive, D1 is cut off, D2 is cut off and the antenna connects to the receiver. During transmit, D1 is forward-biased, as is D2. This connects the receiver input. This causes the input impedance of the transmission line to be high, so little transmitter power reaches the receiver. Although not shown in the schematic, the 12-V supply should have a series resistor of 100 Ω to 2.2 kΩ, depending on diode current, to limit diode current to a safe value.

SERIES/SHUNT PIN-DIODE RF SWITCH

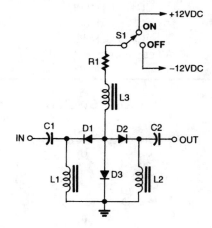

POPULAR ELECTRONICS

Fig. 88-15

A combination of series and shunt switching, like that shown here, results in superior isolation between the input and output when in the off condition.

AUTO-OFF POWER SWITCH

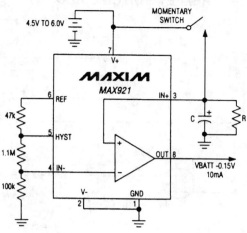

MAXIM

Fig. 88-16

This timed switch circuit can be used where a timed power source is needed. The on-time is approximately $4.6\,RC$.

SWITCH-ON DELAY CIRCUIT

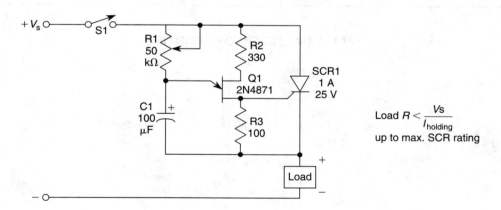

WILLIAM SHEETS

Fig. 88-17

When S1 is turned on, a very small current flows through the load. Almost the entire supply voltage appears across the SCR. When C1 changes up to the firing voltage of Q1 (approximately the standoff ratio of Q1 times V_s usually 0.4 to 0.6 V_s) through R1, Q1 fires, turning on SCR1. This delivers full voltage to load, minus SCR drop (about 1.2 V). Notice that load current must exceed SCR holding current.

HEXFET SWITCH CIRCUITS

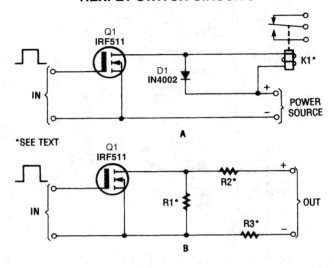

POPULAR ELECTRONICS

Fig. 88-18

The HEXFET can switch dc power to relays (as shown in A), motors, lamps, and numerous other devices. That arrangement can even be used to switch resistors in and out of a circuit, as shown in Fig. 88-18B. R1, R2, and R3 are possible load resistors and represent load configurations that can be used.

ALTERNATING ON/OFF CONTROL

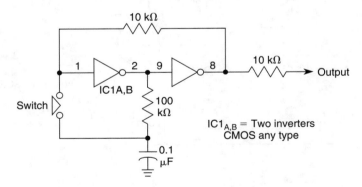

WILLIAM SHEETS

Fig. 88-19

When the switch is closed, it causes a change in the state of pins 1 through 8. This will provide a toggle flip-flop action.

AUDIO-CONTROLLED SWITCH

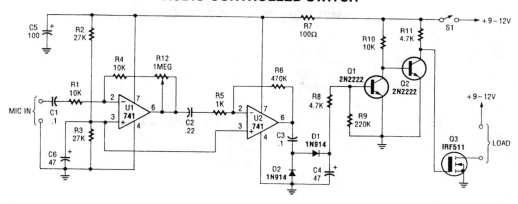

POPULAR ELECTRONICS

Fig. 88-20

The audio-controlled switch combines a pair of 741 op amps, two 2N222 general-purpose transistors, a HEXFET, and a few support components to produce a circuit that can be used to turn on a tape recorder, a transmitter, or just about anything using sound.

SWITCH DEBOUNCERS

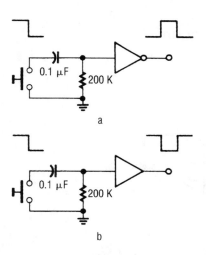

These circuits will cure problems caused by switch-contact bounce. The one shown in Fig. 88-21A provides you a positive output pulse, and the one shown in Fig. 88-21B provides you a negative output pulse.

ELECTRONICS NOW!

Fig. 88-21

SIMPLE SWITCH DEBOUNCER

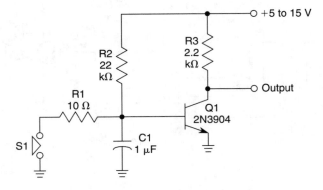

Pressing S1 discharges C1 through R1, causing Q1 to cut off, forcing the output high. Once C1 is discharged below the V_{BE} (ON) of Q1, switch bounce will have no effect on the output.

WILLIAM SHEETS

Fig. 88-22

ANALOG SWITCH CIRCUIT

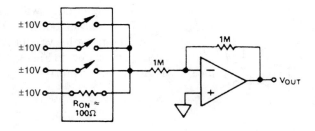

ANALOG DEVICES

Fig. 88-23

89

Sync Circuits

The sources of the following circuits are contained in the Sources section, which begins on page 707. The figure number in the box of each circuit correlates to the entry in the Sources section.

PLD Synchronizes Asynchronous Inputs
Sync Tip dc Restorer
Sync Stretcher Circuit
Synchronizer Circuit

PLD SYNCHRONIZES ASYNCHRONOUS INPUTS

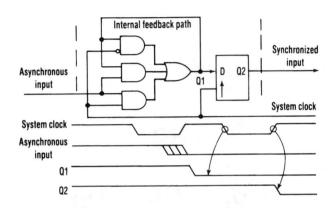

ELECTRONIC DESIGN

Fig. 89-1

A programmable electrically erasable logic (PEEL) device can easily supply the synchronizing function. Digital systems often require synchronization of asynchronous inputs to avoid the potential metastability problems caused by setup-time violations. A common synchronization method uses two rippled 74LS72 D-type flip-flops.

In this circuit, the asynchronous input feeds into the D input of the first flip-flop and its Q output feeds into the D of the second. Because the first flip-flop latches on the falling edge of the system clock, to avoid setup-time violations, the D input signal to the second flip-flop will be stabilized before the rising edge of the clock. Even experienced programmable-logic device designers often resort to such a TTL flip-flop circuit to handle the synchronization function, because of the architectural limitations of standard PLDs.

A programmable electrically erasable logic (PEEL) device, such as the PEEL18CV8 from ICT, however, can easily supply the function. The user-programmable 12-configuration I/O macrocells in the device can internally feed back a signal before the output register. With this feedback arrangement, designing a two-stage input is simple.

A gated-latch internally latches the asynchronous input on the falling edge of the system clock, generating signal Q1. ANDing the input with Q1 through the internal feedback path, eliminates a possible hazard condition during the clock's high-to-low transition time. The latch then holds Q1 stable to ensure meeting the setup-time requirement of the subsequent D flip-flop, which, as before, registers the signal on the next rising system clock edge.

If by chance the input pulse width violates the set-up time of the gated latch, the clock's low time will give more time for settling.

SYNC TIP dc RESTORER

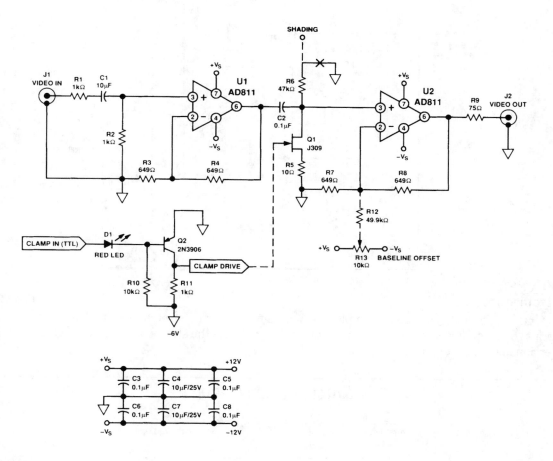

ANALOG DEVICES

Fig. 89-2

The dc restorer shown supplies a video signal with sync tips clamped to a baseline level. Clamp drive signal is supplied from elsewhere, usually a sync generator or a sync separator.

SYNC STRETCHER CIRCUIT

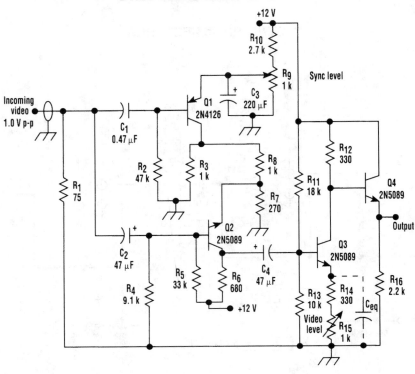

ELECTRONIC DESIGN

Fig. 89-3

Q1, Q2, and Q3 comprise a simple video amplifier and sync stretch circuit. Transistor Q1 sync strips the incoming video, which is amplified and mixed with the stripped sync in Q2. Q3 supplies inversion and video amplitude control.

SYNCHRONIZER CIRCUIT

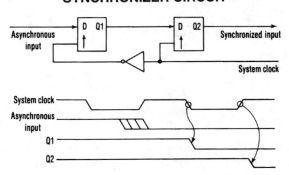

ELECTRONIC DESIGN

Fig. 89-4

This common synchronization method uses two rippled 74LS74 D-type flip-flops.

90

Telephone-Related Circuits

The sources of the following circuits are contained in the Sources section, which begins on page 707. The figure number in the box of each circuit correlates to the entry in the Sources section.

Telephone Line Tester
Caller ID Circuit
Telephone Call Restrictor
Telephone Scrambler
Universal Telephone Hold Circuit
Telephone Ring Amplifier
Bell System 202 Data Encoder
Telephone/Audio Interface
Telephone Recording Circuit
Telephone Bell Amplifier
Phone Line Simulator
Phone Helper
Telephone Ring Signal Detector
Telephone Hold Circuit

TELEPHONE LINE TESTER

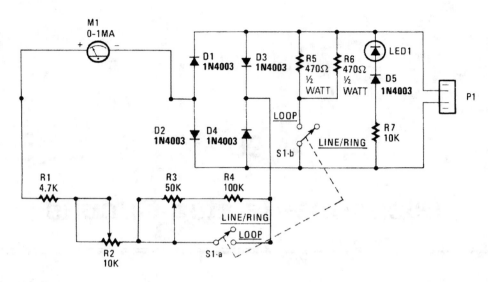

RADIO-ELECTRONICS

Fig. 90-1

The telephone-line tester shown in the figure is connected to the telephone line through modular connector P1. Because the tester's LED polarity indicator is always connected when the tester is plugged in, the instant that the unit is connected, you will have an indication of the polarity. If it is correct—that is, if the green wire is the positive side and the red wire is the negative side—nothing will happen. If the situation is reversed, the LED will light.

With switch S1 set for LINE/RING, both S1-a and S1-b are open and the meter indicates the condition of the line voltage. Any line-voltage reading in the LINE OK range (more on the meter in a moment) indicates a line voltage that is higher than 40 Vdc. If the telephone is caused to ring, either by using a ringback number or by dialing from another phone, the meter will indicate RING OK, and the LED will pulse (indicating ac), if the ringing voltage/current is correct. The actual position of the meter's pointer depends on how many ringers are connected across the line.

When S1 is closed the voltage range of the meter is changed and a nominal load resistance of 230 Ω (R5 and R6) is connected across the line to emulate the off-hook load of the telephone. If the meter indicates LOOP OK, you can be certain that you have sufficient loop voltage for satisfactory telephone operation. If you place another load on the line, perhaps by taking an extension telephone off hook, the meter reading will almost invariably drop below the LOOP OK range. If lifting the handset causes the meter reading to drop, you can at least be certain that the telephone's hook switch is working and that the repeat coil is connected to the line.

CALLER ID CIRCUIT

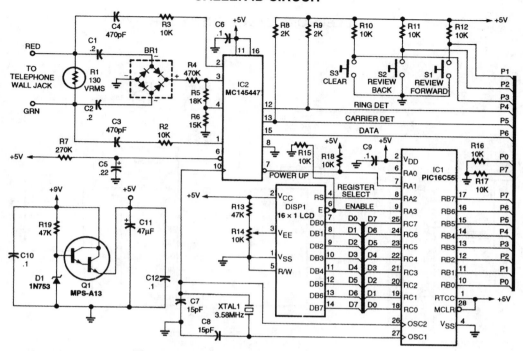

THE HEART OF THE CALLER ID circuit is microcontroller IC1 which processes
the serial data from IC2, outputs ASCII characters to DISP1, and monitors switches
S1–S3.

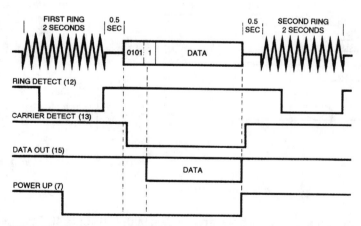

TIMING RELATIONSHIPS between the data present on the phone line (top),
and the output pins of IC2.

Fig. 90-2

This circuit requires programming of the microcontroller. Software information is available from
the reference in the original article.

TELEPHONE CALL RESTRICTOR

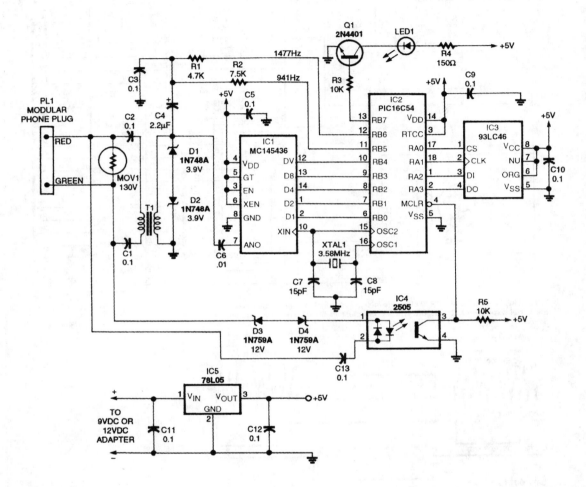

Fig. 90-3

This circuit is designed to restrict phone calls with the area codes: 900, 976, and 540. This device uses a microcontroller to compare the DTMF decoded tones with telephone numbers stored in EEP-ROM (IC3). This device requires a programmed microcontroller. Software and details of programming can be found in the original magazine article.

TELEPHONE SCRAMBLER

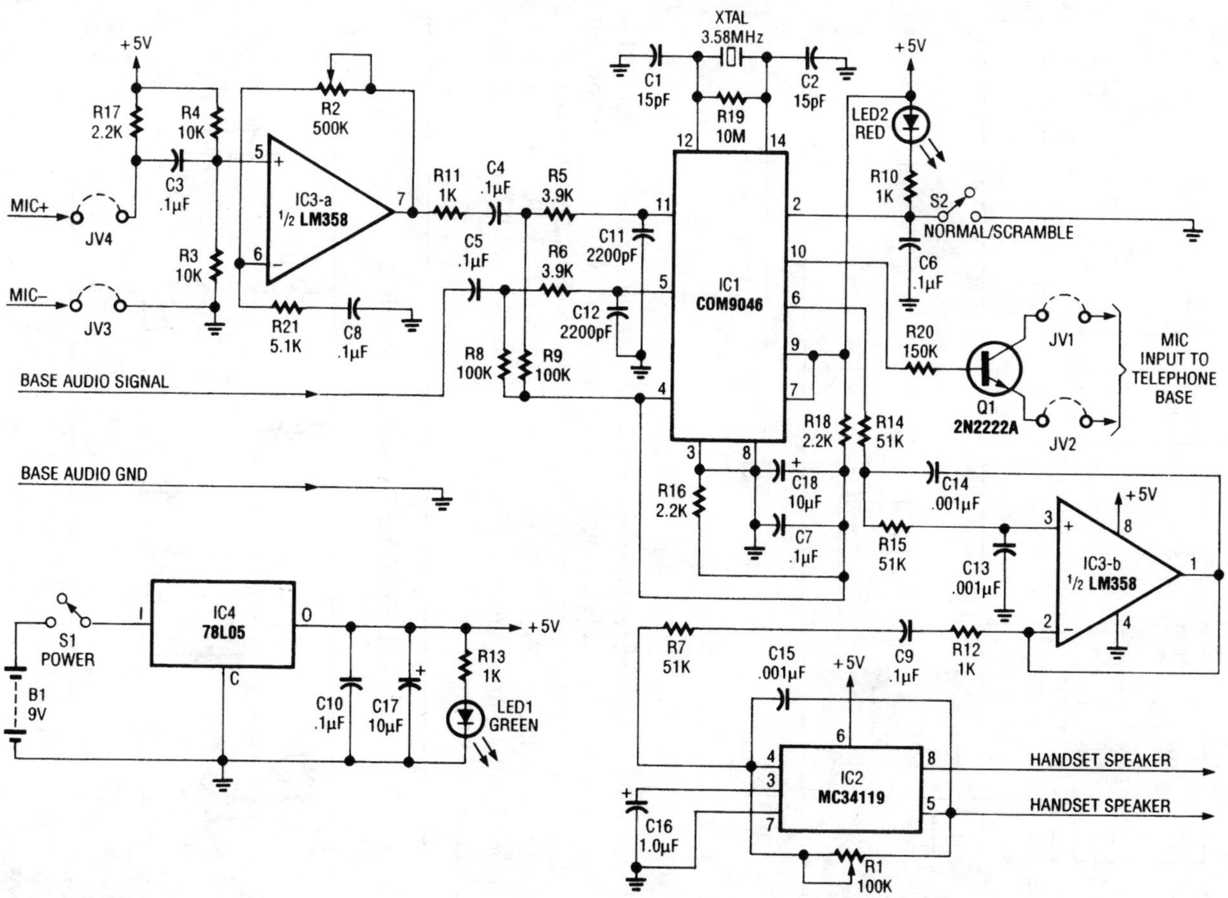

Fig. 90-4

ELECTRONICS EXPERIMENTERS HANDBOOK

This circuit uses the usual speech inversion algorithm, implementing it with a COM9046 ASIC. This unit is designed to fit between the handset and base of a standard telephone. It is powered by a 9-V battery.

623

UNIVERSAL TELEPHONE HOLD CIRCUIT

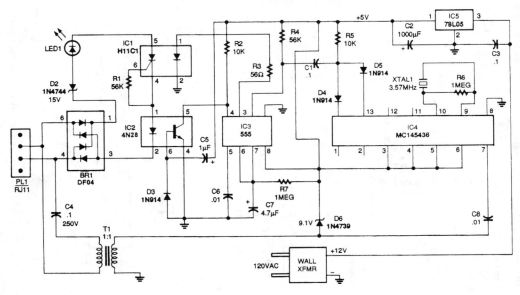

ELECTRONICS NOW

Fig. 90-5

The telephone line is connected to the hold components through bridge rectifier BR1 so that the input is not polarity sensitive. If you have touch-tone telephone service, you can now put a call on hold from any phone in your house by plugging this simple device into any telephone jack. The universal hold circuit works with any phone that has a key pad with a # key. To put a call on hold, press the # key and hang the phone up. A timer extends the #-key function while you hang up phones that have a keypad built into the handset.

TELEPHONE RING AMPLIFIER

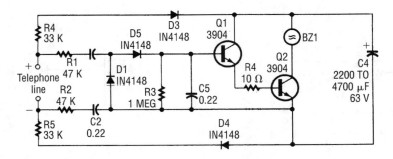

ELECTRONICS NOW

Fig. 90-6

This circuit takes its operating power from the telephone line. BZ1 is a piezoelectric transducer.

BELL SYSTEM 202 DATA ENCODER

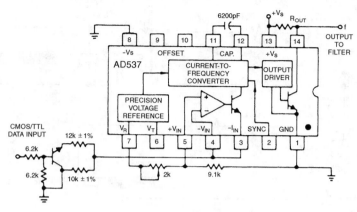

ANALOG DEVICES   *Fig. 90-7*

The AD537 is well-suited for frequency-shift modulator and demodulator applications. Requiring little power, it is especially appropriate for using phone-line power. The Bell-System 202 data encoder shown here delivers the mark frequency of 1.2 kHz with the data input low. When the input goes high, the timing current increases to 165 μA and generates the space frequency of 2.2 kHz. The trim shown provides a ±10% range of frequency adjustment. The output goes to the required bandpass filter before transmission over a public telephone line. A complementary demodulator is easy to implement.

TELEPHONE/AUDIO INTERFACE

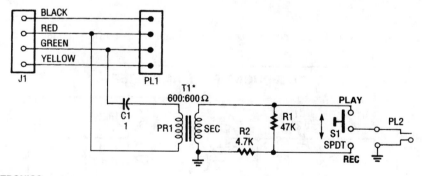

POPULAR ELECTRONICS   *Fig. 90-8*

The telephone audio interface—essentially, a simple isolation/couple circuit—isolates the phone line from any connected audio circuit without presenting any danger to the phone line, the equipment, or the user.

TELEPHONE RECORDING CIRCUIT

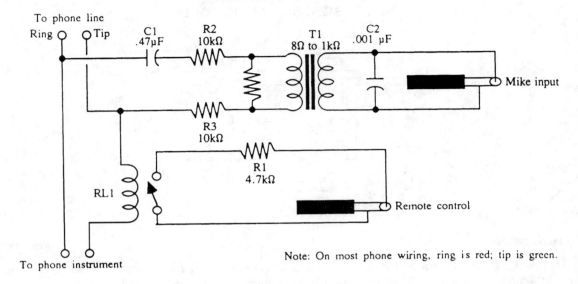

R1   4.7 kΩ
R2,R3   10 kΩ
C1   0.47 μF disc
C2   0.001 μF disc
T1   8 kΩ-to-1 kΩ impedance-matching transformer
RL1   SPST reed relay

All resistors are 5 to 10 percent tolerance, 1/4 watt. All capacitors
are 10 to 20 percent tolerance, rated at 35 volts or more.

McGRAW-HILL   *Fig. 90-9*

This device will automatically record telephone calls. An ordinary cassette recorder can be hooked to it.

TELEPHONE BELL AMPLIFIER

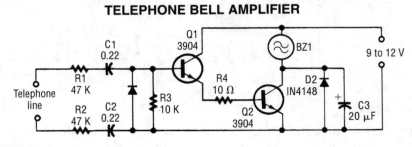

ELECTRONICS NOW   *Fig. 90-10*

Telephone "bell" amplifier circuit will let you hear (or see) an enhanced alarm if you are away from your telephone.

PHONE LINE SIMULATOR

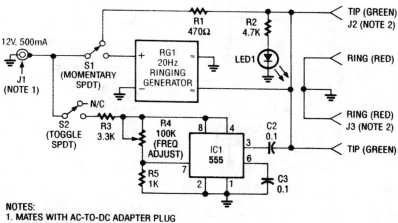

NOTES:
1. MATES WITH AC-TO-DC ADAPTER PLUG
2. LOCATED IN WALL PLATE

ELECTRONICS NOW

Fig. 90-11

This device contains a ringing generator and a tone oscillator. The tone oscillator is set to either 350 or 440 Hz. The ringing generator is a potted module delivering 86 Vac at 20 Hz. It is available from a source listed in the reference.

PHONE HELPER

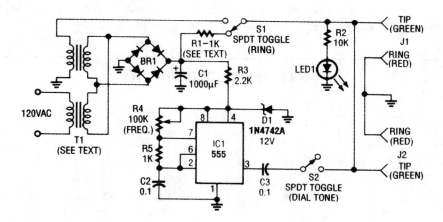

RADIO-ELECTRONICS

Fig. 90-12

This phone-line simulator uses a 60-Hz transformer instead of a ring generator. The dial tone is provided by an NE555 astable oscillator.

TELEPHONE RING SIGNAL DETECTOR

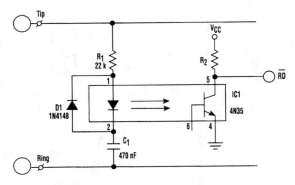

ELECTRONIC DESIGN

Fig. 90-13

Discriminating between telephone-ring signals on a phone line can be accomplished by using dedicated ICs, such as AT&T's LB1006AB or Texas Instruments' TMS1520A. However, if the system already is using a microcontroller, those dedicated chips can be replaced with simpler hardware and a few bytes of code.

Looking at the setup, the ringing voltage pulses the optoisolator's LED, which, in turn, pulses the low-asserted RD line to the microcontroller. The firmware analyzes the pulses to determine whether a valid ringing signal is present.

The frequency limits of a valid signal are 20 to 80 Hz, which is modulated 2 seconds on and 4 seconds off (with distinctive ringing, though, this cadence can vary). Therefore, the simplest analysis is to count down at least 20 pulses of RD in 1 second.

The routine could be expanded to determine what type of ring signal is present in a distinctive ring setting. Such a system could switch the phone line to various output jacks. As a result, several phone devices could use the same line without first picking up the line to determine if it's a voice, fax, or data call.

TELEPHONE HOLD CIRCUIT

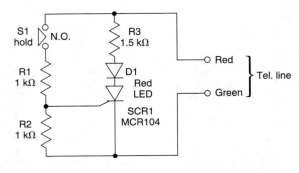

When the hold button S1 is pressed, SCR1 fires via R1 and R2, firing SCR1, and seizing the line via the path through R3, D1, and SCR1.

WILLIAM SHEETS

Fig. 90-14

91

Temperature-Related Circuits

The sources of the following circuits are contained in the Sources section, which begins on page 707. The figure number in the box of each circuit correlates to the entry in the Sources section.

Electronic Thermostat
Temperature Controller
Manual Control for Heater
Proportional Temperature Controller
Eight-Input A/D Converter for Temperature
  Measurements
Cold Junction Compensation for a Grounded
  Thermocouple
Absolute Temperature Log with RS-232
Centigrade Thermometer with Cold-Junction
  Compensation
1.5-V Electronic Thermometer
Two-Wire Temperature Sensor Output
  Referenced to Ground
Two-Wire Remote Temperature Sensor with
  Sensor Grounded
Single-Supply Temperature Sensor (–50 to
  +300°F)
Basic Fahrenheit Temperature Sensor

Temperature-to-Frequency Converter (Celsius)
Temperature-to-Frequency Converter (Kelvin)
Differential Thermometer
Optoelectronic Pyrometer
Bar-Graph Room-Temperature Display
LM3911 Temperature Controller
Thermocouple Amplifier with Cold-Junction
  Compensation
Precision RTD Amplifier Circuit for +5 V
Full-Range Fahrenheit Temperature Sensor
Improved Thermostatic Relay Circuit
Thermocouple Cold-Junction Compensation
Temperature Differential Detector
Thermostatic Relay Application
Temperature Controller
Temperature-to-Digital-Output Converter
Freeze-Up Sensor
Zero-Voltage Switching Temperature Regulator

ELECTRONIC THERMOSTAT

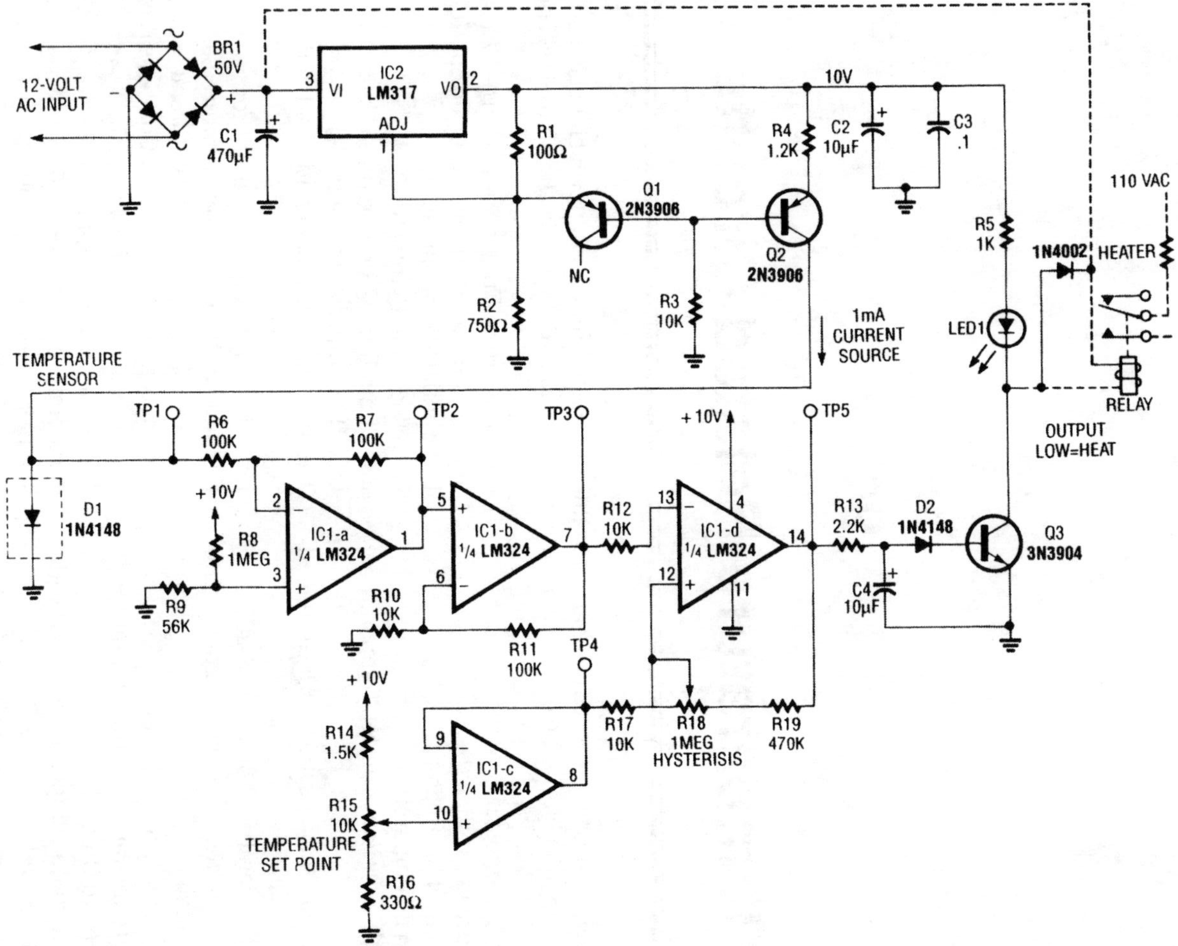

Fig. 91-1

ELECTRONICS EXPERIMENTERS HANDBOOK

Using a 1N914 diode as a temperature sensor, this straightforward circuit has hysteresis and set-point adjustments. Usable range is about –50 to +150°C.

TEMPERATURE CONTROLLER

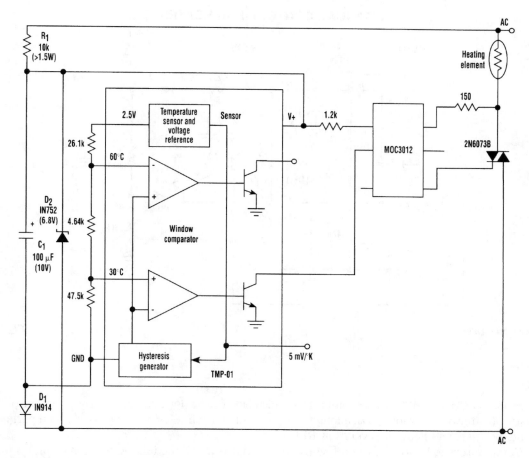

ELECTRONIC DESIGN

Fig. 91-2

The temperature sensor/controller (the TMP-01) is a monolithic device whose low power allows it to operate with a simple half-wave rectified power supply directly from the ac line. Such an arrangement greatly simplifies the power-supply design requirement to the point of only needing a few low-cost components to provide a single +6-Vdc supply.

The TMP-01 is essentially a "thermostat on a chip." It includes a linear temperature sensor (5 mV/K), and also has two comparators that switch at externally determined set points. These set points are established by resistively dividing the internal 2.5-V reference to set appropriate voltages on the inputs to the comparators.

One comparator is used in this circuit to turn on the heating element when the temperature drops below 30°C; it corresponds to a voltage of 1.52 V on the comparator's input.

MANUAL CONTROL FOR HEATER

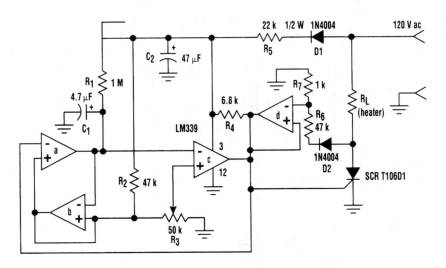

ELECTRONIC DESIGN

Fig. 91-3

Built around an LM339 quad comparator, this circuit provides manual control of the output of a resistive heater or other load with a long time constant. The circuit's design uses minimal parts, thus it's inexpensive, and generates very low RFI.

Comparators a, b, and c form a low-frequency pulse-width modulator. Sections a and b form a sawtooth oscillator (of approximately 0.25 Hz), with capacitor C1 being charged through R1 and discharged through section a's open collector output. R2 and R3 set the upper voltage limit for the sawtooth wave. The hysteresis means that C1 is discharged to nearly 0 V, creating a voltage swing that is identical to the adjustment range of R3.

Comparator c, in conjunction with potentiometer R3, converts the sawtooth wave form to a variable duty-cycle drive for the silicon-controlled rectifier.

Increasing voltage at R3's wiper means increasing the "on" time. Section d holds the SCR gate low if the line voltage is above approximately 3.5 V, preventing turn on at mid-cycle and ensuring low RFI.

The oscillator frequency is roughly determined by $1/0.7R_1C_1$. Resistance R_1 must be greater than $4R_2$ or the oscillator will lock up. Reducing R_2 will increase the lower voltage limit of the sawtooth; increasing it might cause lock-up.

PROPORTIONAL TEMPERATURE CONTROLLER

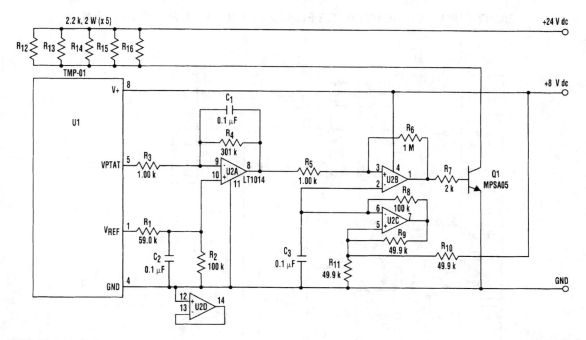

Fig. 91-4

Most temperature-controller circuits use upper and lower trip points to control a heater element, with the heater power full on and full off. Usually, this results in a temperature hysteresis of several degrees. This relatively large temperature hysteresis effect might cause modulation in the output of the circuit that's being controlled.

A proportional temperature controller eliminates this problem by continuously providing the power needed to maintain the "oven" at the desired temperature--within 1°C. From a cold start, maximum power is applied until the temperature is within 2°C of the set point.

The circuit's mechanical construction is important. The five heater resistors (R12 through R16), the temperature-sensor IC (U1), and the circuit being controlled are mounted with thermal epoxy to a small piece of aluminum. This provides excellent heat transfer between the components. The heater resistors must be selected to raise the temperature from ambient to the set point within an acceptable warm-up time.

U1 is Analog Devices' TMP-01 temperature-controller IC. The voltage proportional to absolute temperature (VPTAT) has a temperature coefficient of exactly 5 mV/°C. The set point is determined by the R_1/R_2 ratio. U2 is a Linear Technology LT1014 quad precision op amp. U2C is an oscillator with a 50% duty cycle that supplies a triangle wave between ⅓ and ⅔ of the supply voltage at U2-2.

U2B compares the amplified VPTAT to the triangle wave, which drives Q1 at a duty cycle of 100% or less. Because the triangle wave's peak-to-peak amplitude is 2.7 V, and VPTAT is amplified by a factor of 300, a temperature change of approximately 2 mV moves the duty cycle from 100% to 0%.

EIGHT-INPUT A/D CONVERTER FOR TEMPERATURE MEASUREMENTS

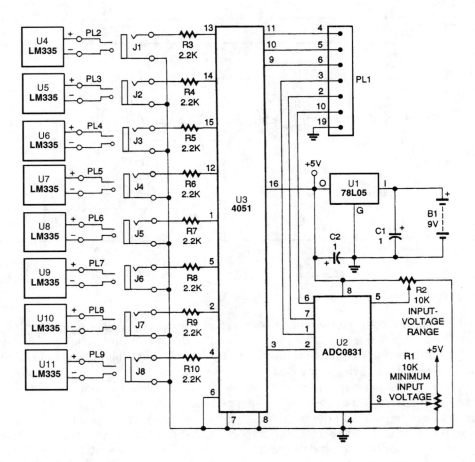

Fig. 91-5

The actual processing circuitry of this A/D converter consists of only four parts: U2, U3, R1 and R2. Eight temperature probes are used with the circuit; however, they can be replaced with other types of sensors, as long as resistors R3 through R10 are removed.

EIGHT-INPUT A/D CONVERTER FOR TEMPERATURE MEASUREMENTS (*Cont.*)

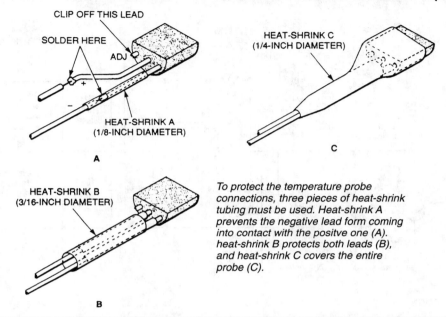

CLIP OFF THIS LEAD

SOLDER HERE

ADJ

+

−

HEAT-SHRINK A
(1/8-INCH DIAMETER)

A

HEAT-SHRINK C
(1/4-INCH DIAMETER)

C

HEAT-SHRINK B
(3/16-INCH DIAMETER)

B

To protect the temperature probe connections, three pieces of heat-shrink tubing must be used. Heat-shrink A prevents the negative lead form coming into contact with the positve one (A). heat-shrink B protects both leads (B), and heat-shrink C covers the entire probe (C).

COLD-JUNCTION COMPENSATION FOR A GROUNDED THERMOCOUPLE

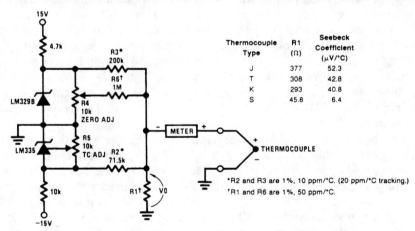

Thermocouple Type	R1 (Ω)	Seebeck Coefficient ($\mu V/°C$)
J	377	52.3
T	308	42.8
K	293	40.8
S	45.8	6.4

15V

4.7k

R3*
200k

R6†
1M

LM329B

R4
10k
ZERO ADJ

R5
10k
TC ADJ

R2*
71.5k

LM335

METER

−

+

+

−

THERMOCOUPLE

10k

R1†

V0

−15V

*R2 and R3 are 1%, 10 ppm/°C. (20 ppm/°C tracking.)
†R1 and R6 are 1%, 50 ppm/°C.

NATIONAL SEMICONDUCTOR

Fig. 91-6

A circuit for use with grounded thermocouples is shown. To trim, short out the LM329B and adjust R5 so that $V° = \alpha T$, where α is the Seebeck coefficient of the thermocouple and T is the absolute temperature. Remove the short and adjust R4 so that $V°$ equals the thermocouple output voltage at ambient. A good grounding system is essential here, for any ground differential will appear in series with the thermocouple output.

ABSOLUTE TEMPERATURE LOG WITH RS-232

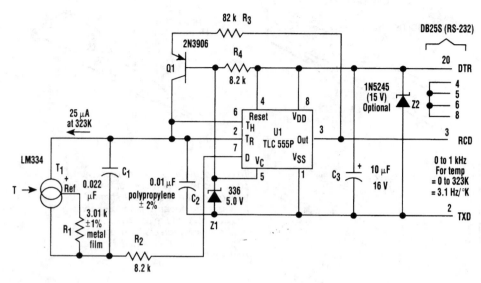

ELECTRONIC DESIGN

Fig. 91-7

In the setup, T1 (an LM334 temperature sensor) generates a constant current that's proportional to absolute temperature, and equal to 25 μA at 323 Kelvin (50°C). R1 sets this constant of proportionality. The current discharges the parallel combination of C1 and C2 connected to the trigger and the threshold pins of U1, which is a CMOS implementation of the venerable 555 analog timer. The negative-going ramp is compared by U1 to an internal 2.5-V trigger level controlled by Z1. When the ramp gets there, U1 triggers. The output pin (3) will go high, presenting a "start" bit to the connected communications port, which causes Q1 to source 100 μA to the timing node. At the same time, the discharge pin (7) will open, allowing R2 and the bottom end of C1 and T1 to be supplied by C1, and isolates it from C2.

Consequently, the current supplied by transistor Q1 will go solely to C2 so that when the resulting positive-going ramp reaches 5 V, exactly 25 nC (2.5 V × C2) will have been deposited in the timing node by the recharge cycle because its threshold level (6) will have been reached. This causes both the output pin to return to the negative rail, restoring the "marking" condition of the R2-232 interface, and Q1 to stop recharging C2. U1's discharge pin (7) now connects R2 to the negative rail, causing the charge that was deposited by T1 on C1 during the recharge interval to rapidly redistribute between C1 and C2. This arrangement creates a very linear (0.01%) relationship between T1 current and pulse output frequency.

While this is happening, the PC's communications port hardware assembles a valid (although meaningless) character because the positive pulse output by U1 looks like the start bit of a character. A simple program running in the PC can then count the frequency of these characters and convert the resulting rate into a direct readout of temperature. Because of the 3.1-Hz/degree slope of frequency versus temperature, a 30-second average suffices for 0.01° resolution.

CENTIGRADE THERMOMETER WITH COLD-JUNCTION COMPENSATION

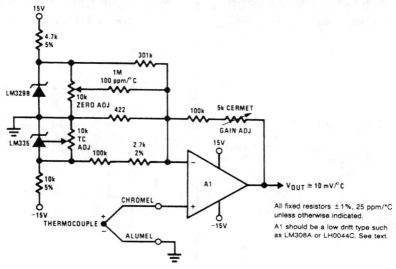

NATIONAL SEMICONDUCTOR

Fig. 91-8

This electronic thermometer has a 10-mV/°C output from 0°C to 1300°C. The trimming procedure is as follows: first short out the LM329B, the LM335 and the thermocouple. Measure the output voltage (equal to the input offset voltage times the voltage gain). Then apply a 50-mV input voltage and adjust the GAIN ADJUST pot until the output voltage is 12.25 V above the previously measured value. Next, short out the thermocouple again and remove the short across the LM335. Adjust the TC ADJUST pot so that the output equals 10 mV/°K times the absolute temperature. Finally, remove the short across the LM329B and adjust the ZERO ADJUST pot so that the output voltage equals 10 mV/°C times the ambient temperature in °C.

1.5-V ELECTRONIC THERMOMETER

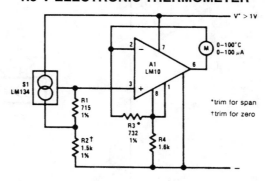

NATIONAL SEMICONDUCTOR

Fig. 91-9

An electronic thermometer design, useful in the range of –55°C to 150°C, is shown. The sensor, S1, develops a current that is proportional to absolute temperature. This is given the required offset and range expansion by the reference and op amp, resulting in a direct readout in either °C or °F.

TWO-WIRE TEMPERATURE SENSOR OUTPUT REFERENCED TO GROUND

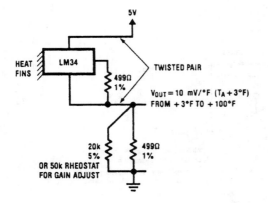

5V

HEAT FINS

LM34

499Ω 1%

TWISTED PAIR

$V_{OUT} = 10$ mV/°F $(T_A + 3°F)$
FROM +3°F TO +100°F

20k 5%
OR 50k RHEOSTAT FOR GAIN ADJUST

499Ω 1%

NATIONAL SEMICONDUCTOR   *Fig. 91-10*

TWO-WIRE REMOTE TEMPERATURE SENSOR WITH SENSOR GROUNDED

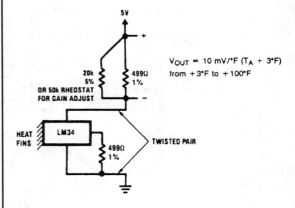

5V

+

20k 5%
OR 50k RHEOSTAT FOR GAIN ADJUST

499Ω 1%

−

$V_{OUT} = 10$ mV/°F $(T_A + 3°F)$
from +3°F to +100°F

HEAT FINS

LM34

499Ω 1%

TWISTED PAIR

NATIONAL SEMICONDUCTOR   *Fig. 91-11*

SINGLE-SUPPLY TEMPERATURE SENSOR (−50 TO +300°F)

−50° to +300°F

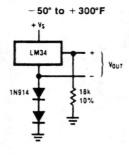

+ Vs

LM34

+
−

V_{OUT}

1N914

18k 10%

NATIONAL SEMICONDUCTOR   *Fig. 91-12*

BASIC FAHRENHEIT TEMPERATURE SENSOR

(+5° to +300°F)

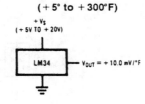

+ Vs
(+5V TO +20V)

LM34

$V_{OUT} = +10.0$ mV/°F

NATIONAL SEMICONDUCTOR   *Fig. 91-13*

TEMPERATURE-TO-FREQUENCY CONVERTER (CELSIUS)

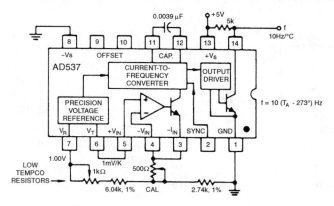

ANALOG DEVICES

Fig. 91-14

The 1.00-V reference output can be combined with the 1-mV/°K output to realize various temperature scales. For the Celsius scale, the lower end of the timing resistor must be offset by 273 mV. This is easily accomplished, and it results in an output from 0 to 1 kHz for temperatures from 0°C to +100°C. Other offsets and scale factors are equally easy to implement.

TEMPERATURE-TO-FREQUENCY CONVERTER (KELVIN)

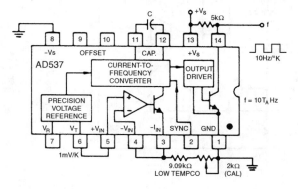

ANALOG DEVICES

Fig. 91-15

This simple connection results in a direct conversion of temperature to frequency. The 1-mV/°K temperature output serves as the input to the buffer amplifier, and the oscillator drive current is scaled to be 298 µA at 298°K (+25°C). Use of a 1000-pF capacitor results in a corresponding frequency of 2.98 kHz. A single-point trim for calibration is normally sufficient to give errors less than ±2°C from –55°C to +125°C. An NPO capacitor is preferred to minimize nonlinearity that results from capacitance drift.

DIFFERENTIAL THERMOMETER

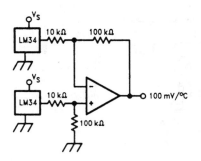

The differential thermometer shown in the figure produces an output voltage which is proportional to the temperature difference between two sensors. This is accomplished by using a difference amplifier to subtract the sensor outputs from one another and then multiplying the difference by a factor of 10 to provide a single-ended output of 100 mV per degree of differential temperature.

NATIONAL SEMICONDUCTOR   *Fig. 91-16*

OPTOELECTRONIC PYROMETER

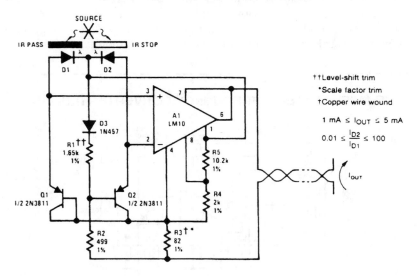

††Level-shift trim

*Scale factor trim

†Copper wire wound

$1 \text{ mA} \leq I_{OUT} \leq 5 \text{ mA}$

$0.01 \leq \dfrac{I_{D2}}{I_{D1}} \leq 100$

NATIONAL SEMICONDUCTOR   *Fig. 91-17*

This setup optically measures the temperature of an incandescent body. It makes use of the shift in the emission spectrum of a black body toward shorter wavelengths, as temperature is increased. Optical filters are used to split the emission spectrum, with one photodiode being illuminated by short wavelengths (visible light) and the other by long (infrared). The photocurrents are converted to logarithms by Q1 and Q2. These are subtracted to generate an output that varies as the log of the ratio of the illumination intensifies. Thus, the circuit is sensitive to changes in spectral distribution, but not intensity.

BAR-GRAPH ROOM TEMPERATURE DISPLAY

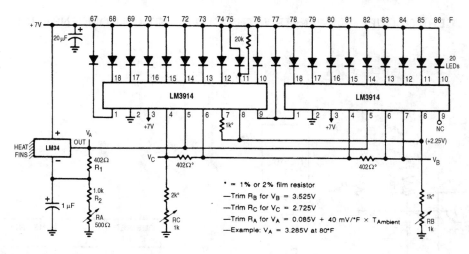

NATIONAL SEMICONDUCTOR

Fig. 91-18

This display shows temperature as a bar graph. The range is 67°F to 86°F.

LM3911 TEMPERATURE CONTROLLER

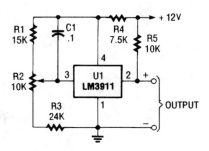

The LM3911 requires only a few resistors and a capacitor to turn into a full-function temperature controller. The LM3911 requires very few external components to implement a full-function temperature controller. The figure illustrates the simplicity of the freeze fighter circuit. The resistor network consisting of R1, R2, and R3 is used to provide the set point voltage for the feedback input of the LM3911.

Resistor R4 limits the current through the internal voltage reference of the LM3911 and can be selected from a wide range of values. 7.5 kΩ is specified in most of the application notes.

Resistor R5 pulls the output of the IC high when the temperature is below the set point. The internal resistor is in series with a diode that allows a switching voltage up to 35 V to be used with the device.

POPULAR ELECTRONICS

Fig. 91-19

THERMOCOUPLE AMPLIFIER WITH COLD-JUNCTION COMPENSATION

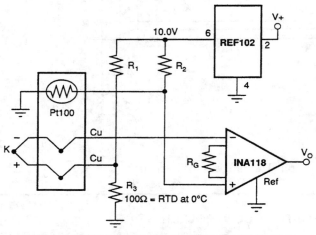

DESIRED GAIN	R_G (Ω)	NEAREST 1% R_G (Ω)
1	NC	NC
2	50.00k	49.9k
5	12.50k	12.4k
10	5.556k	5.62k
20	2.632k	2.61k
50	1.02k	1.02k
100	505.1	511
200	251.3	249
500	100.2	100
1000	50.05	49.9
2000	25.01	24.9
5000	10.00	10
10000	5.001	4.99

ISA TYPE	MATERIAL	SEEBECK COEFFICIENT (μV/°C)	R_1, R_2
E	+ Chromel – Constantan	58.5	66.5kΩ
J	+ Iron – Constantan	50.2	76.8kΩ
K	+ Chromel – Alumel	39.4	97.6kΩ
T	+ Copper – Constantan	38.0	102kΩ

BURR-BROWN

Fig. 91-20

PRECISION RTD AMPLIFIER CIRCUIT FOR +5 V

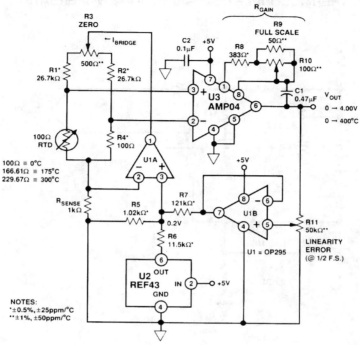

ANALOG DEVICES

Fig. 91-21

This circuit uses a platinum resistance temperature device to sense temperature. It has a range of 0 to 300°C. The RTD bridge is driven with a regulated 200-μA current to minimize self heating of the RTD. A 5-V supply is used.

FULL-RANGE FAHRENHEIT TEMPERATURE SENSOR

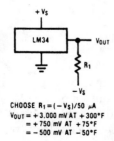

CHOOSE $R_1 = (-V_S)/50 \ \mu A$
$V_{OUT} = +3.000 \ mV \ AT \ +300°F$
$= +750 \ mV \ AT \ +75°F$
$= -500 \ mV \ AT \ -50°F$

NATIONAL SEMICONDUCTOR

Fig. 91-22

IMPROVED THERMOSTATIC RELAY CIRCUIT

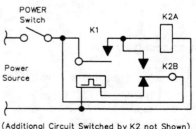

(Additional Circuit Switched by K2 not Shown)

QST

Fig. 91-23

THERMOCOUPLE COLD-JUNCTION COMPENSATION

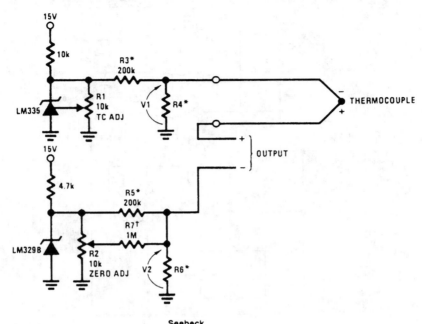

Thermocouple Type	Seebeck Coefficient (μV/°C)	R4 (Ω)	R6 (Ω)
J	52.3	1050	385
T	42.8	856	315
K	40.8	816	300
S	6.4	128	46.3

*R3 thru R6 are 1%, 5 ppm/°C. (10 ppm/°C tracking.)

†R7 is 1%, 25 ppm/°C.

$$\text{choose R4} = \frac{\alpha}{10 \text{ mV/°C}} \cdot \text{R3}$$

$$\text{choose R6} = \frac{-T_0 \cdot \alpha}{V_Z} \cdot (0.9\text{R5})$$

$$\text{choose R7} = 5 \cdot \text{R5}$$

where T_0 is absolute zero (-273.16°C)

V_Z is the reference voltage (6.95V for LM329B)

NATIONAL SEMICONDUCTOR

Fig. 91-24

A single-supply circuit is shown. R3 and R4 divide down the 10-mV/°K output of the LM329B and its associated voltage divider provide a voltage to buck out the 0°C output of the LM335. To calibrate, adjust R_1 so that $V_1 = \alpha T$, where α is the Seebeck coefficient and T is the ambient temperature in degrees Kelvin. Then, adjust R_2 so that $V_1 - V_2$ is equal to the thermocouple output voltage at the known ambient temperature.

TEMPERATURE DIFFERENTIAL DETECTOR

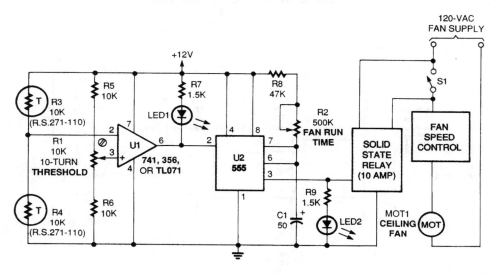

POPULAR ELECTRONICS

Fig. 91-25

This circuit measures temperature differences, not temperature. Once the difference passes a certain threshold, the timer is triggered, activating the solid-state relay.

Op amp U1 is placed in a comparator configuration with two thermistors—one located at the ceiling, one at the floor. The IC senses the temperature difference between the ceiling and floor, but is unaffected by the overall room temperature differential increases. The upper thermistor will decrease in resistance, eventually causing the voltage at pin 2 of U1 to exceed that pin 2 of U1 to go low and trip the timer. This adds hysteresis to the op amp's output, preventing the motor from chattering. When tripped, the timer activates the relay, which activates the fan.

THERMOSTATIC RELAY APPLICATION

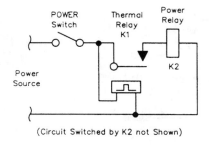

(Circuit Switched by K2 not Shown)

K1, the thermostatic relay, energizes power relay K2, which handles the circuit's power switching. The drawing doesn't show K2's power-switching contacts.

QST

Fig. 91-26

TEMPERATURE CONTROLLER

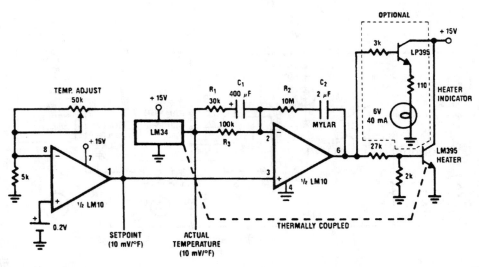

NATIONAL SEMICONDUCTOR

Fig. 91-27

A proportional temperature controller can be made with an LM34 and a few additional parts. The complete circuit is shown. Here, an LM10 serves as both a temperature-setting device and as a driver for the heating unit (an LM395 power transistor). The optional lamp, driven by an LP395 transistor, is for indicating whether or not power is being applied to the heater.

TEMPERATURE-TO-DIGITAL-OUTPUT CONVERTER

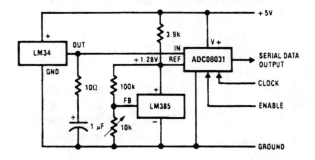

NATIONAL SEMICONDUCTOR

Fig. 91-28

FREEZE-UP SENSOR

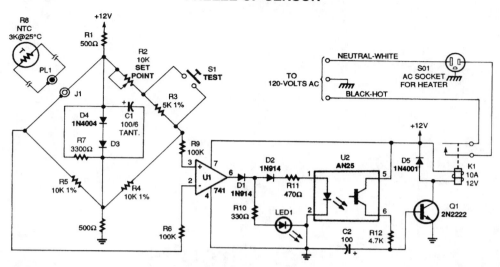

POPULAR ELECTRONICS

Fig. 91-29

Using a bridge circuit to provide an accurate activation temperature, this circuit will turn on a heating unit or other device when the temperature drops below the trip point set by R2. Use a 10-kΩ resistor in place of the thermistor to calibrate it for 32°F activation.

ZERO-VOLTAGE SWITCHING TEMPERATURE REGULATOR

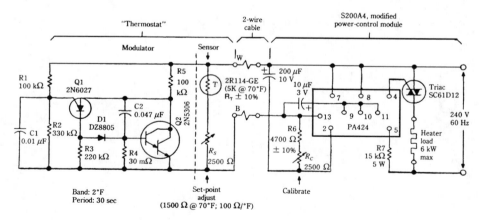

McGRAW-HILL

Fig. 91-30

In this arrangement, an integral number of cycles of ac is fed to the heater. No RFI or EMI is generated with this method. The thermostat uses a thermistor as a sensor. The PA424 (GE) device generates trigger pulses for the triac only at zero crossings of the ac line cycle.

92

Timer Circuits

The sources of the following circuits are contained in the Sources section, which begins on page 707. The figure number in the box of each circuit correlates to the entry in the Sources section.

Alarm Clock Timer
Lamp Timer
Long-Period Timer

ALARM CLOCK TIMER

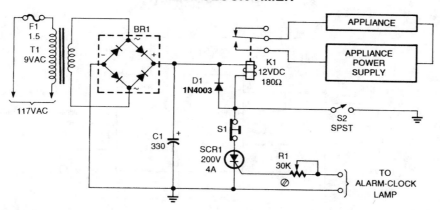

POPULAR ELECTRONICS

Fig. 92-1

Turn your alarm clock into a specialized timer with this simple circuit. The clock used with the circuit should be the kind that turns on a little lamp when the alarm is activated.

LAMP TIMER

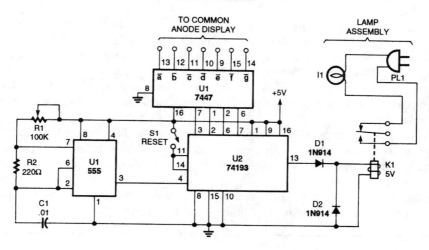

POPULAR ELECTRONICS

Fig. 92-2

A timed switch uses a 555 oscillator/timer wired to operate in the astable mode. The timer supplies a positive pulse to the clock input of a 74193 4-bit binary up/down count every five minutes. Because the 74193 is set to operate in the count-down mode, the output of the 555 is connected to the count-down input of the 74193.

As the binary counter is reset, it starts counting at nine and counts down to zero with each clock pulse. When the counter hits zero, the output from the 74193 goes low, turning off the relay and the light. The light can be turned back on by pressing the reset button again.

LONG-PERIOD TIMER

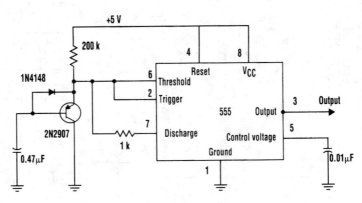

ELECTRONIC DESIGN

Fig. 92-3

Adding a transistor to the 555 timer can create long timer periods, which is a key factor when the timer is operating at low speed. The transistor basically acts as a current divider or capacitance multiplier. The problem with low speed, however, is that the timing resistors and capacitors must be large and the charging current must be small, particularly when the desired timing period is in the range of seconds.

Typically, electrolytic capacitors are used in these situations, but their leakage current tends to aggravate or even prohibit operation at very low charging currents.

This problem can be solved by adding a transistor. In effect, the transistor is used as a current divider or a capacitance multiplier. The normal charging current (emitter current) is divided by the transistor's current gain so that the capacitor charging current (base current) is reduced considerably. For example, 10 μA of emitter current will require approximately 0.1 μA of base current, based on a current gain of 100.

In this circuit, the capacitor will be charged with such a low charging current that timing periods will typically be 100 times longer than usual. This means that substantial time periods can be achieved with film or ceramic capacitors that have much better leakage characteristics and are physically smaller.

The circuit's output period was approximately 6 seconds, compared to 80 ms without the transistor. The transistor multiplied the normal time period by a factor of approximately 75.

93

Tone Control Circuits

The sources of the following circuits are contained in the Sources section, which begins on page 707. The figure number in the box of each circuit correlates to the entry in the Sources section.

TREBLE-CONTROL CIRCUIT

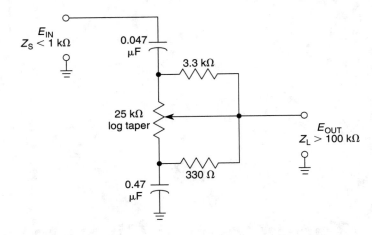

WILLIAM SHEETS

Fig. 93-1

This tone control has an insertion loss of 20 dB at flat setting and is effective above 1 kHz. It has little effect below about 1 kHz.

BASS TONE-CONTROL CIRCUIT

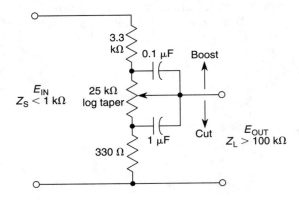

WILLIAM SHEETS

Fig. 93-2

This tone control has an insertion loss of 20 dB at flat setting and is effective below 350 Hz. The control has little effect above this frequency.

COMBINED BASS AND TREBLE CONTROL

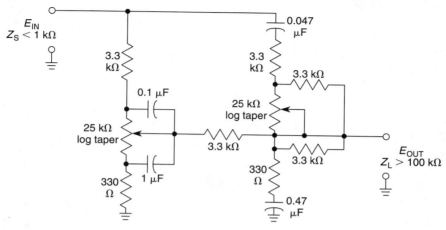

WILLIAM SHEETS

Fig. 93-3

This positive tone control system uses two pots to control bass and treble.

ACTIVE BASS- AND TREBLE-TONE CONTROL

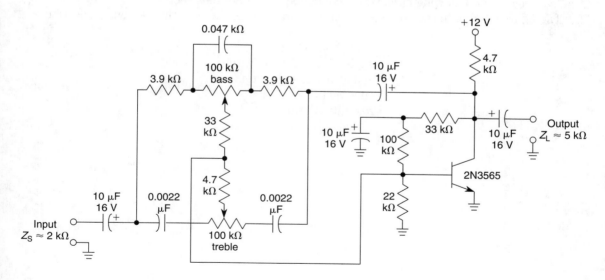

WILLIAM SHEETS

Fig. 93-4

A single transistor used as a feedback amplifier is connected with ac feedback through the tone controls, which determine the frequency response of the stage.

94

Touch/Proximity Control Circuits

The sources of the following circuits are contained in the Sources section, which begins on page 707. The figure number in the box of each circuit correlates to the entry in the Sources section.

Simple Touch Switch
Simple Timed Touch Switch
Capacitive Sensor System
Touch Switch
Proximity Alarm

SIMPLE TOUCH SWITCH

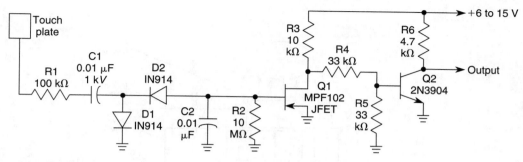

WILLIAM SHEETS

Fig. 94-1

Q2 is held cut off since Q1 normally is conducting. When the touch plate is contacted by a large object (human body, etc.), stray 60-Hz pickup is rectified by D1 and D2, and produces a negative voltage across R2-C2 and the gate of Q1. Q1 cuts off, causes Q2 to conduct, and the output goes low.

SIMPLE TIMED TOUCH SWITCH

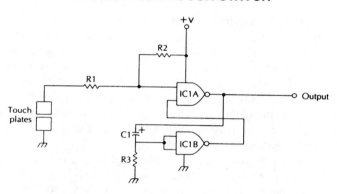

McGRAW-HILL

Fig. 94-2

This circuit produces an output for a time approximately equal to time constant R_3C_1.

IC1, IC2	CD4011 quad NAND gate
IC3	CD4066 quad bilateral switch
C1	47-μF, 25-V electrolytic capacitor
C2	100-μF, 25-V electrolytic capacitor
C3	220-μF, 25-V electrolytic capacitor
C4	470-μF, 25-V electrolytic capacitor
C5, C6, C7	0.1-μF capacitor
R1, R3, R4, R6	100-kΩ, ¼-W 5% resistor
R7, R9, R10, R12	
R2, R5, R8, R11	10-MΩ, ¼-W 5% resistor

CAPACITIVE SENSOR SYSTEM

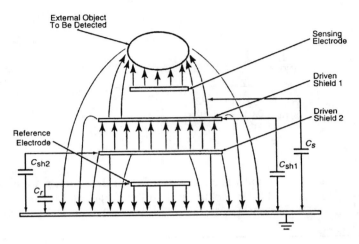

Fig. 94-3

This figure illustrates the electric-field configuration of a capacitive proximity sensor of the "capaciflector" type. It includes a sensing electrode driven by an alternating voltage, which gives rise to an electric field in the vicinity of the electrode; an object that enters the electric field can be detected by its effect on the capacitance between the sensing electrode and electrical ground.

Also, it includes a shielding electrode (in this case, driven shield 1), which is excited via a voltage follower at the same voltage as that applied to the sensing electrode to concentrate more of the electric outward from the sensing electrode, increasing the sensitivity and range of the sensor. Because the shielding electrode is driven via a voltage follower, it does not present a significant electrical load to the source of the alternating voltage.

In this case, the layered electrode structure also includes a reference electrode adjacent to ground, plus a second shielding electrode (driven shield 2), which is excited via a voltage follower at the same voltage as that applied to the reference electrode. Driven shield 2 isolates the reference electrode from the electric field generated by driven shield 1 and the sensing electrode so that a nearby object exerts no capacitive effect on the reference electrode.

The excitation is supplied by a crystal-controlled oscillator and applied to the sensing and reference electrodes via a bridge circuit. Fixed capacitors C1 and C2 (or, alternatively, fixed resistors R1 and R2) are chosen to balance the bridge; that is, to make the magnitude of the voltage at sensing-electrode node S equal the magnitude of the voltage at reference-electrode node R.

The voltages at S and R are peak-detected and fed to a differential amplifier, which puts out voltage V_u proportional to the difference between them. When no object intrudes into the electric field of the sensing electrode, the bridge remains in balance, and $V_u - 0$. When an object intrudes, it changes C_s, unbalances the bridge, and causes V_u to differ from zero. The closer the object comes to the sensing electrode, the larger (V_u) becomes.

An additional output voltage KV_r is available, where K is the amplification and V_r is the voltage on the reference electrode.

TOUCH SWITCH

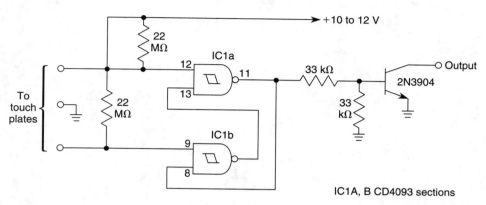

WILLIAM SHEETS

Fig. 94-4

Two NAND Schmitt triggers are used as a flip-flop to produce a bridged touch switch.

PROXIMITY ALARM

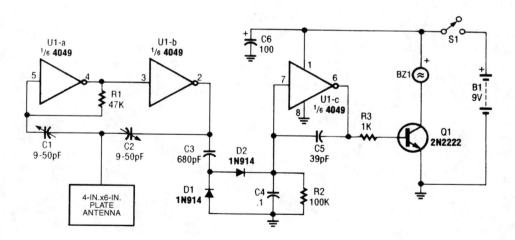

POPULAR ELECTRONICS

Fig. 94-5

95

Tracer Circuits

The sources of the following circuits are contained in the Sources section, which begins on page 707. The figure number in the box of each circuit correlates to the entry in the Sources section.

Wire Tracer
Cable Tracer
Signal Tracer

WIRE TRACER

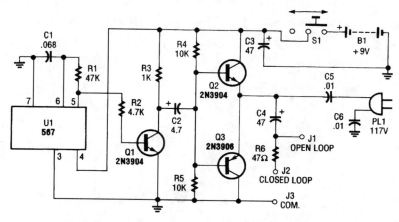

At the heart of the McTrak is a 567 tone decoder, configured as a simple squarewave oscillator, operating at about 250 Hz.

ELECTRONICS HOBBYIST HANDBOOK

Fig. 95-1

This tracer works by placing a square-wave signal on the line to be traced. The square wave is rich in harmonics. A small transistor radio placed close to a wire carrying this signal will buzz. The radio, therefore, is used as a probe to trace out the wire.

CABLE TRACER

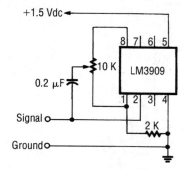

ELECTRONICS NOW

Fig. 95-2

This circuit generates a 1-kHz square wave for cable tracing. Because this circuit is simple and generates from 1.5 V, several can be used at the same time to generate multiple tones for tracing multiconductor cables.

SIGNAL TRACER

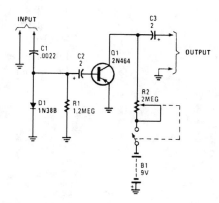

POPULAR ELECTRONICS

Fig. 95-3

This circuit uses a simple detector-audio amplifier. The output can be connected to headphones or another audio amplifier.

96

Transmitter and Transceiver Circuits

The sources of the following circuits are contained in the Sources section, which begins on page 707. The figure number in the box of each circuit correlates to the entry in the Sources section.

WIRELESS GUITAR TRANSMITTER

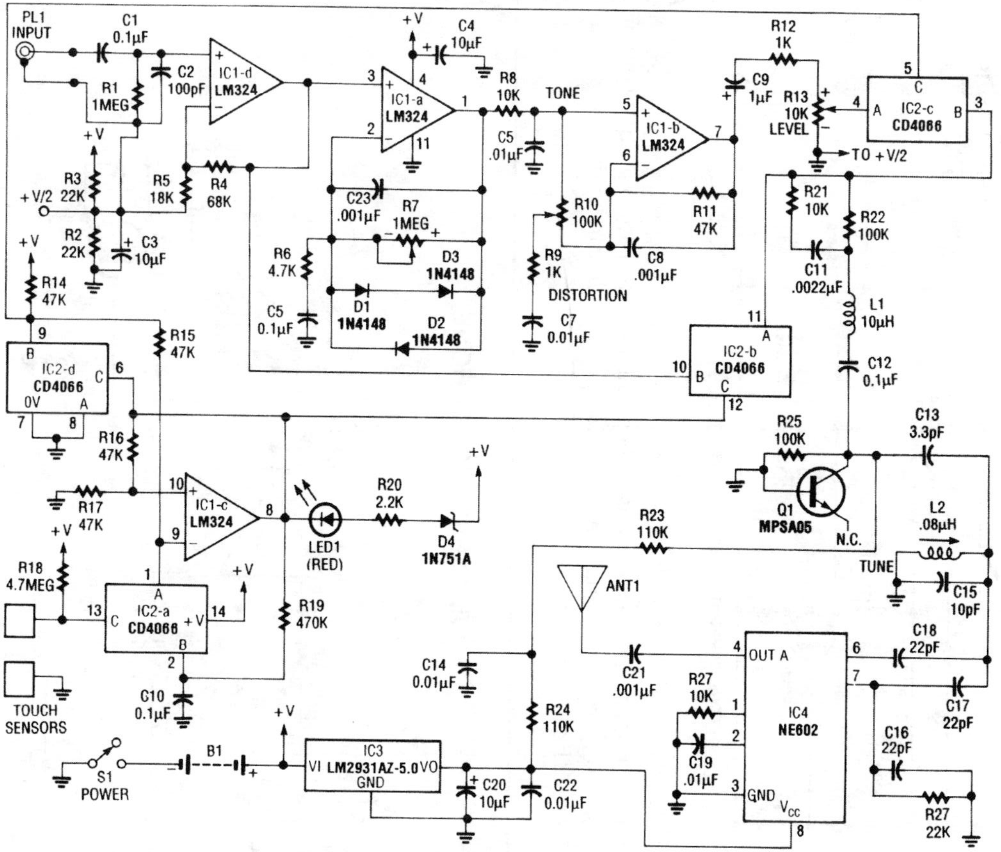

Fig. 96-1

ELECTRONICS NOW

This transmitter has a built-in distortion effects unit and a touch switch to switch effects off and on. The circuit operates from a 9-V battery. IC1-a and IC1-b are used in the effects circuitry. IC1-d is an input preamp and IC2 is a quad analog switch to handle audio switching. Q1 acts as a varactor diode modulator while IC-4 is an 88- to 108-MHz FM oscillator.

MICRO TV TRANSMITTER

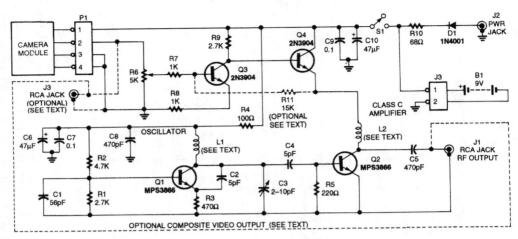

ELECTRONICS NOW

Fig. 96-2

For very low power, noncritical applications, this small TV modulator can be useful as a short-range (50 feet) transmitter for video signals. A small camera module can be used as a source. R11 is used to vary dc offset of the modulator.

WIRELESS MICROPHONE

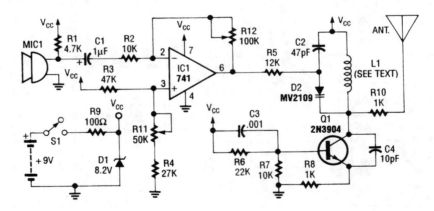

ELECTRONICS EXPERIMENTERS HANDBOOK

Fig. 96-3

An op-amp IC (741) amplifies the audio signal from MIC1, and R12 controls its gain. Audio is fed to the oscillator circuit Q1 and related components. D2 is a varactor diode. Audio fed to D2 causes FM of the oscillator signal. L1 is 2½ turns of #18 wire on a 5⁄16" diameter form. The antenna is a 12" whip.

FM STEREO TRANSMITTER

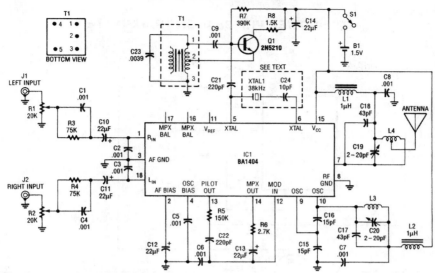

THE HEART OF THE FM TRANSMITTER is a BA1404 FM stereo transmitter IC.
The left input-signal level is adjusted via R1, pre-emphasis is provided by C1 and R3,
and audio is coupled by C10 into the left-channel input. The right-channel input
circuitry is identical.

ELECTRONICS NOW

Fig. 96-4

An FM stereo transmitter can be built around the BA1404 IC. This IC has all the functions necessary to generate an FM MPX signal. A separator oscillator circuit uses a 2N5210 transistor instead of the difficult-to-find 38-kHz crystal that is normally used. T1 is a 455-kHz IF transformer with 0.0039-μF capacitance added across it to enable tuning to 38 kHz. With this circuit, oscillator stability should be adequate.

FM BUG

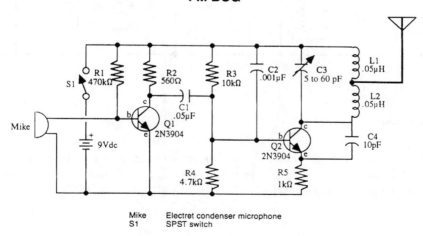

Mike   Electret condenser microphone
S1   SPST switch

McGRAW-HILL

Fig. 96-5

LOW-POWER VHF BEACON TRANSMITTER

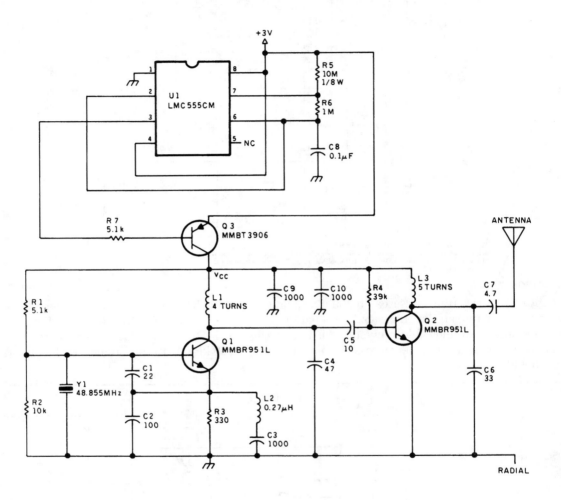

Fig. 96-6

A crystal oscillator and tripler make up the low power beacon transmitter. U1 generates a pulse that keys the transmitter at a 10:1 duty cycle (100 ms on, 1 s off) to conserve battery power. This transmitter was used as a locator beacon.

LOW-COST 6-W, 40-M CW TRANSMITTER

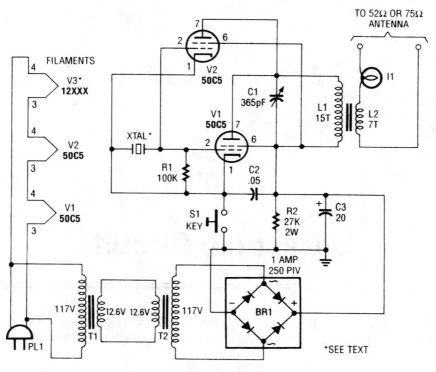

POPULAR ELECTRONICS

Fig. 96-7

L1 and L2 are wound on a ⅞" diameter form. L1 is 15 turns #22 plastic-covered wire, and L2 is 7 turns #22 plastic-covered wire.

97

Ultrasonic Circuits

The sources of the following circuits are contained in the Sources section, which begins on page 707. The figure number in the box of each circuit correlates to the entry in the Sources section.

Ultrasonic Remote-Control Tester
Ultrasonic Motion Detector
Ultrasonic CW Transceiver
Ultrasonic Proximity Sensor
Simple Ultrasonic Generator
Ultrasonic Sound Receiver

ULTRASONIC REMOTE-CONTROL TESTER

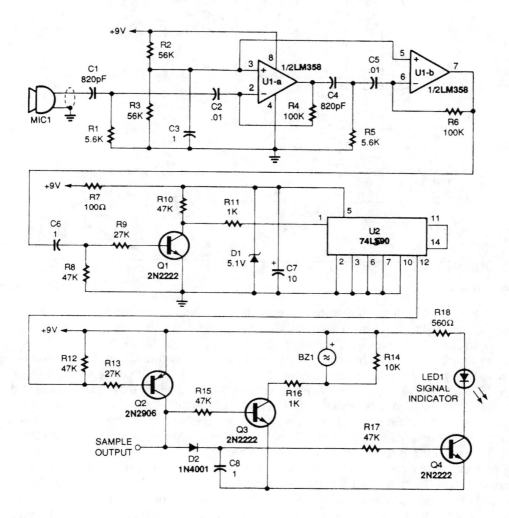

POPULAR ELECTRONICS

Fig. 97-1

This circuit picks up the ultrasonic tone via MIC1, amplifies it, and divides it by 10 in IC U2, a 74LS90. The output of U2 drives an audio amplifier and a piezoelectric element is used as a speaker.

ULTRASONIC MOTION DETECTOR

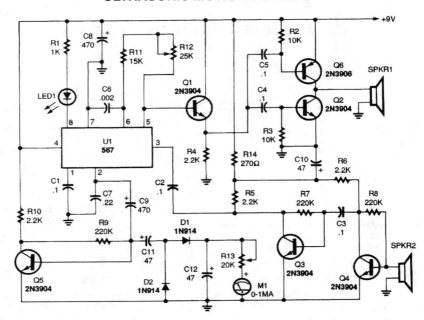

POPULAR ELECTRONICS

Fig. 97-2

A 567 PLL IC operates in a dual-function mode as a signal generator and an FM receiver. The 567's square-wave output at pin 5 is coupled to the base of Q1, and from Q1's emitter to the input of the power amplifier (Q2 and Q6). The output drives the piezo speaker, SPKR1.

The receive portion of the circuit operates as follows: transistors Q3 and Q4 are connected in a two-stage, high-gain, audio-frequency amplifier circuit, with the input connected to a second piezo speaker (SPKR2) operating as a sensitive microphone. The amplifier's output is coupled to the 567's input at pin 3. When an in-band signal is received, the LED lights.

The 567's FM output is coupled from pin 2 to the input of a very-low-frequency single-transistor amplifier, Q5. The amplifier's output at Q5's collector drives a voltage-doubler circuit (C11, D1, D2, and C12). The dc output feeds a 0- to 1-mA analog meter.

By placing the two piezo speakers one foot apart and aiming them in the same direction, toward a nonmoving solid object, the signal from the transmitter's speaker will reflect back into the receiver's speaker, and the frequency at the 567's input will be the same as the one being transmitted.

The ac output at pin 2 is zero when the outgoing and incoming frequencies are the same. However, when the signal is reflected from a moving object, the received frequency will be either lower or higher than the transmitted one. If the object is moving away from the speakers, the received frequency will be lower; if the object is moving toward the speakers, the frequency will be higher.

The pin-2 signal is fed through a 470-μF capacitor to the base of Q5, where the signal is amplified and fed to a voltage doubler, and then on to a meter circuit.

If you wish, the voltage doubler and meter can be removed and replaced with headphones connected between the negative side of C11 and circuit ground. That will allow you to listen to the difference-frequency signal as objects move in front of the speakers.

ULTRASONIC CW TRANSCEIVER

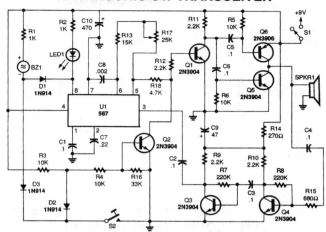

POPULAR ELECTRONICS

Fig. 97-3

With the telegraph key at S2 "up" (open), the 567 PLL's input at pin 3 is coupled to the output of Q3. Transistors Q3 and Q4 are operating in a high-gain, two-stage audio-amplifier circuit. The piezo speaker is coupled to the input of the amplifier though a 0.1-μF capacitor and a 680-Ω isolation resistor.

In the receive mode, the piezo speaker operates as a sensitive microphone. Ultrasonic signals travel from the microphone through the two-stage amplifier to the input of the 567, and, if the signal's frequency is within the IC's bandwidth, the LED will light and piezo-sounder BZ1 will sing out for each "dit" and "dah" received. The receiver can be tuned to the incoming ultrasonic signal by adjusting R17. Of course, adjusting that potentiometer also changes the transmitter's frequency.

The transmitter operates each time the S2 is closed. When the key is closed, diode D3 supplies a path to ground for BZ1, causing that sounder to produce an audible signal for each "dit" and "dah" transmitted. Also, Q2's bias is taken to ground through D2, allowing Q1 to pass the 567's square-wave signal on to the input of the power amplifier and out through the speaker.

ULTRASONIC PROXIMITY SENSOR

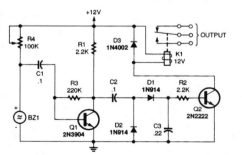

POPULAR ELECTRONICS

Fig. 97-4

A 100-kΩ potentiometer, R4, sets the current fed to the Sonalert sounder. The potentiometer is adjusted to a point where the sounder just begins to make an audible sound. A single-transistor audio amplifier (Q1) is coupled to the positive side of the sounder and its output is fed to a voltage-doubler circuit. The doubler's dc output drives the base of Q2, which, in turn, operates the relay (K1). As long as the Sonalert is producing a sound, the relay stays energized.

When a solid object is moved in close proximity to the front of the sounder, the *Q* of the piezo element is lowered and the Sonalert's internal circuit ceases to operate; as a result, the relay drops out. By carefully adjusting R4, the circuit can be made quite sensitive.

SIMPLE ULTRASONIC GENERATOR

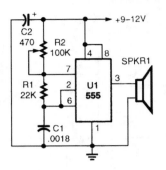

This basic ultrasonic generator can be built easily and quickly. An NE555 drives a speaker. The frequency range is 12 to 50 kHz. SPKR1 is a piezo tweeter, etc.

POPULAR ELECTRONICS

Fig. 97-5

ULTRASONIC SOUND RECEIVER

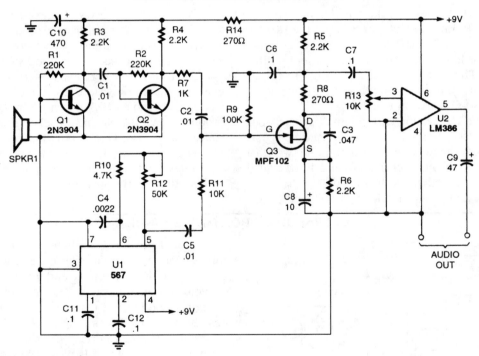

POPULAR ELECTRONICS

Fig. 97-6

You won't be disappointed with the performance of this sensitive ultrasonic receiver. It can let you listen to bugs, bats, engines, and virtually any other source of ultrasonic sounds. This circuit uses a piezo tweeter as an ultrasonic microphone, amplifier stages Q1, Q2, and an LO using a 567 IC. Q3 is a mixer that heterodynes the ultrasonic sounds down to the audible range. U2 is an amplifier that will drive a pair of headphones.

98

Video Circuits

The sources of the following circuits are contained in the Sources section, which begins on page 707. The figure number in the box of each circuit correlates to the entry in the Sources section.

Video Titler
Video Amplifier
RGB Video Amplifier
One-of-Two Video Selector
NTSC-to-RGB Converter
Video IF Amplifier/Detector
Video Cable Driver
Simple NTSC Gray-Scale Video Generator
LM1201 Video Amplifier
Simple Video Gray-Scale Generator European Line Standard
Video Switch
Adjustable Video-Cable Equalizer
Video Summing Amplifier
Video Amplifier
Twisted-Pair Video Driver/Receiver Circuit
250-mA 60-MHz Current-Feedback Amplifier for Video Applications
Video Driver/Amplifier
Video Line Driver

VIDEO TITLER

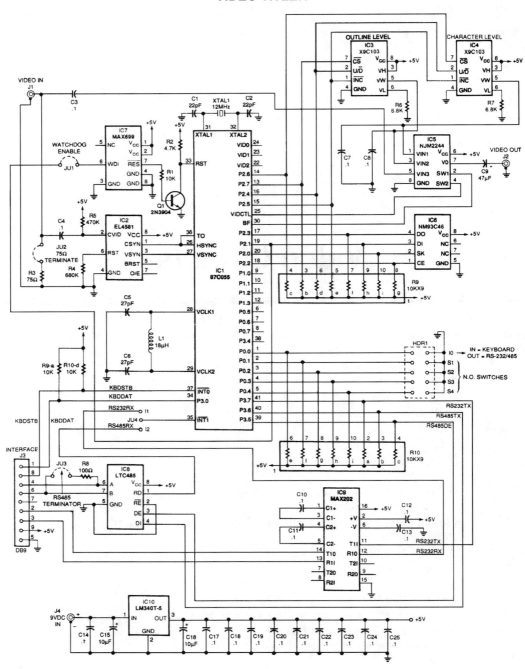

Fig. 98-1

VIDEO TITLER (*Cont.*)

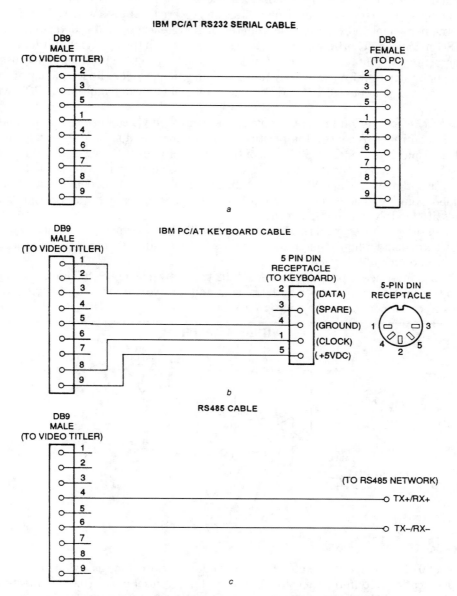

IBM PC/AT RS232 SERIAL CABLE

DB9 MALE (TO VIDEO TITLER)

DB9 FEMALE (TO PC)

2
3
5
1
4
6
7
8
9

a

IBM PC/AT KEYBOARD CABLE

DB9 MALE (TO VIDEO TITLER)

5 PIN DIN RECEPTACLE (TO KEYBOARD)

2 (DATA)
3 (SPARE)
4 (GROUND)
1 (CLOCK)
5 (+5VDC)

5-PIN DIN RECEPTACLE

b

RS485 CABLE

DB9 MALE (TO VIDEO TITLER)

(TO RS485 NETWORK)

TX+/RX+

TX–/RX–

c

There is a different cable for each interface. The RS-232 serial cable is shown in a, the PC/AT keyboard cable is shown in b, and the RS-485 cable is shown in c.

VIDEO TITLER (*Cont.*)

The figure shows the schematic of the video titler circuit. The power-on reset function is generated by IC7, a Maxim MAX699 reset, and watchdog pulse generator. That device supplies a reset pulse of 140 to 500 ms at power-up. This is accomplished with some external parts, as well as the OSD controller in IC1. First, the horizontal and vertical sync from the composite video input is detected by IC2, which is set for NTSC specification horizontal and vertical synchronization timing via resistor R4.

The detected horizontal and vertical sync is fed to IC1. The OSC controller in IC1 uses these signals to internally synchronize the overlay text to the incoming video. The frequency of the dot clock is controlled by components L1, C5, and C6. Text is overlaid by video multiplexer IC5, which is controlled by IC1.

The overlay character outline and intensity are controlled via solid-state potentiometers, allowing the microcontroller to control the position of their wipers and store the settings in an onboard EEPROM. The microcontroller's OSC logic controls the multiplexer timing from the BF (IC1, pin 30) and VIDCTL (IC1, pin 25) signals. The BF signal switches the video multiplexer between character and character-outline video, and VIDCTL switches the multiplexer between the input video and the overlay video from IC1. The dc levels from IC3 and IC4 set the character and outline intensity, and these levels are fed to video multiplexer IC5.

The video titler can store and recall text from EEPROM IC6, which has enough capacity to store one overlay screen and other required data, such as network address, horizontal and vertical overlay fine position, and type of interface.

The RS-232 interface is provided by a MAX202 transceiver. The RS-485 interface is provided by an LTC485 transceiver that provides both transmit and receive functions. The keyboard interface is basically a direct connection to the microprocessor.

VIDEO AMPLIFIER

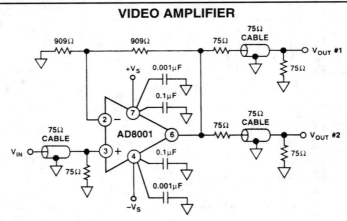

ANALOG DEVICES   *Fig. 98-2*

The AD8001 has been designed to offer outstanding performance as a video line driver. The important specification of differential gain (0.01%) and differential phase (0.025°) meet the most exacting HDTV demands for driving one video load. The AD8001 also drives up to two back-terminated loads with equally impressive performance (0.01%, 0.07°). Another important consideration is isolation between loads in a multiple-load application. The AD8001 has more than 40 dB of isolation at 5 MHz when driving two 75-Ω terminated loads.

RGB VIDEO AMPLIFIER

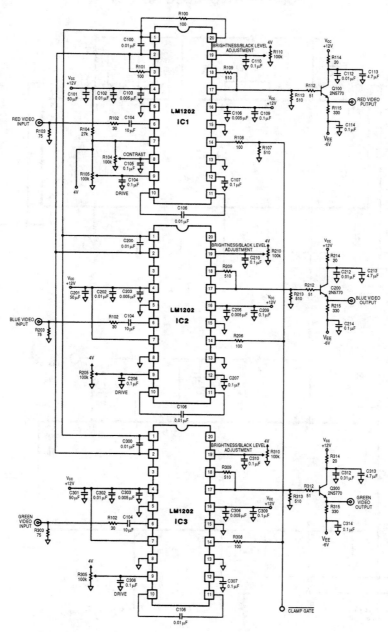

NATIONAL SEMICONDUCTOR

Fig. 98-3

This circuit is a three-channel RGB video amplifier with individual brightness, black level and drive controls.

ONE-OF-TWO VIDEO SELECTOR

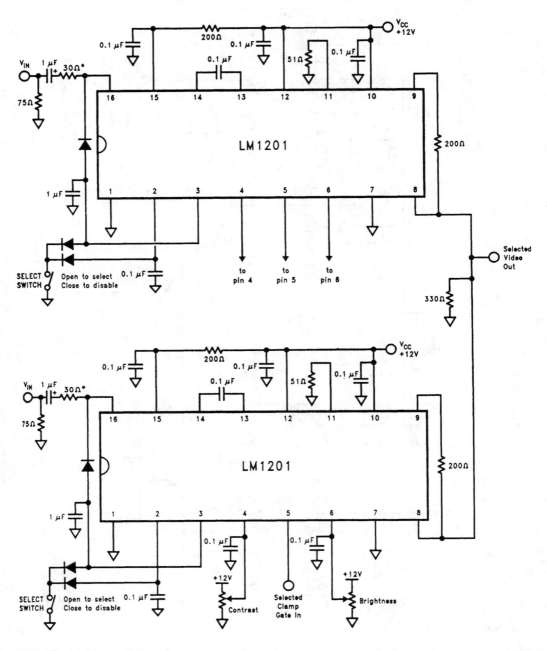

NTSC-TO-RGB CONVERTER

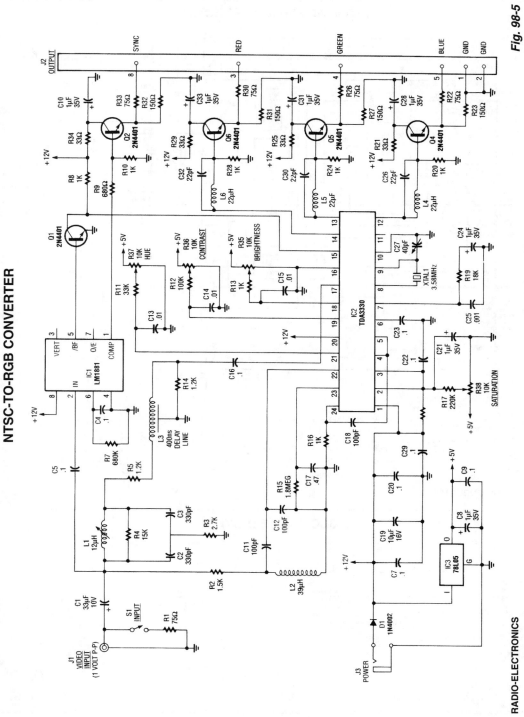

Fig. 98-5

RADIO-ELECTRONICS

This circuit takes baseband NTSC video, decodes it, and derives RGB video suitable for driving a color multisync computer monitor. This enables the user to take advantage of the generally better resolution of computer monitors.

VIDEO IF AMPLIFIER/DETECTOR

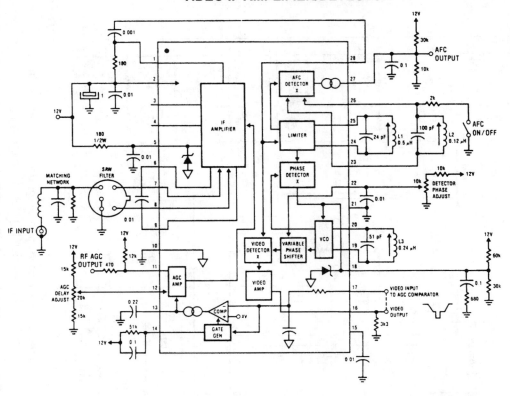

SAW Filter-MuRata SAF45MC/MA

L1-9 ½T   #22 wire

L2-4 ½T   } on 3.16" form with

L3-6 ½T   HF core, shielded

All caps in uF unless noted

NATIONAL SEMICONDUCTOR   *Fig. 98-6*

VIDEO CABLE DRIVER

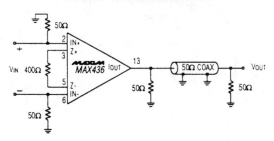

MAXIM   *Fig. 98-7*

This is a MAX436 coaxial-cable driving circuit.

678

SIMPLE NTSC GRAY-SCALE VIDEO GENERATOR

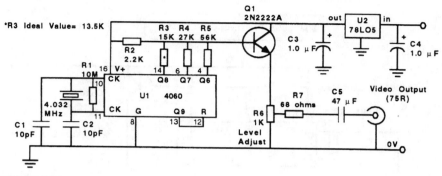

73 AMATEUR RADIO TODAY

Fig. 98-8

A 4.032-MHz crystal oscillator (256 × horizontal line scan rate) drives a BCD counter. The binary outputs of the counter are fed to R2 through R5, a simple weighting network for D/A conversion, resulting in a staircase video output with a rep rate of 15.75 kHz. This circuit should be useful for amateur TV linearity testing and setup purposes.

LM1201 VIDEO AMPLIFIER

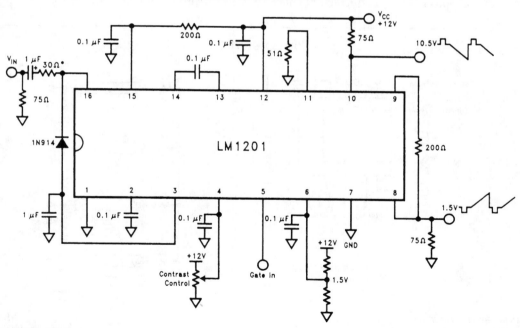

NATIONAL SEMICONDUCTOR

Fig. 98-9

This video amplifier has 75-Ω bi-phase outputs.

SIMPLE VIDEO GRAY-SCALE GENERATOR EUROPEAN LINE STANDARD

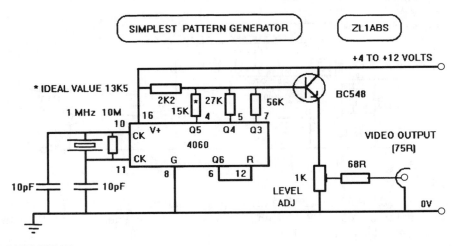

73 AMATEUR RADIO TODAY

Fig. 98-10

A simple gray-scale generator (staircase waveform) can be obtained with a CD4060 counter, a 1-MHz crystal oscillator, and several resistors to act as an elementary D/A converter to convert the binary count output to analog equivalent. This circuit is for European (PAL) standards.

VIDEO SWITCH

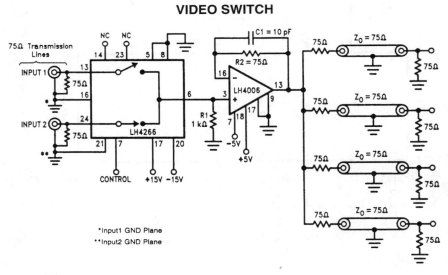

NATIONAL SEMICONDUCTOR

Fig. 98-11

Using National Semiconductor LH4266 and LH4006, this circuit switches one of two inputs to four output (75 Ω) lines.

ADJUSTABLE VIDEO-CABLE EQUALIZER

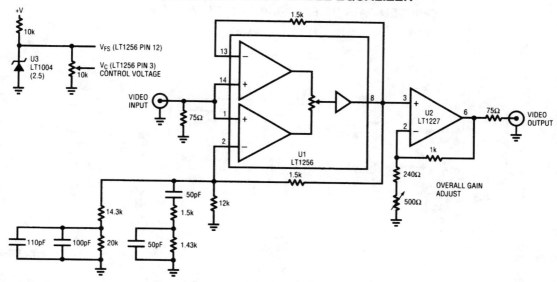

LINEAR TECHNOLOGY

Fig. 98-12

The figure is a complete schematic of the cable equalizer. The LT1256 (U1) is a two-input/one-output 40-MHz current feedback amplifier with a linear control circuit that sets the amount that each input contributes to the output. One amplifier (input pins 13 and 14) of the LT1256 is configured as a gain of one with no frequency equalization. The other amplifier (input pins 1 and 2) has frequency equalizing components in parallel with the 12-kΩ gain resistor. An additional amplifier (U2, LT1227) is used to set the overall gain. Two amplifiers were used here to make setting the gain a single adjustment, but in a production circuit, the LT1256 can be configured to have the necessary gain and the whole function can be done with one chip.

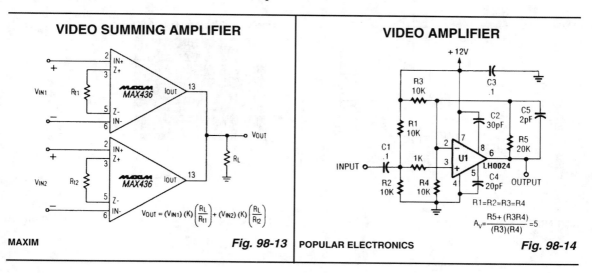

VIDEO SUMMING AMPLIFIER

$$V_{OUT} = (V_{IN1})(K)\left(\frac{R_L}{R_{t1}}\right) + (V_{IN2})(K)\left(\frac{R_L}{R_{t2}}\right)$$

MAXIM

Fig. 98-13

VIDEO AMPLIFIER

$$R1 = R2 = R3 = R4$$

$$A_V = \frac{R5 + (R3R4)}{(R3)(R4)} = 5$$

POPULAR ELECTRONICS

Fig. 98-14

TWISTED-PAIR VIDEO DRIVER/RECEIVER CIRCUIT

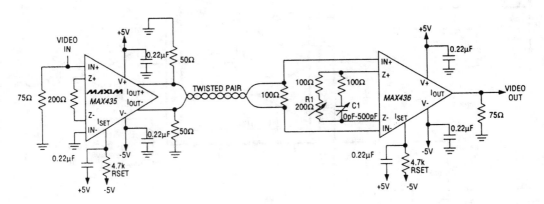

MAXIM

Fig. 98-15

This circuit should be useful where a twisted-pair video line is to be used. R1 is adjusted for proper gain (monitor brightness and contrast) and C1 is adjusted for best color.

250-mA 60-MHz CURRENT-FEEDBACK AMPLIFIER FOR VIDEO APPLICATIONS

Noninverting Amplifier with Shutdown

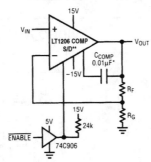

*OPTIONAL, USE WITH CAPACITIVE LOADS
**GROUND SHUTDOWN PIN FOR NORMAL OPERATION

Large-Signal Response, C_L = 10,000pF

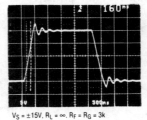

$V_S = \pm15V$, $R_L = \infty$, $R_F = R_G = 3k$

LINEAR TECHNOLOGY

Fig. 98-16

VIDEO DRIVER/AMPLIFIER

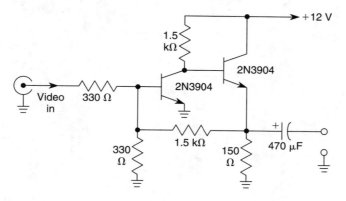

WILLIAM SHEETS

Fig. 98-17

This simple circuit has a voltage gain of about 5× and will drive low-impedance loads (75 Ω) to 1.5 V p-p or better.

VIDEO LINE DRIVER

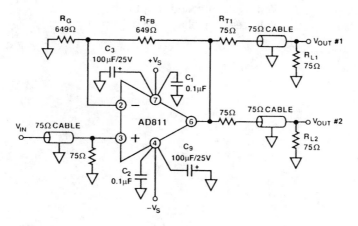

ANALOG DEVICES

Fig. 98-18

This video buffer/line driver operates at a gain of +2 and drives a pair of 75-Ω lines with 75-Ω back terminations. The overall terminated gain is unity.

99

Voltage-Controlled Amplifier Circuits

The sources of the following circuits are contained in the Sources section, which begins on page 707. The figure number in the box of each circuit correlates to the entry in the Sources section.

Voltage-Controlled Amplifier
Voltage-Controlled Audio Amplifier

VOLTAGE-CONTROLLED AMPLIFIER

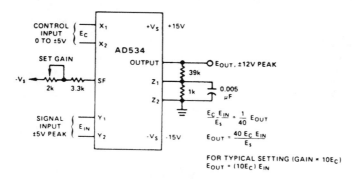

ANALOG DEVICES

Fig. 99-1

A constant or varying signal applied to the X input, E_c, controls the gain for a constant or variable signal applied to the Y input, E_{in}. The inputs could be interchanged.

For this circuit, the "set gain" potentiometer is typically adjusted to provide a calibration for gain of Z 10 per-V-of-E_c. The bandwidth is dc to 30 kHz, independent of the gain. The wideband noise (10 Hz to 30 kHz) is 3 mV rms, typically, corresponding to full-scale signal-to-noise of 70 dB. Noise, referred to the signal input ($E_c = \pm 5$ V) is 60 μV rms, typically.

VOLTAGE-CONTROLLED AUDIO AMPLIFIER

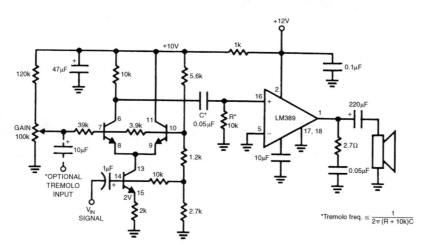

NATIONAL SEMICONDUCTOR

Fig. 99-2

The LM389 has internal transistors used in this circuit.

100

Voltage-Controlled Oscillator Circuits

The sources of the following circuits are contained in the Sources section, which begins on page 707. The figure number in the box of each circuit correlates to the entry in the Sources section.

Three-Decade VCO
Voltage-Controlled Two-Phase Oscillator

THREE-DECADE VCO

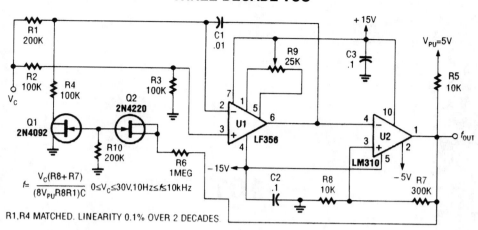

$$f = \frac{V_c(R8+R7)}{(8V_{PU}R8R1)C} \quad 0 \leq V_c \leq 30V, 10Hz \leq f \leq 10kHz$$

R1,R4 MATCHED. LINEARITY 0.1% OVER 2 DECADES.

POPULAR ELECTRONICS

Fig. 100-1

A range of 10 Hz to 10 kHz is covered by this circuit.

VOLTAGE-CONTROLLED TWO-PHASE OSCILLATOR

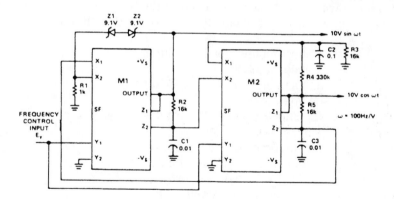

ANALOG DEVICES

Fig. 100-2

This circuit uses two multipliers for integration-with-controllable-time-constants in a feedback loop. R2 and R5 will be recognized in the AD534 voltage-to-current configuration; the currents are integrated in C1 and C3, and the voltages they develop are connected at high impedance in proper polarity to the X inputs of the "next" AD534. The frequency-control input, EY, varies the integrator gains, with a sensitivity of 100 Hz/V, and frequency error typically less than 0.1% of full scale from 0.1 V to 10 V (10 Hz to 1 kHz). C2 (proportional to C1 and C3), R3, R4 provide regenerative damping to start and maintain oscillation. Z_1 and Z_2 stabilize the amplitude at low distortion by degenerative damping above ±10 V.

101

Voltage-Measuring Circuits

The sources of the following circuits are contained in the Sources section, which begins on page 706. The figure number in the box of each circuit correlates to the entry in the Sources section.

METER AMPLIFIER FOR 1.5-V SUPPLY

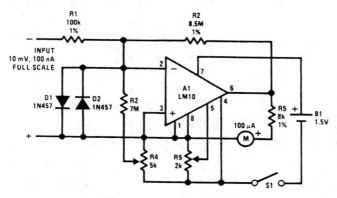

NATIONAL SEMICONDUCTOR

Fig. 101-1

An LM10 is used as a meter amplifier. Accuracy can be maintained over a 15°C to 55°C range for a full-scale sensitivity of 10 mV and 100 nA. The offset voltage error is nulled with R5, and the bias current can be balanced out with R4. The zeroing circuits operate from the reference output and are essentially unaffected by changes in battery voltage, so frequent adjustments should not be necessary. Total current drain is under 0.5 mA, giving an approximate life of 3 to 6 months with an "AA" cell and over a year with a "D" cell. With these lifetimes, an ON/OFF switch might be unnecessary. A test switch that converts to a battery-test mode might be of greater value.

VOLTAGE MONITOR

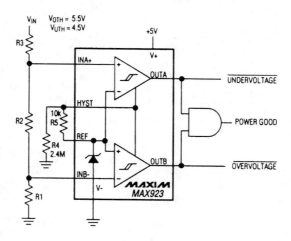

MAXIM

Fig. 101-2

A MAX923 dual comparator is used as a window detector. For a threshold of 4.5 V and 5.5 V, $R_1 = 1.068$ MΩ, $R_2 = 61.9$ kΩ, and $R_3 = 1$ MΩ.

ac VOLTMETER HAS UNIQUE FEATURES

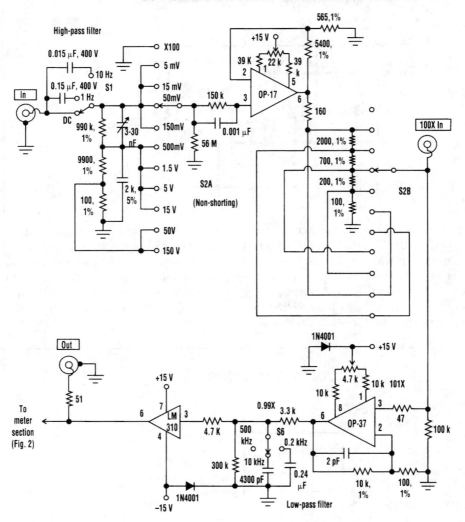

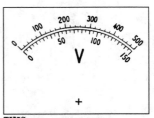

THIS meter scale can be enlarged
on a copier and attached to a meter face
for the dual-scale ac voltmeter.

Fig. 101-3

ac VOLTMETER HAS UNIQUE FEATURES (*Cont.*)

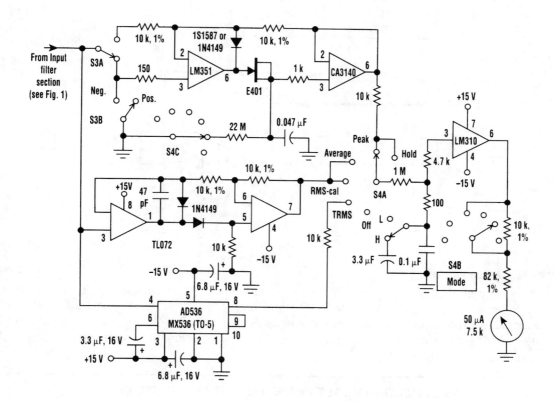

Though it's built with standard components, this ac voltmeter contains many features not typically found in commercial meters; the most unusual is a selection of rectification modes. The meter responses available include true RMS (TRMS), average, RMS-calibrated average responding, positive peak, negative peak, positive-peak hold, and negative-peak hold.

High- and low-pass filters (S1 and S6, respectively) allow the –3-dB-passband to be varied from as little as 10 Hz to 200 Hz, to as wide as dc to 500 kHz. The low-pass filter also is effective in the 100× amplifier mode, where the input equivalent noise level is only 0.3 µV, with 10-kHz roll-off.

dc VOLTMETER

Resistance Values for a DC Voltmeter

V FULL SCALE	R_V [Ω]	R_f [Ω]	R'_f [Ω]
10 mV	100k	1.5M	1.5M
100 mV	1M	1.5M	1.5M
1V	10M	1.M	1.5M
10V	10M	300k	0
100V	10M	30k	0

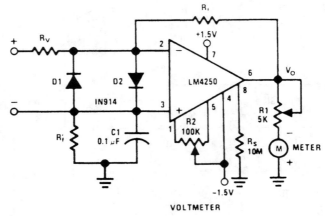

NATIONAL SEMICONDUCTOR

Fig. 101-4

A wide-range voltmeter circuit. This inverting amplifier has a gain varying from –30 for the 10-mV full-scale range to –0.003 for the 100-V full-scale range. Diodes D1 and D2 provide complete amplifier protection for input overvoltages as high as 500 V on the 10-mV range, but if overvoltages of this magnitude are expected under continuous operation, the power rating of R_v should be adjusted accordingly.

EXPANDED-SCALE dc METER FOR 12-V SYSTEMS

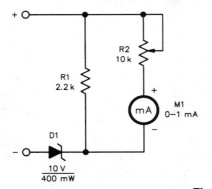

This circuit can be used to monitor a 12-V system with a meter reading 10 V to another voltage. Expect 10 to 20 V, depending on the setting of R2. Depending on the characteristics of D1, R1 might be increased or eliminated entirely.

QST

Fig. 101-5

SIMPLE 3-DIGIT DVM

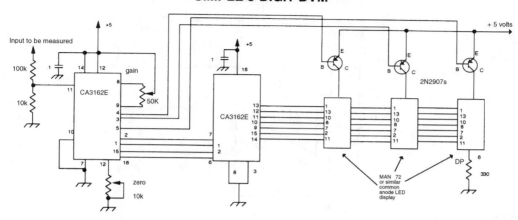

73 AMATEUR RADIO TODAY

Fig. 101-6

A CA3162ZE A-D converter drives a CA3161 BCD decoder/driver and LED display to form a simple DVM circuit. The 50-kΩ gain control and 100-kΩ/10-kΩ voltage divider determine full-scale range.

INEXPENSIVE VOLTAGE CALIBRATOR

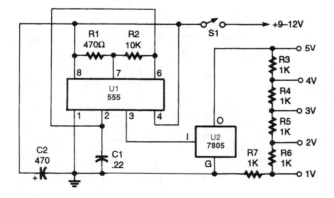

POPULAR ELECTRONICS

Fig. 101-7

In the voltage calibrator, two low-cost ICs—a 555 oscillator/timer and a 78055 5-V 1.5-A voltage regulator—along with a precision voltage divider network are used to provide outputs of 1- to 5-V peak-to-peak.

DOUBLE-ENDED VOLTAGE MONITOR

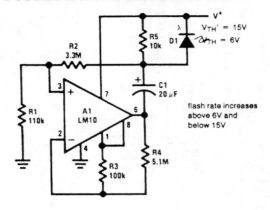

NATIONAL SEMICONDUCTOR *Fig. 101-8*

This circuit has the added feature that it can sense an over-voltage condition. The lower activation threshold is given by equation (1), but above a threshold,

$$V_{TH} = \frac{R4(R_1 + R_2)\, V_{REF}}{R_1(R_3 + R_4) - R3(R_1 + R_2)}$$

oscillation again ceases. Below V_{TH}, the op amp output is saturated negative, but above V_{TH}, it is saturated positive. The flash rate approaches zero near either limit.

AUDIBLE VOLTAGE INDICATOR

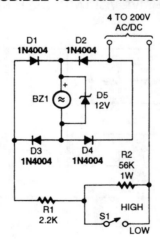

POPULAR ELECTRONICS *Fig. 101-9*

The audible voltmeter can be used to test for ac or dc voltages in a circuit. With S1 closed, the circuit can be used to test for voltages between 4 and 24 V, and when S1 is open, it can be used to check for the presence voltages of up to 200 V.

LOW-DRAIN METER AMPLIFIER

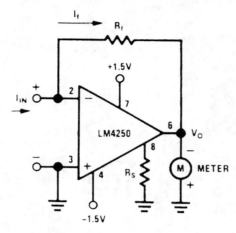

NATIONAL SEMICONDUCTOR   *Fig. 101-10*

   Meter amplifiers normally require one or two 9-V transistor batteries. Because of the heavy current drain on these supplies, the meters must be switched to the OFF position when not in use. The meter circuit described here operates on two 1.5-V flashlight batteries and has a quiescent power drain so low that no on/off switch is needed. A pair of Eveready No. 950 "D" cells will serve for a minimum of one year without replacement. As a dc ammeter, the circuit will provide current ranges as low as 100 nA full-scale.

   The basic meter amplifier circuit shown is a current-to-voltage converter. Negative feedback around the amplifier ensures that currents I_{IN} and I_f are always equal, and the high gain of the op amp ensures that the input voltage between pins 2 and 3 is in the microvolt region. Output voltage V_o is therefore equal to $-I_f R_f$. Considering the ±1.5-V sources (±1.2 V end of life) a practical value of V_o for full-scale meter deflection is 300 mV. With the master bias-current setting resistor (R_s) set at 10 MΩ, the total quiescent current drain of the circuit is 0.6 μA for a total power supply drain of 1.8 μW. The input bias current, required by the amplifier at this low level of quiescent current, is in the range of 600 pA.

102

Waveform Generator Circuits

The sources of the following circuits are contained in the Sources section, which begins on page 707. The figure number in the box of each circuit correlates to the entry in the Sources section.

SIMPLE TRIANGLE WAVEFORM GENERATOR

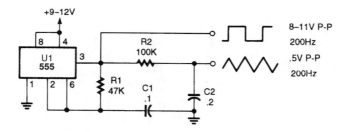

POPULAR ELECTRONICS

Fig. 102-1

The circuit is a triangle waveform-generator circuit that uses as few parts as possible. A 555 timer IC, two resistors, and two capacitors make the triangle waveform. The IC is connected in a 50% duty-cycle astable square-wave oscillator circuit. The square-wave output is fed from pin 3 of the IC to an RC shaping circuit.

When the 555's square-wave output goes high, C2 begins to charge through R2 and the voltage across C2 increases as long as the output remains high. When the IC's output goes low again, C2 begins to discharge through R2 reducing the voltage across C2 as long as the output remains low. The resulting waveform across C2 takes the shape of a triangle. The best waveform linearity is obtained when R_2 and C_2 are made as large as possible. With the component values shown, the peak-to-peak output is 0.5 V at a frequency of about 200 Hz.

TRIANGLE-WAVE GENERATOR

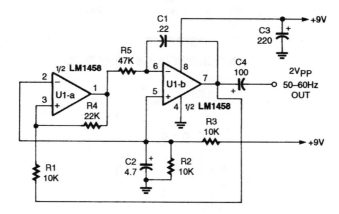

POPULAR ELECTRONICS

Fig. 102-2

This oscillator, which is built around an LM1458 dual op amp and a few inexpensive components, produces a 2-V peak-to-peak, triangular waveform.

697

GENERATE ACCURATE PWM SIGNALS

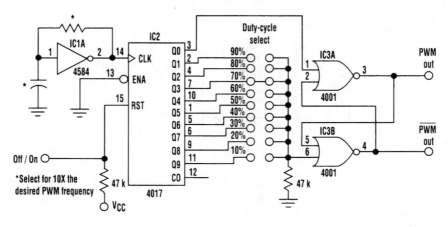

ELECTRONIC DESIGN

Fig. 102-3

Accurate 10 to 90% duty-cycle PWM signals can be generated using this simple circuit setup. The desired duty cycle is selected by a single jumper block. PWM clock IC1 runs at 10× the desired pulse drive frequency. IC2, a 4017 divide-by-10 counter, decodes the clock pulses into one of 10 outputs. Output 0 resets IC3, the PWM latch. The latch stays reset until the desired duty-cycle output set by the jumper block is reached. At this point, the PWM latch is set, and the PWM output line remains high until output 0 is decoded again.

By calling IC2's output (0) the "reset" line for the latch, the PWM output is forced inactive if the jumper strap is removed to change duty cycles without first powering down.

Using the zero-state reset allows IC2's reset pin to be used as an on/off control line for the circuit. The complementary PWM output could be used in a full bridge design.

TRIANGLE-WAVE GENERATOR

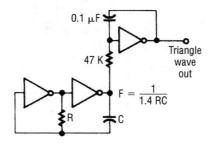

$$F = \frac{1}{1.4\,RC}$$

ELECTRONICS NOW

Fig. 102-4

The first two gates are set up as a square-wave oscillator, and the last one makes the conversion to triangle waves.

LOW-FREQUENCY PULSE GENERATOR

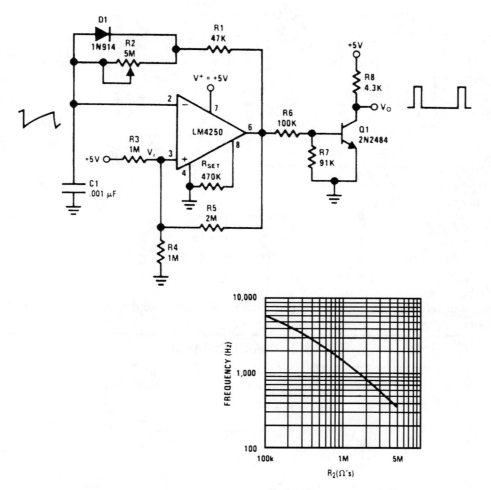

Pulse frequency vs. R_2

NATIONAL SEMICONDUCTOR

Fig. 102-5

SINE/COSINE AUDIO GENERATOR FOR GALVANOMETER EXPERIMENTS

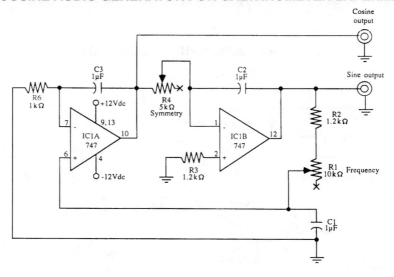

C1 – C3   1 μF polarized electrolytic
IC1   LM747 dual op amp IC
J1,J2   1/8-inch miniature phone jack

All resistors are 5 to 10 percent tolerance, 1/4 watt.
All capacitors are 10 to 20 percent tolerance, rated
at 35 volts or more.

McGRAW-HILL   **Fig. 102-6**

This circuit shows how to implement a sine/cosine audio generator for operating two galvanometers.

BASIC 555 MONOSTABLE

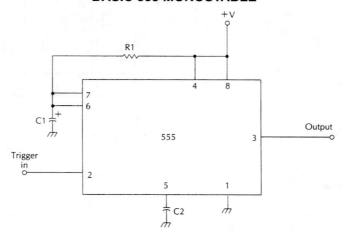

McGRAW-HILL   **Fig. 102-7**

$$T + 1.1\,R_1 C_1$$

OP-AMP SAWTOOTH GENERATOR

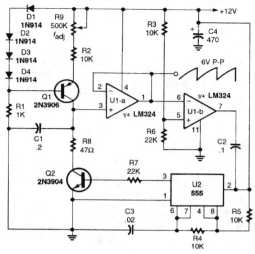

POPULAR ELECTRONICS
Fig. 102-8

The sawtooth generator circuit shown is reset at the end of each cycle. The result is a constant peak-to-peak output throughout the circuit's frequency range.

The constant-current generator circuit, the voltage-follower circuit, and the comparator circuit produce the waveform. A 555 timer IC (U2) is configured as a one-shot multivibrator that's triggered by the comparator's negative output pulse.

DIGITAL SINE-WAVE GENERATOR

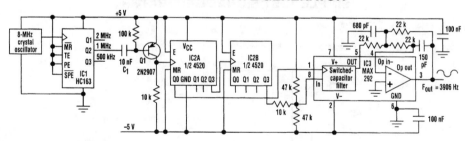

ELECTRONIC DESIGN
Fig. 102-9

The sine-wave generator starts with an 8-MHz signal and divides it by eight to obtain 1 MHz at C1 (IC1's 2-MHz and 500-kHz outputs can serve as alternate drive signals). Q1 level-shifts the 1-MHz pulses so that they can drive the bipolar circuitry necessary for producing a bipolar output. Synchronous counter IC2 divides 1 MHz by 256 to give the desired output frequency (3906 Hz), and IC3 filters the harmonic frequencies.

The filter's clock is taken from the first divide-by-2 tap of IC2 to assure a 50% duty cycle. IC2 further divides this signal by 128 to ensure that the filter's input signal (1 MHz/256) falls within the flat portion of the filter response.

The output of the switched-capacitor filter resembles a sampled sine wave. It can be smoothed by building a first- or second-order low-pass filter around the otherwise uncommitted output op amp.

SIGNAL SOURCE FOR AUDIO AMPLIFIER/INVERTER

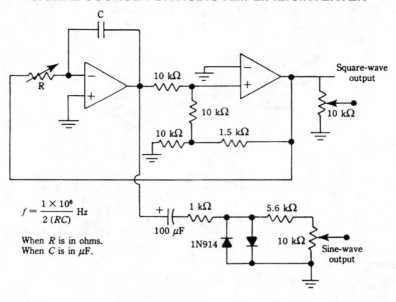

$$f = \frac{1 \times 10^6}{2\,(RC)}\ \text{Hz}$$

When R is in ohms.
When C is in μF.

McGRAW-HILL   **Fig. 102-10**

Two op amps (741, etc..) are used in this oscillator circuit. A square wave is available and a sine wave, obtained by shaping the triangle waveform, is also provided.

SIMPLE TEST SIGNAL GENERATOR

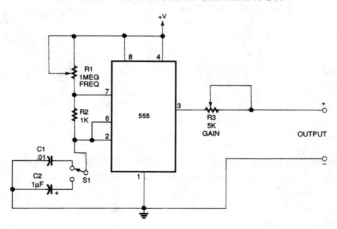

ELECTRONICS NOW   **Fig. 102-11**

An NE555 generates signals for test purposes. Frequency range is from 20 Hz to 10 kHz, depending on the setting of S1. $+V$ is 5 V.

103

Waveguide Circuits

The sources of the following circuits are contained in the Sources section, which begins on page 707. The figure number in the box of each circuit correlates to the entry in the Sources section.

10-GHz Waveguide Detector for Amateur Radio Use
10-GHz Waveguide Transition for Amateur Radio Use

10-GHz WAVEGUIDE DETECTOR FOR AMATEUR RADIO USE

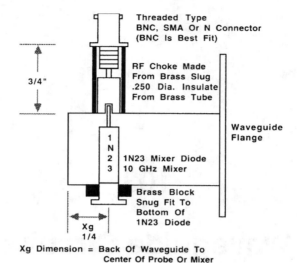

Fig. 103-1

This shows the construction of a waveguide detector for use at the 10-GHz amateur radio frequencies.

10-GHz WAVEGUIDE TRANSITION FOR AMATEUR RADIO USE

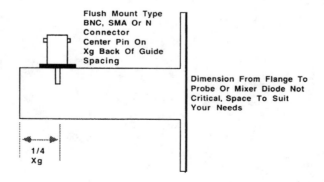

Fig. 103-2

A transistor adapts the waveguide to coaxial cable or other types of transmission lines.

104

White-Noise Generator Circuits

The sources of the following circuits are contained in the Sources section, which begins on page 707. The figure number in the box of each circuit correlates to the entry in the Sources section.

Zener-Diode White-Noise Generator
White-Noise Generator

ZENER-DIODE WHITE-NOISE GENERATOR

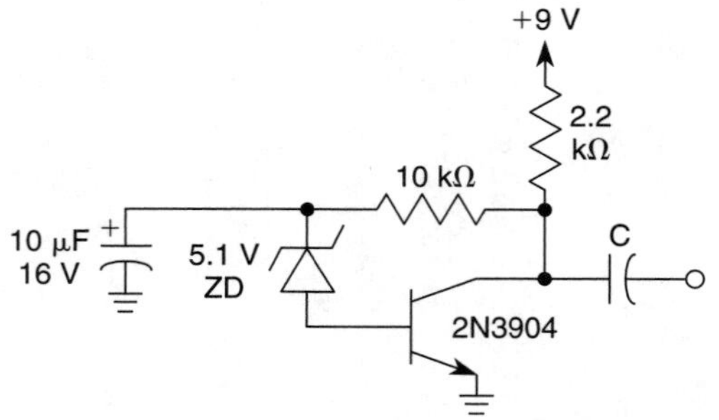

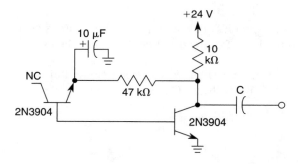

WILLIAM SHEETS

Fig. 104-1

This circuit uses a Zener diode as a noise source. C is chosen to pass the lowest-desired frequency components of the noise.

WHITE-NOISE GENERATOR

WILLIAM SHEETS

Fig. 104-2

Here, a 2N3904 E-B junction is used as a noise generator, reversed bias. C is chosen to pass the lowest-desired frequency components of the noise.

Sources

Chapter 1

Fig. 1-1. QST, 2/95, p. 58.

Fig. 1-2. Analog Devices, Analog Dialogue, Vol. 26, No. 2, p. 5.

Fig. 1-3. Reprinted with permission from Electronic Design, 2/95, p. 108. Copyright 1995, Penton Publishing, Inc.

Fig. 1-4. 1994 Experimenters Handbook, p. 113.

Chapter 2

Fig. 2-1. Reprinted with permission from Electronic Design, 10/94, p. 90. Copyright 1994, Penton Publishing, Inc.

Fig. 2-2. Reprinted with permission from Popular Electronics, 1/94, p. 37. (C) Copyright Gernsback Publications, Inc., 1994.

Fig. 2-3. Reprinted with permission from Popular Electronics, Fact Card No. 270. (C) Copyright Gernsback Publications, Inc.

Chapter 3

Fig. 3-1. Reprinted with permission from Popular Electronics, 9/94, p. 26. (C) Copyright Gernsback Publications, Inc., 1994.

Fig. 3-2. Reprinted with permission from Popular Electronics, 10/94, p. 48. (C) Copyright Gernsback Publications, Inc., 1994.

Fig. 3-3. Reprinted with permission from Popular Electronics, 10/94, p. 35. (C) Copyright Gernsback Publications, Inc., 1994.

Fig. 3-4. Reprinted with permission from Popular Electronics, 10/94, p. 49. (C) Copyright Gernsback Publications, Inc., 1994.

Fig. 3-5. Reprinted with permission from Popular Electronics, 5/95, pp. 69–70. (C) Copyright Gernsback Publications, Inc., 1995.

Fig. 3-6. Reprinted with permission from Popular Electronics, 9/94, p. 24. (C) Copyright Gernsback Publications, Inc., 1994.

Fig. 3-7. Reprinted with permission from Popular Electronics, 2/94, p. 37. (C) Copyright Gernsback Publications, Inc., 1994.

Fig. 3-8. Reprinted with permission from Popular Electronics, 5/95, p. 70. (C) Copyright Gernsback Publications, Inc., 1995.

Fig. 3-9. Reprinted with permission from Popular Electronics, 9/94, p. 25 (C)

Copyright Gernsback Publications, Inc., 1994.

Fig. 3-10. Reprinted with permission from Popular Electronics, 4/95, p. 70. (C) Copyright Gernsback Publications, Inc., 1995.

Fig. 3-11. Reprinted with permission from Popular Electronics, 5/95, p. 71. (C) Copyright Gernsback Publications, Inc., 1995.

Chapter 4

Fig. 4-1. QST, 12/94, p. 36.

Fig. 4-2. QST, 12/94, p. 35.

Fig. 4-3. QST, 12/94, p. 34.

Fig. 4-4. QST, 5/89, pp. 25–27.

Fig. 4-5. 73 Amateur Radio Today, 5/94, pp. 58–59.

Fig. 4-6. 73 Amateur Radio Today, 8/89, p. 60.

Fig. 4-7. QST, 4/95, p. 56.

Fig. 4-8. QST, 7/94, p. 24.

Fig. 4-9. QST, 3/95, p. 28.

Fig. 4-10. QST, 10/94, p. 65.

Fig. 4-11. 73 Amateur Radio Today, 1/95, p. 32.

Fig. 4-12. 73 Amateur Radio Today, 6/83, p. 99.

Fig. 4-13. 73 Amateur Radio Today, 10/94, p. 14.

Fig. 4-14. 73 Amateur Radio Today, 5/94, p. 10.

Fig. 4-15. William Sheets.

Fig. 4-16. QST, 4/95, p. 61.

Fig. 4-17. QST, 10/94, p. 42.

Fig. 4-18. 73 Amateur Radio Today, 6/94, p. 48.

Fig. 4-19. 73 Amateur Radio Today, 6/94, pp. 32–34.

Chapter 5

Fig. 5-1. Reprinted with permission from Electronics Now, 7/94, p. 34. (C) Copyright Gernsback Publications, Inc., 1994.

Fig. 5-2. Reprinted with permission from Electronics Now, 7/94, p. 39. (C) Copyright Gernsback Publications, Inc., 1994.

Fig. 5-3. Reprinted with permission from Electronics Now, 7/94, p. 36. (C) Copyright Gernsback Publications, Inc., 1994.

Fig. 5-4. William Sheets.

Fig. 5-5. 73 Amateur Radio Today, 9/94, p. 62.

Fig. 5-6. QST, 11/94, p. 23.

Fig. 5-7. QST, 12/94, p. 28.

Fig. 5-8. Rudolf F. Graf and William Sheets.

Fig. 5-9. Reprinted with permission from Radio-Electronics, September 1992, p. 79. (C) Copyright Gernsback Publications, Inc., 1994.

Chapter 6

Fig. 6-1. Reprinted with permission from Electronic Design, 11/93, p. 102. Copyright 1993, Penton Publishing, Inc.

Fig. 6-2. Reprinted with permission from Radio-Electronics, Experimenters Handbook, p. 4. (C) Copyright Gernsback Publications, Inc., 1994.

Fig. 6-3. Maxim, Vol. III, New Releases Data Book, 1994, p. 8-17.

Fig. 6-4. Maxim, Vol. III, New Releases Data Book, 1994, p. 8-17.

Fig. 6-5. Reprinted with permission from Electronics Now, 2/94, p. 75. (C) Copyright Gernsback Publications, Inc., 1994.

Fig. 6-6. Reprinted with permission of National Semiconductor Corporation, National Semiconductor Linear Applications Handbook 1991, p. 492.

Fig. 6-7. Amplifiers, Waveform Generators & Other Low-Cost IC Projects, McGraw-Hill, pp. 104–105.

Fig. 6-8. Amplifiers, Waveform Generators & Other Low-Cost IC Projects, McGraw-Hill, p. 107.

Fig. 6-9. Maxim, Vol. III, New Releases Data Book, 1994, p. 3-30.

Fig. 6-10. Maxim, Vol. III, New Releases Data Book, 1994, p. 8-18.

Fig. 6-11. Linear Technology, 2/95.

Fig. 6-12. Maxim, Vol. III, New Releases Data Book, 1994, p. 8-18.

Fig. 6-13. Linear Technology, 2/95.

Fig. 6-14. Reprinted with permission of National Semiconductor Corporation, National Semiconductor Linear Applications Handbook 1991, p. 594.

Fig. 6-15. Power Supplies, Switching Regulators, Inverters, and Converters, McGraw-Hill, pp. 127–128.

Fig. 6-16. Reprinted with permission of National Semiconductor Corporation, National Semiconductor Linear Applications Handbook 1991, p. 501.

Fig. 6-17. Reprinted with permission from Electronics Now, 6/94, p. 12. (C) Copyright Gernsback Publications, Inc., 1994.

Fig. 6-18. Maxim, Vol. III, New Releases Data Book, 1994, p. 8-19.

Chapter 7

Fig. 7-1. Analog Devices, AD8001 Data Sheet.

Fig. 7-2. Reprinted with permission from Popular Electronics, 4/94, p. 47. (C) Copyright Gernsback Publications, Inc., 1994.

Chapter 8

Fig. 8-1. 73 Amateur Radio Today, 12/93, p. 32.

Fig. 8-2. Reprinted with permission from Electronic Design, 2/95, p. 107. Copyright 1995, Penton Publishing, Inc.

Fig. 8-3. Reprinted with permission from Electronics Now, 11/93, p. 31. (C) Copyright Gernsback Publications, Inc., 1993.

Fig. 8-4. Reprinted with permission from Electronics Now, 5/94, p. 49. (C) Copyright Gernsback Publications, Inc., 1994.

Fig. 8-5. QST, 5/95, p. 35.

Fig. 8-6. QST, 5/95, p. 35.

Fig. 8-7. William Sheets.

Chapter 9

Fig. 9-1. QST, 10/94, p. 47.

Fig. 9-2. Reprinted with permission from Popular Electronics, 12/94, p. 88. (C) Copyright Gernsback Publications, Inc., 1994.

Chapter 10

Fig. 10-1. Reprinted with permission from Electronics Now, 6/94, p. 34. (C) Copyright Gernsback Publications, Inc., 1994.

Fig. 10-2. Reprinted with permission from Electronics Now, 5/93, p. 65. (C) Copyright Gernsback Publications, Inc., 1993.

Fig. 10-3. William Sheets.

Fig. 10-4. William Sheets.

Fig. 10-5. Reprinted with permission from Popular Electronics, 12/94, p. 31. (C) Copyright Gernsback Publications, Inc., 1994.

Fig. 10-6. William Sheets.

Fig. 10-7. William Sheets.

Fig. 10-8. Reprinted with permission from Electronics Now, 8/94, p. 12. (C) Copyright Gernsback Publications, Inc., 1994.

Fig. 10-9. Analog Devices, The Best of Analog Dialogue, 1967–1992, p. 180.

Fig. 10-10. Analog Devices, Analog Dialogue, Vol. 27, No. 2 (1993), p. 22.

Fig. 10-11. Reprinted with permission from Popular Electronics, Fact Card No. 229. (C) Copyright Gernsback Publications, Inc.

Fig. 10-12. William Sheets.

Fig. 10-13. William Sheets.

Fig. 10-14. William Sheets.

Fig. 10-15. William Sheets.

Fig. 10-16. Analog Devices, The Best of Analog Dialogue, 1967–1991, p. 180.

Fig. 10-17. William Sheets.

Fig. 10-18. William Sheets.

Fig. 10-19. Gordon McComb's Gadgeteer's Goldmine, McGraw-Hill, p. 329.

Chapter 11

Fig. 11-1. William Sheets.

Fig. 11-2. Reprinted with permission of National Semiconductor Corporation, National Semiconductor Linear Applications Handbook 1991, p. 1058.

Fig. 11-3. William Sheets.

Fig. 11-4. Reprinted with permission of National Semiconductor Corporation, National Semiconductor Linear Applications Handbook 1991, p. 1060.

Fig. 11-5. Reprinted with permission of National Semiconductor Corporation, National Semiconductor Linear Applications Handbook 1991, p. 253.

Fig. 11-6. William Sheets.

Fig. 11-7. Reprinted with permission from Popular Electronics, 1/95, p. 54. (C) Copyright Gernsback Publications, Inc., 1995.

Fig. 11-8. Reprinted with permission from

Popular Electronics, 1/95, p. 54. (C) Copyright Gernsback Publications, Inc., 1995.

Fig. 11-9. Reprinted with permission of National Semiconductor Corporation, National Semiconductor Linear Application Specific IC's Databook 1993, p. 1-14.

Fig. 11-10. William Sheets.

Fig. 11-11. William Sheets.

Fig. 11-12. Reprinted with permission of National Semiconductor Corporation, National Semiconductor Linear Applications Handbook 1991, p. 199.

Fig. 11-13. William Sheets.

Fig. 11-14. Reprinted with permission of National Semiconductor Corporation, National Semiconductor Linear Applications Handbook, 1991, p. 198.

Fig. 11-15. Reprinted with permission of National Semiconductor Corporation, National Semiconductor Linear Edge, #8, p. 14.

Fig. 11-16. William Sheets.

Fig. 11-17. QST, 12/94, p. 44.

Fig. 11-18. Reprinted with permission of National Semiconductor Corporation, National Semiconductor Linear Applications Handbook 1991, p. 1059.

Fig. 11-19. Reprinted with permission of National Semiconductor Corporation, National Semiconductor Linear Applications Handbook 1991, p. 1059.

Fig. 11-20. Reprinted with permission from Popular Electronics, 5/94, p. 80. (C) Copyright Gernsback Publications, Inc., 1994.

Chapter 12

Fig. 12-1. Reprinted with permission from Popular Electronics, 4/95, p. 47. (C) Copyright Gernsback Publications, Inc., 1995.

Fig. 12-2. Reprinted with permission from Popular Electronics, 3/95, p. 67. (C) Copyright Gernsback Publications, Inc., 1995.

Fig. 12-3. William Sheets.

Fig. 12-4. William Sheets.

Fig. 12-5. Reprinted with permission from Popular Electronics, 4/95, p. 70. (C) Copyright Gernsback Publications, Inc., 1995.

Fig. 12-6. Electronics Now, 4/94, p. 25.

Fig. 12-7. Reprinted with permission from Popular Electronics, 5/95, p. 69. (C) Copyright Gernsback Publications, Inc., 1995.

Fig. 12-8. Reprinted with permission from Radio-Electronics, June 1984, p. 39. (C) Copyright Gernsback Publications, Inc., 1984.

Fig. 12-9. Reprinted with permission from Radio-Electronics, 1994 Electronics Experimenters Handbook, p. 37. (C) Copyright Gernsback Publications, Inc., 1994.

Fig. 12-10. William Sheets.

Fig. 12-11. Power Supplies, Switching Regulators, Inverters, and Converters, McGraw-Hill, p. 104.

Fig. 12-12. Reprinted with permission from Popular Electronics, 6/95, p. 32. (C) Copyright Gernsback Publications, Inc., 1995.

Fig. 12-13. Reprinted with permission from Popular Electronics, 4/95, p. 68. (C) Copyright Gernsback Publications, Inc., 1995.

Fig. 12-14. Reprinted with permission from Popular Electronics, 4/95, p. 68. (C) Copyright Gernsback Publications, Inc., 1995.

Fig. 12-15. Electronics Now, 4/95, p. 18.

Fig. 12-16. Reprinted with permission from Popular Electronics, 6/95, p. 30. (C) Copyright Gernsback Publications, Inc., 1995.

Chapter 13

Fig. 13-1. Reprinted with permission from Electronics Now, 5/95, p. 65. (C) Copyright Gernsback Publications, Inc., 1995.

Fig. 13-2. Reprinted with permission from Electronics Now, 10/94, pp. 65–66. (C) Copyright Gernsback Publications, Inc., 1994.

Fig. 13-3. William Sheets.

Fig. 13-4. Reprinted with permission from

Electronic Design, 6/94, pp. 42–43. Copyright 1994, Penton Publishing, Inc.

Fig. 13-5. Reprinted with permission from Popular Electronics, 6/95, p. 32. (C) Copyright Gernsback Publications, Inc., 1995.

Fig. 13-6. Reprinted with permission from Electronic Design, 6/94, p. 42. Copyright 1994, Penton Publishing, Inc.

Fig. 13-7. Linear Technology, Design Note #98.

Fig. 13-8. 73 Amateur Radio Today, 6/93, p. 41.

Fig. 13-9. 73 Amateur Radio Today, 6/93, p. 35.

Fig. 13-10. Reprinted with permission from Electronic Design, 6/94, p. 42. Copyright 1994, Penton Publishing, Inc.

Fig. 13-11. Reprinted with permission from Popular Electronics, 6/93, p. 76. (C) Copyright Gernsback Publications, Inc., 1993.

Fig. 13-12. 73 Amateur Radio Today, 5/93, p. 69.

Chapter 14

Fig. 14-1. Reprinted with permission from Electronic Design, 1/95, pp. 81–82. Copyright 1995, Penton Publishing, Inc.

Fig. 14-2. Spring 1994 Electronics Hobbyist Handbook.

Fig. 14-3. Reprinted with permission of National Semiconductor Corporation, National Semiconductor Linear Applications Handbook 1991, p. 499.

Fig. 14-4. Reprinted with permission from Popular Electronics, 3/92, p. 74. (C) Copyright Gernsback Publications, Inc., 1992.

Fig. 14-5. Maxim, Vol. III, New Releases Data Book, 1994, p. 529.

Fig. 14-6. Reprinted with permission from Radio-Electronics, June 1984, p. 90.

Fig. 14-7. Reprinted with permission of National Semiconductor Corporation, National Semiconductor Linear Applications Handbook 1991, p. 498.

Fig. 14-8. QST, 10/94, p. 75.

Fig. 14-9. Laser Cookbook, McGraw-Hill, p. 183.

Chapter 15

Fig. 15-1. Reprinted with permission of Burr-Brown Corporation, Burr-Brown Data Sheet INA118, (C) 1989–1995 Burr-Brown Corporation.

Fig. 15-2. Reprinted with permission from Popular Electronics, 3/94, p. 83. (C) Copyright Gernsback Publications, Inc., 1994.

Fig. 15-3. Reprinted with permission from Popular Electronics, Fact Card No. 248. (C) Copyright Gernsback Publications, Inc.

Chapter 16

Fig. 16-1. Analog Devices, The Best of Analog Dialogue, 1967–1991, p. 180.

Fig. 16-2. William Sheets.

Fig. 16-3. Analog Devices, Analog Dialogue, Vol. 26, No. 2, 1992, p. 17.

Fig. 16-4. Analog Devices, Analog Dialogue, Vol. 26, No. 1, 1992, p. 12.

Fig. 16-5. Reprinted with permission of National Semiconductor Corporation, National Semiconductor Linear Applications Handbook 1991, p. 542.

Fig. 16-6. Analog Devices, The Best of Analog Dialogue, 1967–1991, p. 117.

Fig. 16-7. Reprinted with permission of National Semiconductor Corporation, National Semiconductor Linear Applications Handbook 1991, p. 543.

Fig. 16-8. Reprinted with permission from Electronic Design, 8/94, p. 104. Copyright 1994, Penton Publishing, Inc.

Fig. 16-9. Linear Technology, Design Note #89.

Chapter 17

Fig. 17-1. Reprinted with permission from Electronic Design, 11/94, p. 115. Copyright 1994, Penton Publishing, Inc.

Fig. 17-2. Reprinted with permission from Electronics Now, 3/95, p. 8. (C) Copyright Gernsback Publications, Inc., 1995.

Chapter 18

Fig. 18-1. Reprinted with permission from Electronic Design, 4/89, p. 108. Copyright 1989, Penton Publishing, Inc.

Fig. 18-2. Reprinted with permission from Electronic Design, 10/94, pp. 107–108. Copyright 1994, Penton Publishing, Inc.

Fig. 18-3. 1994 Electronic Experimenters Handbook, p. 99.

Fig. 18-4. Reprinted with permission from Electronic Design, 12/94, p. 115. Copyright 1994, Penton Publishing, Inc.

Fig. 18-5. Maxim, Vol. III, New Releases Data Book, 1994, p. 5-17.

Fig. 18-6. Reprinted with permission from Electronic Design, 10/94, p. 101. Copyright 1994, Penton Publishing, Inc.

Fig. 18-7. Maxim, Vol. III, New Releases Data Book, 1994, p. 4-107.

Fig. 18-8. Reprinted with permission from Electronic Design, 1/84, p. 440. Copyright 1984, Penton Publishing, Inc.

Fig. 18-9. Reprinted with permission from Electronics Now, 5/95, p. 44. (C) Copyright Gernsback Publications, Inc., 1994.

Chapter 19

Fig. 19-1. Reprinted with permission from Electronic Design, 3/95, p. 111. Copyright 1995, Penton Publishing, Inc.

Fig. 19-2. Reprinted with permission from Electronics Now, 8/94, p. 8. (C) Copyright Gernsback Publications, Inc., 1994.

Fig. 19-3. Reprinted with permission from Electronics Now, 4/95, p. 83. (C) Copyright Gernsback Publications, Inc., 1995.

Fig. 19-4. Sound Light and Music, Delton T. Horn, McGraw-Hill, p. 168.

Chapter 20

Fig. 20-1. Radio Craft, 1993, p. 64.

Fig. 20-2. William Sheets.

Fig. 20-3. Power Supplies, Switching Regulators, Inverters, and Converters, McGraw-Hill, p. 97.

Fig. 20-4. Analog Devices, Analog Dialogue, Vol. 27, No. 2 (1993), p. 21.

Fig. 20-5. Power Supplies, Switching Regulators, Inverters, and Converters, McGraw-Hill, p. 103.

Fig. 20-6. William Sheets.

Fig. 20-7. Power Supplies, Switching Regulators, Inverters, and Converters, McGraw-Hill, pp. 170–173.

Fig. 20-8. Maxim, Vol. III, New Releases Data Book, 1994, p. 4-79.

Fig. 20-9. Reprinted with permission from Electronic Design, 11/93, p. 89. Copyright 1995, Penton Publishing, Inc.

Fig. 20-10. Reprinted with permission of Burr-Brown Corporation, Burr-Brown Data Sheet INA118, (C) 1989–1995 Burr-Brown Corporation.

Fig. 20-11. Reprinted with permission from Radio-Electronics Experimenters Handbook, p. 76. (C) Copyright Gernsback Publications, Inc., 1994.

Fig. 20-12. Reprinted with permission from Radio-Electronics Experimenters Handbook, p. 76. (C) Copyright Gernsback Publications, Inc., 1994.

Fig. 20-13. Amplifiers, Waveform Generators & Other Low-Cost IC Projects, McGraw-Hill, pp. 65–66.

Fig. 20-14. Reprinted with permission from Electronic Design, 3/95, p. 96. Copyright 1995, Penton Publishing, Inc.

Fig. 20-15. Reprinted with permission from Electronic Design, 3/95, pp. 94–96. Copyright 1995, Penton Publishing, Inc.

Fig. 20-16. Analog Devices, The Best of Analog Dialogue, 1967–1991, p. 75.

Chapter 21

Fig. 21-1. Reprinted with permission from Popular Electronics, 2/94, p. 80. (C) Copyright Gernsback Publications, Inc., 1994.

Fig. 21-2. Reprinted with permission from Popular Electronics, 2/94, p. 81. (C) Copyright Gernsback Publications, Inc., 1994.

Fig. 21-3. Reprinted with permission from Popular Electronics, 2/94, p. 81. (C) Copyright Gernsback Publications, Inc., 1994.

Fig. 21-4. Reprinted with permission from Popular Electronics, 2/94, p. 80. (C) Copyright Gernsback Publications, Inc., 1994.

Fig. 21-5. William Sheets.

Fig. 21-6. Reprinted with permission from Popular Electronics, 2/94, p. 90. (C) Copyright Gernsback Publications, Inc., 1994.

Fig. 21-7. William Sheets.

Fig. 21-8. Reprinted with permission from Radio-Electronics Experimenters Handbook, p. 75. (C) Copyright Gernsback Publications, Inc., 1994.

Fig. 21-9. Reprinted with permission from Radio-Electronics Experimenters Handbook, p. 76. (C) Copyright Gernsback Publications, Inc., 1994.

Fig. 21-10. Radio Craft, 1993, p. 63.

Fig. 21-11. Reprinted with permission from Radio-Electronics Experimenters Handbook, p. 76. (C) Copyright Gernsback Publications, Inc., 1994.

Chapter 22

Fig. 22-1. Reprinted with permission from Popular Electronics, 5/95, pp. 30–31. (C) Copyright Gernsback Publications, Inc., 1995.

Fig. 22-2. Reprinted with permission of National Semiconductor Corporation, National Semiconductor Linear Applications Handbook 1991, p. 245.

Fig. 22-3. Reprinted with permission of National Semiconductor Corporation, National Semiconductor Linear Applications Handbook 1991, p. 245.

Fig. 22-4. Reprinted with permission of National Semiconductor Corporation, National Semiconductor Linear Applications Handbook, 1991, p. 245.

Chapter 23

Fig. 23-1. 73 Amateur Radio Today, 3/95, p. 62.

Fig. 23-2. William Sheets.

Fig. 23-3. Linear Technology, Design Note #86.

Fig. 23-4. Maxim, Vol. III, New Releases Data Book, 1994, p. 4-92.

Fig. 23-5. Reprinted with permission from Popular Electronics, 1/94, p. 73. (C) Copyright Gernsback Publications, Inc., 1994.

Chapter 24

Fig. 24-1. Reprinted with permission from Electronics Now, 11/93, p. 53. (C) Copyright Gernsback Publications, Inc., 1993.

Fig. 24-2. Reprinted with permission of National Semiconductor Corporation, National Semiconductor Linear Applications Handbook 1991, p. 300.

Fig. 24-3. William Sheets.

Fig. 24-4. Reprinted with permission from Electronics Now, 3/95, p. 86. (C) Copyright Gernsback Publications, Inc., 1995.

Fig. 24-5. Reprinted with permission from Electronics Now, 3/95, p. 86. (C) Copyright Gernsback Publications, Inc., 1995.

Chapter 25

Fig. 25-1. Reprinted with permission of National Semiconductor Corporation, National Semiconductor Linear Applications Handbook 1991, p. 267.

Fig. 25-2. Reprinted with permission from Electronics Now, 8/93, p. 12. (C) Copyright Gernsback Publications, Inc., 1993.

Chapter 26

Fig. 26-1. Reprinted with permission of National Semiconductor Corporation, National Semiconductor Linear Applications Handbook, 1991, p. 541.

Fig. 26-2. Reprinted with permission from Electronic Design, 12/94, p. 129. Copyright 1994, Penton Publishing, Inc.

Fig. 26-3. 1994 Analog Application Issue, Electronic Design, June 27, 1994.

Fig. 26-4. Reprinted with permission from Popular Electronics, 10/94, p. 82. (C) Copyright Gernsback Publications, Inc., 1994.

Fig. 26-5. Reprinted with permission from Popular Electronics, 12/94, p. 30. (C) Copyright Gernsback Publications, Inc., 1994.

Fig. 26-6. Reprinted with permission from Electronics Now, 3/94, p. 68. (C) Copyright Gernsback Publications, Inc., 1994.

Fig. 26-7. Reprinted with permission of National Semiconductor Corporation, National Semiconductor Linear Applications Handbook, 1991, p. 271.

Fig. 26-8. Reprinted with permission from Popular Electronics, Fact Card No. 249. (C) Copyright Gernsback Publications, Inc.

Fig. 26-9. Reprinted with permission of National Semiconductor Corporation, National Semiconductor Linear Applications Handbook, 1991, p. 271.

Fig. 26-10. Reprinted with permission from Electronic Design, 9/94, p. 136. Copyright 1994, Penton Publishing, Inc.

Fig. 26-11. Reprinted with permission from Electronic Design, 1/95, p. 81. Copyright 1995, Penton Publishing, Inc.

Fig. 26-12. Reprinted with permission from Electronics Now, 11/93, p. 8. (C) Copyright Gernsback Publications, Inc., 1993.

Fig. 26-13. Reprinted with permission from Electronics Now, 3/94, p. 67. (C) Copyright Gernsback Publications, Inc., 1994.

Fig. 26-14. Reprinted with permission from Electronic Design, 5/90, p. 79. Copyright 1990, Penton Publishing, Inc.

Fig. 26-15. Reprinted with permission from Popular Electronics, Fact Card No. 270. (C) Copyright Gernsback Publications, Inc.

Fig. 26-16. Reprinted with permission from Popular Electronics, Fact Card No. 270. (C) Copyright Gernsback Publications, Inc.

Fig. 26-17. Reprinted with permission from Popular Electronics, Fact Card No. 269. (C) Copyright Gernsback Publications, Inc.

Fig. 26-18. Reprinted with permission from Popular Electronics, Fact Card No. 269. (C) Copyright Gernsback Publications, Inc.

Fig. 26-19. Linear Technology, Design Note 88.

Fig. 26-20. Reprinted with permission from Popular Electronics, 10/94, p. 82. (C) Copyright Gernsback Publications, Inc., 1994.

Chapter 27

Fig. 27-1. Amplifiers, Waveform Generators & Other Low-Cost IC Projects, McGraw-Hill, p. 37.

Fig. 27-2. Maxim, Vol. III, New Releases Data Book, 1994, p. 3-45.

Chapter 28

Fig. 28-1. Reprinted with permission from Electronics Now, 3/95, p. 64. (C) Copyright Gernsback Publications, Inc., 1995.

Fig. 28-2. Reprinted with permission from Popular Electronics, 1/94, p. 24. (C) Copyright Gernsback Publications, Inc., 1994.

Fig. 28-3. Reprinted with permission from Popular Electronics, 1/94, p. 25. (C) Copyright Gernsback Publications, Inc., 1994.

Fig. 28-4. Linear Technology, Advertisement.

Fig. 28-5. Maxim, Vol. III, New Releases Data Book, 1994, p. 3-127.

Fig. 28-6. Sound Light and Music, Delton T. Horn, McGraw-Hill, pp. 132–135.

Fig. 28-7. Reprinted with permission from Popular Electronics, 2/94, p. 26. (C) Copyright Gernsback Publications, Inc., 1994.

Fig. 28-8. Reprinted with permission from Electronic Design, 12/93, p. 74. Copyright 1993, Penton Publishing, Inc.

Fig. 28-9. Reprinted with permission from Electronics Now, 3/95, p. 61. (C) Copyright Gernsback Publications, Inc., 1994.

Chapter 29

Fig. 29-1. Linear Technology, 2/95.

Fig. 29-2. Analog Devices, Analog Dialogue, Vol. 26, No. 2, 1992, p. 18.

Fig. 29-3. Reprinted with permission from Popular Electronics, 10/94, p. 24. (C) Copyright Gernsback Publications, Inc., 1994.

Fig. 29-4. Analog Devices, Analog Dialogue, Vol. 26, No. 2, 1992, p. 12.

Fig. 29-5. Reprinted with permission of National Semiconductor Corporation, National Semiconductor Linear Applications Handbook, 1991, p. 542.

Fig. 29-6. Reprinted with permission from Popular Electronics, 1/94, p. 24. (C) Copyright Gernsback Publications, Inc., 1994.

Fig. 29-7. 73 Amateur Radio Today, 3/95, p. 62.

Fig. 29-8. Reprinted with permission from Popular Electronics, 11/94, p. 31. (C) Copyright Gernsback Publications, Inc., 1994.

Fig. 29-9. Analog Devices, Analog Dialogue, Vol. 27, No. 1 (1993), p. 17.

Fig. 29-10. Reprinted with permission from Electronic Design, 7/94, p. 62. Copyright 1994, Penton Publishing, Inc.

Fig. 29-11. Reprinted with permission of National Semiconductor Corporation, National Semiconductor Linear Applications Handbook, 1991, p. 541.

Fig. 29-12. Analog Devices, Analog Dialogue, Vol. 27, No. 1 (1993), p. 16.

Fig. 29-13. Reprinted with permission from Electronic Design, 9/94, p. 135. Copyright 1994, Penton Publishing, Inc.

Chapter 30

Fig. 30-1. Reprinted with permission from Popular Electronics, 4/95, pp. 29–30. (C) Copyright Gernsback Publications, Inc., 1995.

Fig. 30-2. Reprinted with permission from Popular Electronics, 4/95, p. 31. (C) Copyright Gernsback Publications, Inc., 1995.

Fig. 30-3. Reprinted with permission from Popular Electronics, 4/95, p. 30. (C) Copyright Gernsback Publications, Inc., 1995.

Chapter 31

Fig. 31-1. Laser Cookbook, McGraw-Hill, p. 231.

Fig. 31-2. Laser Cookbook, McGraw-Hill, p. 231.

Chapter 32

Fig. 32-1. Reprinted with permission from Electronic Design, 2/95, p. 115. Copyright 1995, Penton Publishing, Inc.

Fig. 32-2. Amplifiers, Waveform Generators, & Other Low-Cost IC Projects, McGraw-Hill, p. 186.

Fig. 32-3. Amplifiers, Waveform Generators & Other Low-Cost IC Projects, McGraw-Hill, p. 193.

Fig. 32-4. Radio-Electronics Experimenters Handbook, p. 66.

Fig. 32-5. Reprinted with permission from Electronic Design, 12/94, p. 134. Copyright 1994, Penton Publishing, Inc.

Fig. 32-6. Reprinted with permission from Electronic Design, 9/94, p. 79. Copyright 1994, Penton Publishing, Inc.

Fig. 32-7. William Sheets.

Fig. 32-8. Reprinted with permission from Popular Electronics, 12/94, p. 42. (C) Copyright Gernsback Publications, Inc., 1994.

Fig. 32-9. Reprinted with permission of National Semiconductor Corporation, National Semiconductor Linear Applications Handbook, 1991, p. 1202.

Fig. 32-10. Radio Receiver Projects You Can Build, McGraw-Hill, p. 165.

Fig. 32-11. Radio Receiver Projects You Can Build, McGraw-Hill, p. 291.

Fig. 32-12. Reprinted with permission from Electronic Design, 8/94, p. 102. Copyright 1994, Penton Publishing, Inc.

Fig. 32-13. Amplifiers, Waveform, Generators & Other Low-Cost IC Projects, McGraw-Hill, p. 178.

Fig. 32-14. Reprinted with permission of National Semiconductor Corporation, National Semiconductor Linear Applications Handbook, 1991, p. 229.

Fig. 32-15. Linear Technology, 2/95.

Fig. 32-16. Reprinted with permission of National Semiconductor Corporation, National Semiconductor Linear Applications Handbook, 1991, p. 1187.

Fig. 32-17. Reprinted with permission of National Semiconductor Corporation, National Semiconductor Linear Applications Handbook, 1991, p. 1187.

Fig. 32-18. Amplifiers, Waveform Generators & Other Low-Cost IC Projects, McGraw-Hill, p. 222.

Fig. 32-19. Analog Devices, The Best of Analog Dialogue, 1967–1991, p. 79.

Fig. 32-20. Reprinted with permission of National Semiconductor Corporation, National Semiconductor Linear Applications Handbook, 1991, p. 1032.

Fig. 32-21. Linear Technology, Design Note #84.

Fig. 32-22. Reprinted with permission of National Semiconductor Corporation, National Semiconductor Linear Applications Handbook, 1991, p. 1202.

Fig. 32-23. Amplifiers, Waveform Generators, & Other Low-Cost IC Projects, McGraw-Hill, p. 186.

Fig. 32-24. Amplifiers, Waveform Generators, & Other Low-Cost IC Projects, McGraw-Hill, p. 185.

Fig. 32-25. Amplifiers, Waveform Generators & Other Low-Cost IC Projects, McGraw-Hill, p. 201.

Fig. 32-26. Linear Technology, Design Note #89.

Fig. 32-27. Reprinted with permission of National Semiconductor Corporation, National Semiconductor Linear Applications Handbook, 1991, p. 1015.

Fig. 32-28. Reprinted with permission of National Semiconductor Corporation, National Semiconductor Linear Applications Handbook, 1991, p. 228.

Chapter 33

Fig. 33-1. Reprinted with permission from Popular Electronics, 5/95, pp. 57–59. (C) Copyright Gernsback Publications, Inc., 1995.

Fig. 33-2. Reprinted with permission from Popular Electronics, 6/95, p. 77. (C) Copyright Gernsback Publications, Inc., 1995.

Fig. 33-3. Reprinted with permission from Popular Electronics, 6/95, p. 78. (C) Copyright Gernsback Publications, Inc., 1995.

Fig. 33-4. Reprinted with permission from Popular Electronics, 1/94, p. 73. (C) Copyright Gernsback Publications, Inc., 1994.

Fig. 33-5. Sound Light and Music, Delton T. Horn, McGraw-Hill, p. 34.

Fig. 33-6. Reprinted with permission from Popular Electronics, 5/93, p. 70. (C) Copyright Gernsback Publications, Inc., 1993.

Fig. 33-7. Reprinted with permission from Popular Electronics, 5/93, p. 70. (C) Copyright Gernsback Publications, Inc., 1993.

Fig. 33-8. Reprinted with permission from Popular Electronics, 5/93, p. 72. (C) Copyright Gernsback Publications, Inc., 1993.

Fig. 33-9. Reprinted with permission from Popular Electronics, 2/94, p. 25. (C) Copyright Gernsback Publications, Inc., 1994.

Chapter 34

Fig. 34-1. Reprinted with permission of National Semiconductor Corporation, National Semiconductor Linear Applications Handbook, 1991, p. 241.

Fig. 34-2. Reprinted with permission of National Semiconductor Corporation, National Semiconductor Linear Applications Handbook, 1991, p. 241.

Chapter 35

Fig. 35-1. Reprinted with permission from Electronic Design, 2/94, p. 115. Copyright 1994, Penton Publishing, Inc.

Fig. 35-2. Analog Devices, The Best of Analog Dialogue, 1967–1991, p. 75.

Chapter 36

Fig. 36-1. Reprinted with permission from Electronics Now, 4/94, pp. 41–45. (C) Copyright Gernsback Publications, Inc., 1994.

Fig. 36-2. Reprinted with permission from Electronics Now, 5/95, pp. 53–55. (C) Copyright Gernsback Publications, Inc., 1995.

Fig. 36-3. Reprinted with permission from Popular Electronics, Fact Card No. 249. (C) Copyright Gernsback Publications, Inc.

Fig. 36-4. Reprinted with permission from Electronic Design, 10/94, p. 110. Copyright 1994, Penton Publishing, Inc.

Fig. 36-5. Reprinted with permission of National Semiconductor Corporation, National Semiconductor Linear Applications Handbook, 1991, p. 368.

Chapter 37

Fig. 37-1. Reprinted with permission from Popular Electronics, 9/94, p. 81. (C) Copyright Gernsback Publications, Inc., 1994.

Fig. 37-2. Reprinted with permission from Popular Electronics, 9/94, p. 83. (C) Copyright Gernsback Publications, Inc., 1994.

Fig. 37-3. Reprinted with permission from Popular Electronics, 4/94, p. 79. (C) Copyright Gernsback Publications, Inc., 1994.

Fig. 37-4. Reprinted with permission from Popular Electronics, 12/94, pp. 44–47. (C) Copyright Gernsback Publications, Inc., 1994.

Fig. 37-5. Reprinted with permission from Popular Electronics, 4/94, p. 80. (C) Copyright Gernsback Publications, Inc., 1994.

Fig. 37-6. Reprinted with permission from Popular Electronics, 9/94, p. 81. (C) Copyright Gernsback Publications, Inc., 1994.

Fig. 37-7. Reprinted with permission from Popular Electronics, 9/94, pp. 83 and 92. (C) Copyright Gernsback Publications, Inc., 1994.

Fig. 37-8. Reprinted with permission from Popular Electronics, 5/93, p. 62. (C) Copyright Gernsback Publications, Inc., 1993.

Fig. 37-9. Reprinted with permission from Popular Electronics, 4/94, p. 78. (C) Copyright Gernsback Publications, Inc., 1994.

Fig. 37-10. Reprinted with permission from Popular Electronics, 4/94, p. 78. (C) Copyright Gernsback Publications, Inc., 1994.

Fig. 37-11. Reprinted with permission from Popular Electronics Hobbyist Handbook, 1991, p. 4. (C) Copyright Gernsback Publications, Inc., 1991.

Chapter 38

Fig. 38-1. Reprinted with permission from 1987 Radio-Electronics Experimenters Handbook, p. 63. (C) Copyright Gernsback Publications, Inc., 1987.

Fig. 38-2. Reprinted with permission from Popular Electronics, 1/94, p. 62. (C) Copyright Gernsback Publications, Inc., 1994.

Chapter 39

Fig. 39-1. Reprinted with permission from Popular Electronics, 6/93, p. 74. (C) Copyright Gernsback Publications, Inc., 1993.

Fig. 39-2. William Sheets.

Fig. 39-3. Analog Devices, Analog Dialogue, Vol. 27, No. 1 (1993), p. 21.

Chapter 40

Fig. 40-1. Reprinted with permission from Popular Electronics, 4/90, p. 102. (C) Copyright Gernsback Publications, Inc., 1990.

Fig. 40-2. Reprinted with permission from Popular Electronics, 6/95, p. 76. (C) Copyright Gernsback Publications, Inc., 1995.

Fig. 40-3. Reprinted with permission from Popular Electronics, 12/93, p. 31. (C) Copyright Gernsback Publications, Inc., 1993.

Fig. 40-4. Reprinted with permission from Popular Electronics, 5/94, p. 69. (C) Copyright Gernsback Publications, Inc., 1994.

Fig. 40-5. Reprinted with permission from Popular Electronics, 4/90, p. 90. (C) Copyright Gernsback Publications, Inc., 1990.

Fig. 40-6. Reprinted with permission from Popular Electronics, 4/94, p. 34. (C)

Copyright Gernsback Publications, Inc., 1994.

Fig. 40-7. Maxim, Vol. III, New Releases Data Book, 1994, p. 3-100.

Fig. 40-8. Reprinted with permission from Electronics Now, 2/94, p. 38. (C) Copyright Gernsback Publications, Inc., 1994.

Fig. 40-9. Reprinted with permission from Popular Electronics, 6/95, p. 76. (C) Copyright Gernsback Publications, Inc., 1995.

Fig. 40-10. Sound Light and Music, Delton T. Horn, McGraw-Hill, pp. 140–142.

Fig. 40-11. Reprinted with permission from Popular Electronics, 4/90, p. 103. (C) Copyright Gernsback Publications, Inc., 1990.

Fig. 40-12. Sound Light and Music, Delton T. Horn, McGraw-Hill, pp. 144–146.

Fig. 40-13. Reprinted with permission from Popular Electronics, 12/93, p. 32. (C) Copyright Gernsback Publications, Inc., 1993.

Fig. 40-14. Laser Cookbook, McGraw-Hill, p. 187.

Fig. 40-15. Reprinted with permission from Popular Electronics, 4/90, p. 91. (C) Copyright Gernsback Publications, Inc., 1990.

Fig. 40-16. Sound Light and Music, McGraw-Hill.

Fig. 40-17. Reprinted with permission of National Semiconductor Corporation, National Semiconductor Linear Applications Handbook, 1991, p. 209.

Chapter 41

Fig. 41-1.

Fig. 41-2. Linear Technology, 2/95.

Fig. 41-3. Reprinted with permission from Popular Electronics, Fact Card No. 248. (C) Copyright Gernsback Publications, Inc.

Fig. 41-4. Analog Devices, Analog Dialogue, Vol. 27, No. 2 (1993), p. 17.

Fig. 41-5. Reprinted with permission from Popular Electronics, Fact Card No. 248. (C) Copyright Gernsback Publications, Inc.

Fig. 41-6. Reprinted with permission from Popular Electronics, Fact Card No. 247. (C) Copyright Gernsback Publications, Inc.

Fig. 41-7. Reprinted with permission of Burr-Brown Corporation, Burr-Brown Data Sheet INA118, (C) 1989-1995 Burr-Brown Corporation.

Fig. 41-8. Maxim, Vol. III, New Releases Data Book, 1994, p. 3-33.

Fig. 41-9. Maxim, Vol. III, New Releases Data Book, 1994, p. 3-69.

Fig. 41-10. Analog Devices, Analog Dialogue, Vol. 27, No. 1 (1993), p. 20.

Fig. 41-11. Linear Technology, 2/95.

Chapter 42

Fig. 42-1. Reprinted with permission from Popular Electronics, Fact Card No. 247. (C) Copyright Gernsback Publications, Inc.

Fig. 42-2. Reprinted with permission from Popular Electronics, Fact Card No. 249. (C) Copyright Gernsback Publications, Inc.

Fig. 42-3. William Sheets.

Chapter 43

Fig. 43-1. Reprinted with permission from Popular Electronics, 4/90, p. 78. (C) Copyright Gernsback Publications, Inc., 1990.

Fig. 43-2. Analog Devices, The Best of Analog Dialogue, 1967–1991, p. 75.

Chapter 44

Fig. 44-1. Power Supplies, Switching Regulators, Inverters, and Converters, McGraw-Hill, pp. 173–174.

Fig. 44-2. Power Supplies, Switching Regulators, Inverters, and Converters, McGraw-Hill, p. 126.

Fig. 44-3. Power Supplies, Switching Regulators, Inverters, and Converters, McGraw-Hill, p. 128.

Fig. 44-4. Maxim, Vol. III, New Releases Data Book, 1994, p. 4-92.

Fig. 44-5. Power Supplies, Switching Regulators, Inverters, and Converters, McGraw-Hill, pp. 152–153.

Chapter 45

Fig. 45-1. Fantastic Electronics, McGraw-Hill, pp. 36–38.

Fig. 45-2. 73 Amateur Radio Today, 7/82, p. 53.

Fig. 45-3. 73 Amateur Radio Today, 7/82, p. 53.

Fig. 45-4. Spring 1994 Electronic Hobbyists Handbook.

Chapter 46

Fig. 46-1. Laser Cookbook, McGraw-Hill, pp. 165–167.

Fig. 46-2. Laser Cookbook, McGraw-Hill, pp. 157–158.

Fig. 46-3. Gordon McComb's Gadgeteer's Goldmine, McGraw-Hill, p. 125.

Fig. 46-4. Laser Cookbook, McGraw-Hill, p. 201.

Fig. 46-5. Laser Cookbook, McGraw-Hill, p. 190.

Fig. 46-6. Laser Cookbook, McGraw-Hill, p. 200.

Fig. 46-7. Laser Cookbook, McGraw-Hill, p. 170.

Fig. 46-8. Reprinted with permission from Popular Electronics, 6/93, p. 78. (C) Copyright Gernsback Publications, Inc., 1993.

Fig. 46-9. Laser Cookbook, McGraw-Hill, p. 169.

Fig. 46-10. Laser Cookbook, McGraw-Hill, p. 168.

Chapter 47

Fig. 47-1. Reprinted with permission from Electronics Now, 5/94, p. 12. (C) Copyright Gernsback Publications, Inc., 1994.

Fig. 47-2. William Sheets.

Fig. 47-3. Reprinted with permission from Popular Electronics, 6/95, p. 54. (C) Copyright Gernsback Publications, Inc., 1995.

Fig. 47-4. Laser Cookbook, McGraw-Hill, pp. 140–141.

Fig. 47-5. Linear Technology, 2/95.

Fig. 47-6. Reprinted with permission from Electronic Design, 12/93, p. 75. Copyright 1993, Penton Publishing, Inc.

Fig. 47-7. Reprinted with permission from Popular Electronics, 9/94, p. 24. (C) Copyright Gernsback Publications, Inc., 1994.

Fig. 47-8. Reprinted with permission from Electronic Design, 10/94, p. 92. Copyright 1994, Penton Publishing, Inc.

Fig. 47-9. Reprinted with permission from Popular Electronics, 4/90, p. 77. (C) Copyright Gernsback Publications, Inc., 1990.

Fig. 47-10. Reprinted with permission from Popular Electronics, 9/94, p. 83. (C) Copyright Gernsback Publications, Inc., 1994.

Fig. 47-11. Amplifiers, Waveform Generators & Other Low-Cost IC Projects, McGraw-Hill, pp. 82–83.

Fig. 47-12. Reprinted with permission from Electronics Now, 5/94, p. 45. (C) Copyright Gernsback Publications, Inc., 1994.

Fig. 47-13. Reprinted with permission from Popular Electronics, 1995, p. 77. (C) Copyright Gernsback Publications, Inc., 1995.

Fig. 47-14. Reprinted with permission of National Semiconductor Corporation, National Semiconductor Linear Applications Handbook, 1991, pp. 589–590.

Fig. 47-15. Gordon McComb's Gadgeteer's Goldmine, McGraw-Hill, p. 218.

Fig. 47-16. Gordon McComb's Gadgeteer's Goldmine, McGraw-Hill, p. 219.

Fig. 47-17. Spring 1994 Electronics Hobbyist Handbook.

Fig. 47-18. Spring 1994 Electronics Hobbyist Handbook.

Fig. 47-19. Reprinted with permission of National Semiconductor Corporation, National Semiconductor Linear Applications Handbook, 1991, p. 493.

Fig. 47-20. Reprinted with permission from Popular Electronics, 6/95, p. 56. (C) Copyright Gernsback Publications, Inc., 1995.

Fig. 47-21. Reprinted with permission from Electronics Now, 9/94, p. 66. (C) Copyright Gernsback Publications, Inc., 1994.

Chapter 48

Fig. 48-1. NASA Tech Briefs, Spring 1977.

Fig. 48-2. Reprinted with permission from Electronic Design, 12/93, p. 73. Copyright 1993, Penton Publishing, Inc.

Fig. 48-3. Reprinted with permission from Popular Electronics, Fact Card No. 268. (C) Copyright Gernsback Publications, Inc.

Fig. 48-4. William Sheets.

Fig. 48-5. William Sheets.

Chapter 49

Fig. 49-1. Reprinted with permission of National Semiconductor Corporation, National Semiconductor Linear Applications Handbook, 1991, p. 520.

Fig. 49-2. Reprinted with permission of National Semiconductor Corporation, National Semiconductor Linear Applications Handbook, 1991, pp. 519–520.

Fig. 49-3. Reprinted with permission of National Semiconductor Corporation, National Semiconductor Linear Applications Handbook, 1991, p. 519.

Fig. 49-4. Analog Devices, The Best of Analog Dialogue, 1967–1991, p. 78.

Fig. 49-5. Analog Devices, The Best of Analog Dialogue, 1967–1991, p. 79.

Fig. 49-6. Analog Dialogue, Analog Devices, Vol. 26, No. 1, pp. 14–15.

Fig. 49-7. Analog Devices, The Best of Analog Dialogue, 1967–1991, p. 78.

Fig. 49-8. Analog Devices, The Best of Analog Dialogue, 1967–1991, p. 78.

Fig. 49-9. Reprinted with permission from Electronic Design, 3/95, p. 116. Copyright 1995, Penton Publishing, Inc.

Fig. 49-10. Reprinted with permission of National Semiconductor Corporation, National Semiconductor Linear Applications Handbook, 1991, p. 27.

Fig. 49-11. Analog Devices, The Best of Analog Dialogue, 1967–1991, p. 77.

Chapter 50

Fig. 50-1. Reprinted with permission from Popular Electronics, 4/95, p. 54. (C) Copyright Gernsback Publications, Inc., 1995.

Fig. 50-2. 1994 Electronic Experimenters Handbook, p. 89.

Fig. 50-3. 73 Amateur Radio Today, 7/93, p. 34.

Fig. 50-4. NASA Tech Briefs, November 1993, p. 56.

Fig. 50-5. Reprinted with permission of National Semiconductor Corporation, National Semiconductor Linear Applications Handbook, 1991, p. 588.

Fig. 50-6. Reprinted with permission of National Semiconductor Corporation, National Semiconductor Linear Applications Handbook, pp. 519–520.

Fig. 50-7. Reprinted with permission from Electronics Now, 3/95, p. 42. (C) Copyright Gernsback Publications, Inc., 1995.

Fig. 50-8. Reprinted with permission from Radio-Electronics, February 1989, p. 64. (C) Copyright Gernsback Publications, Inc., 1989.

Fig. 50-9. Reprinted with permission from Electronics Now, 5/95, p. 10. (C) Copyright Gernsback Publications, Inc., 1995.

Fig. 50-10. Reprinted with permission from Electronics Now, 5/95, p. 10. (C) Copyright Gernsback Publications, Inc., 1995.

Fig. 50-11. Reprinted with permission from Popular Electronics, Fact Card No. 221. (C) Copyright Gernsback Publications, Inc.

Fig. 50-12. Fantastic Electronics, McGraw-Hill, pp. 52–60.

Fig. 50-13. Analog Devices, Analog Dialogue, Vol. 27, No. 2, p. 20.

Fig. 50-14. Electronics Now, 9/94, pp. 73–74.

Fig. 50-15. Reprinted with permission from Popular Electronics, 1/94, pp. 31–36. (C) Copyright Gernsback Publications, Inc., 1994.

Fig. 50-16. Electronics Design, June 27, 1994, p. 33.

Fig. 50-17. Reprinted with permission from Popular Electronics, 10/94, p. 90. (C) Copyright Gernsback Publications, Inc., 1994.

Fig. 50-18. Reprinted with permission from Popular Electronics, 11/94, p. 62. (C) Copyright Gernsback Publications, Inc., 1994.

Fig. 50-19. Radio-Electronics, Feb. 1989, p. 64.

Fig. 50-20. Reprinted with permission from Popular Electronics, 11/94, p. 91. (C) Copyright Gernsback Publications, Inc., 1994.

Fig. 50-21. Reprinted with permission from Electronic Design, 10/94, pp. 102–104. Copyright 1994, Penton Publishing, Inc.

Fig. 50-22. Reprinted with permission from Electronic Design, 12/94, p. 132. Copyright 1994, Penton Publishing, Inc.

Fig. 50-23. Reprinted with permission from Electronics Now, 3/95, p. 63. (C) Copyright Gernsback Publications, Inc., 1995.

Fig. 50-24. Reprinted with permission from Popular Electronics, 5/95, pp. 47–48. (C) Copyright Gernsback Publications, Inc., 1995.

Fig. 50-25. Reprinted with permission from Popular Electronics, 5/95, pp. 47–48. (C) Copyright Gernsback Publications, Inc., 1995.

Fig. 50-26. Reprinted with permission from Popular Electronics, 11/94, p. 38. (C) Copyright Gernsback Publications, Inc., 1994.

Fig. 50-27. Reprinted with permission from Electronic Design, 10/94, pp. 92–94. Copyright 1994, Penton Publishing, Inc.

Fig. 50-28. RF Design, August 1994, p. 78.

Fig. 50-29. Reprinted with permission of National Semiconductor Corporation, National Semiconductor Linear Applications Handbook, 1991, p. 206.

Fig. 50-30. Reprinted with permission of National Semiconductor Corporation, National Semiconductor Linear Applications Handbook, 1991, p. 500.

Fig. 50-31. Reprinted with permission from Electronic Design, 11/94, pp. 116–118. Copyright 1994, Penton Publishing, Inc.

Fig. 50-32. Reprinted with permission from Electronic Design, 10/94, pp. 107. Copyright 1994, Penton Publishing, Inc.

Fig. 50-33. QST, 12/94, p. 27.

Fig. 50-34. Reprinted with permission from Electronic Design, 8/94, pp. 109. Copyright 1994, Penton Publishing, Inc.

Fig. 50-35. Reprinted with permission of Burr-Brown Corporation, Burr-Brown Data Sheet INA118, (C) 1989–1995 Burr-Brown Corporation.

Fig. 50-36. William Sheets.

Fig. 50-37. Linear Technology, Design Note 96.

Fig. 50-38. Reprinted with permission from Popular Electronics, 11/94, p. 31. (C) Copyright Gernsback Publications, Inc., 1994.

Fig. 50-39. Reprinted with permission of XICOR, XICOR Data Sheet, p. 7. (C) Copyright XICOR, Inc.

Fig. 50-40. Reprinted with permission from Popular Electronics, Fact Card No. 221. (C) Copyright Gernsback Publications, Inc.

Fig. 50-41. Reprinted with permission from Popular Electronics, 11/94, p. 30. (C) Copyright Gernsback Publications, Inc., 1994.

Fig. 50-42. William Sheets.

Fig. 50-43. QST, 10/94, P. 75.

Fig. 50-44. Reprinted with permission from Popular Electronics, 5/95, p. 69. (C) Copyright Gernsback Publications, Inc., 1995.

Fig. 50-45. Reprinted with permission from Popular Electronics, 11/94, p. 33. (C) Copyright Gernsback Publications, Inc., 1994.

Fig. 50-46. Reprinted with permission from Radio-Electronics, February 1989, pp. 63–65. (C) Copyright Gernsback Publications, Inc., 1989.

Fig. 50-47. Amplifiers, Waveform Generators & Other Low-Cost IC Projects, McGraw-Hill, p. 45.

Fig. 50-48. Amplifiers, Waveform Generators & Other Low-Cost IC Projects, McGraw-Hill, pp. 40–41.

Chapter 51

Fig. 51-1. Sound Light and Music, Delton T. Horn, McGraw-Hill, p. 123.

Fig. 51-2. Sound Light and Music, Delton T. Horn, McGraw-Hill, p. 120.

Chapter 52

Fig. 52-1. Reprinted with permission from Electronic Design, 5/94, pp. 79–80. Copyright 1994, Penton Publishing, Inc.

Fig. 52-2. Reprinted with permission from Popular Electronics, 1/94, p. 26. (C) Copyright Gernsback Publications, Inc., 1994.

Fig. 52-3. Analog Devices, Analog Dialogue, Vol. 26, No. 1, 1992, p. 15.

Fig. 52-4. Reprinted with permission from Popular Electronics, 8/89, pp. 33–35. (C) Copyright Gernsback Publications, Inc., 1989.

Fig. 52-5. NASA Tech Briefs, March 1995, pp. 39–40.

Fig. 52-6. Reprinted with permission from Radio-Electronics Experimenters Handbook, pp. 98–99. (C) Copyright Gernsback Publications, Inc., 1994.

Fig. 52-7. Reprinted with permission from Radio-Electronics Experimenters Handbook, p. 99. (C) Copyright Gernsback Publications, Inc., 1994.

Fig. 52-8. 1994 Electronics Experimenters Handbook, p. 57.

Fig. 52-9. Power Supplies, Switching Regulators, Inverters, and Converters, Mc-Graw-Hill, pp. 119–123.

Fig. 52-10. Amplifiers, Waveform Generators & Other Low-Cost IC Projects, Mc-Graw-Hill, pp. 121–122.

Fig. 52-11. Reprinted with permission from Radio-Electronics Experimenters Handbook, pp. 36–38. (C) Copyright Gernsback Publications, Inc., 1994.

Fig. 52-12. Reprinted with permission from Popular Electronics, 4/95, p. 70. (C) Copyright Gernsback Publications, Inc., 1995.

Fig. 52-13. Reprinted with permission from Popular Electronics, 4/90, p. 37. (C) Copyright Gernsback Publications, Inc., 1990.

Fig. 52-14. Power Supplies, Switching Regulators, Inverters, and Converters, Mc-Graw-Hill, pp. 175–177.

Fig. 52-15. Linear Technology, Design Note 88.

Fig. 52-16. Fantastic Electronics, McGraw-Hill, p. 59.

Fig. 52-17. 73 Amateur Radio Today, 6/83, p. 99.

Fig. 52-18. Spring 1994 Electronic Hobbyists Handbook.

Fig. 52-19. Reprinted with permission from Electronic Design, 1/95, p. 78. Copyright 1995, Penton Publishing, Inc.

Fig. 52-20. Sound Light and Music, Delton T. Horn, McGraw-Hill, p. 123.

Fig. 52-21. Power Supplies, Switching Regulators, Inverters, and Converters, Mc-Graw-Hill, pp. 140–145.

Fig. 52-22. Reprinted with permission from Electronic Design, 3/95, pp. 111–112. Copyright 1995, Penton Publishing, Inc.

Fig. 52-23. Reprinted with permission from Electronic Design, 3/95, p. 94. Copyright 1995, Penton Publishing, Inc.

Fig. 52-24. Reprinted with permission from Electronic Design, 4/89, p. 107. Copyright 1989, Penton Publishing, Inc.

Fig. 52-25. Reprinted with permission of National Semiconductor Corporation, National Semiconductor Linear Applications Handbook, 1991, p. 367.

Fig. 52-26. 73 Amateur Radio Today, 6/83, p. 99.

Fig. 52-27. Reprinted with permission from Popular Electronics, 4/94, p. 52. (C) Copyright Gernsback Publications, Inc., 1994.

Fig. 52-28. Reprinted with permission from Electronics Now, 2/94, p. 16. (C) Copyright Gernsback Publications, Inc., 1994.

Fig. 52-29. Reprinted with permission from Electronic Design, 3/95, p. 91. Copyright 1995, Penton Publishing, Inc.

Fig. 52-30. William Sheets.

Fig. 52-31. William Sheets.

Fig. 52-32. Reprinted with permission from Electronic Design, 7/94, p. 96. Copyright 1994, Penton Publishing, Inc.

Fig. 52-33. Reprinted with permission from Electronics Now, 4/94, p. 11. (C) Copyright Gernsback Publications, Inc., 1994.

Fig. 52-34. Reprinted with permission from Popular Electronics, 5/95, p. 30. (C)

Copyright Gernsback Publications, Inc., 1994.

Fig. 52-35. Reprinted with permission from Electronic Design, 12/94, pp. 115–116. Copyright 1994, Penton Publishing, Inc.

Chapter 53

Fig. 53-1. William Sheets.
Fig. 53-2. William Sheets.

Chapter 54

Fig. 54-1. Reprinted with permission from Popular Electronics, 10/94, p. 28. (C) Copyright Gernsback Publications, Inc., 1994.

Fig. 54-2. Reprinted with permission from Popular Electronics, 10/94, p. 26. (C) Copyright Gernsback Publications, Inc., 1994.

Chapter 55

Fig. 55-1. Reprinted with permission from Electronic Design, 10/94, pp. 94–96. Copyright 1994, Penton Publishing, Inc.

Fig. 55-2. Reprinted with permission of National Semiconductor Corporation, National Semiconductor Linear Applications Handbook, 1991, p. 1033.

Fig. 55-3. NASA Tech Briefs, January 1995, pp. 28–29.

Fig. 55-4. Linear Technology, 2/95.

Fig. 55-5. Reprinted with permission from Popular Electronics, Fact Card No. 249. (C) Copyright Gernsback Publications, Inc.

Fig. 55-6. Analog Devices, The Best of Analog Dialogue, 1967–1991, p. 75.

Fig. 55-7. William Sheets.

Chapter 56

Fig. 56-1. 73 Amateur Radio Today, 5/94, p. 12.

Fig. 56-2. 73 Amateur Radio Today, 6/94, pp. 46–48.

Fig. 56-3. Reprinted with permission from Popular Electronics, 5/93, p. 71. (C) Copyright Gernsback Publications, Inc., 1993.

Chapter 57

Fig. 57-1. Power Supplies, Switching Regulators, Inverters, and Converters, McGraw-Hill, pp. 167–169.

Fig. 57-2. Reprinted with permission from Electronic Design, 10/94, pp. 104–105. Copyright 1994, Penton Publishing, Inc.

Fig. 57-3. Gordon McComb's Gadgeteer's Goldmine, McGraw-Hill, p. 355.

Fig. 57-4. Electronics Now, 5/95, p. 8.

Fig. 57-5. Power Supplies, Switching Regulators, Inverters, and Converters, McGraw-Hill, pp. 166–167.

Fig. 57-6. Power Supplies, Switching Regulators, Inverters, and Converters, McGraw-Hill, pp. 163–166.

Fig. 57-7. Gordon McComb's Gadgeteer's Goldmine, McGraw-Hill, p. 48.

Fig. 57-8. Reprinted with permission from Electronics Now, 5/94, p. 10. (C) Copyright Gernsback Publications, Inc., 1994.

Fig. 57-9. Gordon McComb's Gadgeteer's Goldmine, McGraw-Hill, p. 357.

Chapter 58

Fig. 58-1. William Sheets.

Fig. 58-2. Reprinted with permission of National Semiconductor Corporation, National Semiconductor Linear Applications Handbook, 1991, p. 240.

Fig. 58-3. Reprinted with permission from Popular Electronics, 1/94, p. 72. (C) Copyright Gernsback Publications, Inc., 1994.

Fig. 58-4. Reprinted with permission from Popular Electronics, 1/94, p. 73. (C) Copyright Gernsback Publications, Inc., 1994.

Fig. 58-5. Reprinted with permission from Popular Electronics, Fact Card No. 268. (C) Copyright Gernsback Publications, Inc.

Fig. 58-6. Amplifiers, Waveform Generators & Other Low-Cost IC Projects, McGraw-Hill, p. 24.

Fig. 58-7. William Sheets.

Chapter 59

Fig. 59-1. Reprinted with permission of National Semiconductor Corporation, National Semiconductor Linear Application Specific IC's Databook 1993, p. 1-38.

Fig. 59-2. Reprinted with permission of National Semiconductor Corporation, National Semiconductor NSLAH 1991, p. 998.

Fig. 59-3. QST, 10/92, p. 22.

Fig. 59-4. William Sheets.

Fig. 59-5. William Sheets.

Chapter 60

Fig. 60-1. Reprinted with permission from Radio-Electronics, June 1987, p. 69. (C) Copyright Gernsback Publications, Inc., 1987.

Fig. 60-2. Reprinted with permission of National Semiconductor Corporation, National Semiconductor Linear Application Handbook, 1991, p. 578.

Fig. 60-3. Reprinted with permission from Radio-Electronics, June 1987, p. 70. (C) Copyright Gernsback Publications, Inc., 1987.

Fig. 60-4. Reprinted with permission from Radio-Electronics, June 1987, p. 75. (C) Copyright Gernsback Publications, Inc., 1987.

Fig. 60-5. Reprinted with permission of National Semiconductor Corporation, National Semiconductor Linear Applications Handbook, 1991, P. 576.

Fig. 60-6. Reprinted with permission from Electronic Design, 12/94, p. 118. Copyright 1994, Penton Publishing, Inc.

Fig. 60-7. Reprinted with permission of National Semiconductor Corporation, National Semiconductor Linear Application Handbook, 1991, p. 1058.

Fig. 60-8. Reprinted with permission from Radio-Electronics, June 1987, p. 75. (C) Copyright Gernsback Publications, Inc., 1987.

Fig. 60-9. Reprinted with permission of National Semiconductor Corporation, National Semiconductor Linear Applications Handbook, 1991, p. 491.

Fig. 60-10. Reprinted with permission of National Semiconductor Corporation,

National Semiconductor Linear Applications Handbook, 1991, p. 578.

Fig. 60-11. Analog Devices, Analog Dialogue, Vol. 27, No. 2 (1993), p. 17.

Chapter 61

Fig. 61-1. Reprinted with permission of National Semiconductor Corporation, National Semiconductor Linear Applications Handbook, p. 1209.

Fig. 61-2. William Sheets.

Fig. 61-3. William Sheets.

Fig. 61-4. Amplifiers, Waveform Generators & Other Low-Cost IC Projects, McGraw-Hill, p. 17.

Fig. 61-5. Reprinted with permission of National Semiconductor Corporation, National Semiconductor Linear Applications Handbook, 1991, P. 1210.

Fig. 61-6. William Sheets.

Fig. 61-7. Reprinted with permission from Popular Electronics, 11/94, p. 31. (C) Copyright Gernsback Publications, Inc., 1994.

Fig. 61-8. William Sheets.

Fig. 61-9. Reprinted with permission from Popular Electronics, 5/93, p. 71. (C) Copyright Gernsback Publications, Inc., 1993.

Fig. 61-10. William Sheets.

Fig. 61-11. Reprinted with permission from Popular Electronics, 3/94, p. 83. (C) Copyright Gernsback Publications, Inc., 1994.

Fig. 61-12. Reprinted with permission from Popular Electronics, 12/93, p. 71. (C) Copyright Gernsback Publications, Inc., 1993.

Chapter 62

Fig. 62-1. Reprinted with permission from Popular Electronics, 12/93, p. 70. (C) Copyright Gernsback Publications, Inc., 1993.

Fig. 62-2. Reprinted with permission from Popular Electronics, 12/93, p. 68. (C) Copyright Gernsback Publications, Inc., 1993.

Fig. 62-3. Reprinted with permission from

Popular Electronics, 12/93, p. 71. (C) Copyright Gernsback Publications, Inc., 1993.

Fig. 62-4. Reprinted with permission from Popular Electronics, 12/93, p. 70. (C) Copyright Gernsback Publications, Inc., 1993.

Fig. 62-5. Reprinted with permission from Electronic Design, 11/94, pp. 130–132. Copyright 1994, Penton Publishing, Inc.

Fig. 62-6. Analog Devices, The Best of Analog Dialogue, 1967–1991, p. 79.

Fig. 62-7. Amplifiers, Waveform Generators & Other Low-Cost IC Projects, McGraw-Hill, p. 116.

Fig. 62-8. Reprinted with permission from Electronics Now, 12/93, p. 16. (C) Copyright Gernsback Publications, Inc., 1993.

Fig. 62-9. Reprinted with permission from Electronic Design, 7/94, p. 94. Copyright 1994, Penton Publishing, Inc.

Fig. 62-10. QST, 5/95, p. 50.

Chapter 63

Fig. 63-1. William Sheets.

Fig. 63-2. Reprinted with permission from Popular Electronics, 3/94, p. 84. (C) Copyright Gernsback Publications, Inc., 1993.

Fig. 63-3. William Sheets.

Fig. 63-4. William Sheets.

Fig. 63-5. QST, 4/95, pp. 38–39.

Fig. 63-6. William Sheets.

Fig. 63-7. Reprinted with permission from Radio-Electronics Experimenters Handbook, p. 75. (C) Copyright Gernsback Publications, Inc., 1994.

Fig. 63-8. Radio Receiver Projects You Can Build, McGraw-Hill, p. 237.

Fig. 63-9. William Sheets.

Fig. 63-10. Reprinted with permission from Popular Electronics, 2/94, p. 90. (C) Copyright Gernsback Publications, Inc., 1994.

Fig. 63-11. William Sheets.

Fig. 63-12. Reprinted with permission from Popular Electronics, 11/94, p. 42. (C) Copyright Gernsback Publications, Inc., 1994.

Fig. 63-13. Reprinted with permission from Popular Electronics, 2/94, p. 90. (C) Copyright Gernsback Publications, Inc., 1994.

Fig. 63-14. Reprinted with permission from Electronic Design, 10/94, p. 96. Copyright 1994, Penton Publishing, Inc.

Fig. 63-15. William Sheets.

Fig. 63-16. 73 Amateur Radio Today, 5/94, p. 66.

Fig. 63-17. Radio Craft, 1993, p. 63.

Fig. 63-18. Radio-Electronics Experimenters Handbook, 1992, p. 76.

Fig. 63-19. William Sheets.

Fig. 63-20. William Sheets.

Fig. 63-21. William Sheets.

Chapter 64

Fig. 64-1. Reprinted with permission from Electronics Now, 5/95, pp. 65–66. (C) Copyright Gernsback Publications, Inc., 1995.

Fig. 64-2. Reprinted with permission from Electronic Design, 11/94, pp. 118–120. Copyright 1994, Penton Publishing, Inc.

Fig. 64-3. Reprinted with permission from Electronic Design, 9/94, p. 84. Copyright 1994, Penton Publishing, Inc.

Fig. 64-4. 73 Amateur Radio Today, 7/94, p. 39.

Chapter 65

Fig. 65-1. Power Supplies, Switching Regulators, Inverters, and Converters, McGraw-Hill, pp. 107–108.

Fig. 65-2. Reprinted with permission from Electronics Now, 5/95, p. 91. (C) Copyright Gernsback Publications, Inc., 1995.

Fig. 65-3. 1994 Electronics Experimenters Handbook, p. 67.

Chapter 66

Fig. 66-1. Reprinted with permission from Popular Electronics, 3/94, p. 70. (C) Copyright Gernsback Publications, Inc., 1994.

Fig. 66-2. Gordon McComb's Gadgeteer's Goldmine, McGraw-Hill, p. 94.

Chapter 67

Fig. 67-1. Reprinted with permission from Electronics Now, 5/95, pp. 67–68. (C) Copyright Gernsback Publications, Inc., 1995.

Fig. 67-2. Reprinted with permission from Popular Electronics, 10/94, p. 84. (C) Copyright Gernsback Publications, Inc., 1994.

Fig. 67-3. Reprinted with permission from Electronics Now, 3/94, p. 59. (C) Copyright Gernsback Publications, Inc., 1994.

Fig. 67-4. Reprinted with permission from Popular Electronics, 12/94, pp. 57–59. (C) Copyright Gernsback Publications, Inc., 1994.

Chapter 68

Fig. 68-1. Reprinted with permission from Popular Electronics, 4/95, p. 60. (C) Copyright Gernsback Publications, Inc., 1995.

Fig. 68-2. Reprinted with permission from Popular Electronics, 4/95, p. 60. (C) Copyright Gernsback Publications, Inc., 1995.

Fig. 68-3. Reprinted with permission from Electronic Design, 11/94, p. 133. Copyright 1994, Penton Publishing, Inc.

Fig. 68-4. Reprinted with permission from Popular Electronics, 4/90, pp. 45–46. (C) Copyright Gernsback Publications, Inc., 1990.

Fig. 68-5. Maxim, Vol. III, New Releases Data Book, 1994, p. 4-131.

Fig. 68-6. Power Supplies, Switching Regulators, Inverters, and Converters, McGraw-Hill, p. 133.

Fig. 68-7. Linear Technology, Design Note #87.

Fig. 68-8. 1994 Electronics Experimenters Handbook, p. 39.

Fig. 68-9. Reprinted with permission of National Semiconductor Corporation, National Semiconductor Linear Applications Handbook, 1991, p. 449.

Fig. 68-10. Linear Technology, Design Note #74.

Fig. 68-11. Reprinted with permission from Radio-Electronics Experimenters Handbook 1992, p. 74. (C) Copyright Gernsback Publications, Inc., 1992.

Fig. 68-12. William Sheets.

Fig. 68-13. Reprinted with permission of National Semiconductor Corporation, National Semiconductor Linear Applications Handbook, 1991, p. 30.

Fig. 68-14. Reprinted with permission from Electronic Design, 7/94, p. 34. Copyright 1994, Penton Publishing, Inc.

Fig. 68-15. Reprinted with permission from Electronic Design, 6/94, p. 32. Copyright 1994, Penton Publishing, Inc.

Fig. 68-16. Power Supplies, Switching Regulators, Inverters, and Converters, McGraw-Hill, pp. 161–163.

Fig. 68-17. NASA Tech Briefs, August 1994, p. 39.

Fig. 68-18. Linear Technology, Design Note #87.

Fig. 68-19. Reprinted with permission of National Semiconductor Corporation, National Semiconductor Linear Applications Handbook, 1991, p. 367.

Fig. 68-20. Reprinted with permission from Electronic Design, 12/94, p. 130. Copyright 1994, Penton Publishing, Inc.

Fig. 68-21. Reprinted with permission from Electronic Design, 10/94, pp. 108–110. Copyright 1994, Penton Publishing, Inc.

Fig. 68-22. Reprinted with permission of National Semiconductor Corporation, National Semiconductor Linear Applications Handbook, 1991, p. 1063.

Fig. 68-23. Reprinted with permission from Popular Electronics, 6/95, p. 78. (C) Copyright Gernsback Publications, Inc., 1995.

Fig. 68-24. Linear Technology, Design Note #74.

Fig. 68-25. Reprinted with permission from Electronic Design, 9/94, p. 140. Copyright 1994, Penton Publishing, Inc.

Fig. 68-26. Reprinted with permission of National Semiconductor Corporation, National Semiconductor Linear Applications Handbook, 1991, p. 1063.

Fig. 68-27. Reprinted with permission from

Electronic Design, 7/94, p. 30. Copyright 1994, Penton Publishing, Inc.

Fig. 68-28. Reprinted with permission of National Semiconductor Corporation, National Semiconductor Linear Applications Handbook, 1991, p. 450.

Fig. 68-29. William Sheets.

Fig. 68-30. Maxim, Vol. III, New Releases Data Book, 1994, p. 4-131.

Fig. 68-31. Laser Cookbook, McGraw-Hill, p. 172.

Fig. 68-32. Analog Devices, Analog Dialogue, Vol. 27, No. 2, p. 19.

Fig. 68-33. Reprinted with permission of National Semiconductor Corporation, National Semiconductor Linear Applications Handbook, 1991, p. 450.

Fig. 68-34. Laser Cookbook, McGraw-Hill, p. 172.

Fig. 68-35. Reprinted with permission from Popular Electronics, 5/95, p. 94. (C) Copyright Gernsback Publications, Inc., 1995.

Fig. 68-36. 73 Amateur Radio Today, 5/93, p. 51.

Fig. 68-37. Reprinted with permission from Electronics Now, 12/93, p. 14. (C) Copyright Gernsback Publications, Inc., 1993.

Chapter 69

Fig. 69-1. Reprinted with permission from Popular Electronics, 6/93, p. 77. (C) Copyright Gernsback Publications, Inc., 1993.

Fig. 69-2. Reprinted with permission of National Semiconductor Corporation, National Semiconductor Linear Applications Handbook, 1991, p. 351.

Fig. 69-3. Reprinted with permission from Electronics Now, 10/94, pp. 58–61. (C) Copyright Gernsback Publications, Inc., 1994.

Fig. 69-4. William Sheets.

Fig. 69-5. Reprinted with permission from Popular Electronics, 6/93, p. 77. (C) Copyright Gernsback Publications, Inc., 1993.

Fig. 69-6. Laser Cookbook, McGraw-Hill, p. 163.

Fig. 69-7. Reprinted with permission of National Semiconductor Corporation, National Semiconductor Linear Applications Handbook, 1991, p. 498.

Fig. 69-8. Fantastic Electronics, McGraw-Hill, p. 177.

Fig. 69-9. Gordon McComb's Gadgeteer's Goldmine, p. 269.

Fig. 69-10. Reprinted with permission of National Semiconductor Corporation, National Semiconductor Linear Applications Handbook, 1991, p. 497.

Fig. 69-11. Laser Cookbook, McGraw-Hill, pp. 160–161.

Chapter 70

Fig. 70-1. 73 Amateur Radio Today, 5/93, p. 32.

Fig. 70-2. Laser Cookbook, McGraw-Hill, p. 172.

Fig. 70-3. Reprinted with permission from Electronic Design, 5/90, p. 80. Copyright 1990, Penton Publishing, Inc.

Fig. 70-4. Linear Technology, Design Note #99.

Fig. 70-5. Radio Receiver Projects You Can Build, McGraw-Hill 4256, p. 241.

Fig. 70-6. Reprinted with permission from Popular Electronics, 11/94, p. 41. (C) Copyright Gernsback Publications, Inc., 1994.

Fig. 70-7. Linear Technology, Design Note #78.

Fig. 70-8. Reprinted with permission from Electronics Now, 2/94, p. 83. (C) Copyright Gernsback Publications, Inc., 1994.

Fig. 70-9. Reprinted with permission from Radio-Electronics, June 1987, p. 75. (C) Copyright Gernsback Publications, Inc., 1987.

Chapter 71

Fig. 71-1. Reprinted with permission from Popular Electronics, 4/95, p. 96. (C) Copyright Gernsback Publications, Inc., 1995.

Fig. 71-2. Reprinted with permission from Electronic Design, 6/94, p. 62. Copyright 1994, Penton Publishing, Inc.

Fig. 71-3. QST, 10/92, p. 50.

Fig. 71-4. William Sheets.

Fig. 71-5. Reprinted with permission from Popular Electronics, 4/95, p. 96. (C) Copyright Gernsback Publications, Inc., 1995.

Fig. 71-6. Reprinted with permission of National Semiconductor Corporation, National Semiconductor Linear Applications Handbook, 1991, p. 450.

Fig. 71-7. Reprinted with permission of National Semiconductor Corporation, National Semiconductor Linear Applications Handbook, 1991, p. 495.

Fig. 71-8. William Sheets.

Chapter 72

Fig. 72-1. Reprinted with permission from Electronic Design, 1/95, p. 133. Copyright 1995, Penton Publishing, Inc.

Fig. 72-2. Reprinted with permission from Electronics Now, 3/95, p. 64. (C) Copyright Gernsback Publications, Inc., 1995.

Fig. 72-3. 73 Amateur Radio Today, 6/83, p. 99.

Fig. 72-4. Reprinted with permission from Popular Electronics, 2/94, p. 58. (C) Copyright Gernsback Publications, Inc., 1994.

Fig. 72-5. Analog Devices, Analog Dialogue, Vol. 27, No. 1 (1993), p. 20.

Fig. 72-6. Reprinted with permission from Radio-Electronics Experimenters Handbook, pp. 77–82. (C) Copyright Gernsback Publications, Inc., 1994.

Chapter 73

Fig. 73-1. Reprinted with permission from Popular Electronics, 11/94, pp. 31 and 91. (C) Copyright Gernsback Publications, Inc., 1994.

Fig. 73-2. William Sheets.

Chapter 74

Fig. 74-1. Reprinted with permission from Popular Electronics, 6/95, p. 38. (C) Copyright Gernsback Publications, Inc., 1995.

Fig. 74-2. Reprinted with permission from Popular Electronics, 6/95, p. 39. (C) Copyright Gernsback Publications, Inc., 1995.

Chapter 75

Fig. 75-1. Fantastic Electronics, McGraw-Hill, pp. 67–73.

Fig. 75-2. Reprinted with permission from Electronics Now, 1/94, p. 61. (C) Copyright Gernsback Publications, Inc., 1994.

Fig. 75-3. Reprinted with permission from Electronics Now, 1/94, p. 61. (C) Copyright Gernsback Publications, Inc., 1994.

Fig. 75-4. Reprinted with permission from Electronics Now, 1/94, p. 59. (C) Copyright Gernsback Publications, Inc., 1994.

Fig. 75-5. Reprinted with permission from Popular Electronics, 5/93, p. 40. (C) Copyright Gernsback Publications, Inc., 1993.

Chapter 76

Fig. 76-1. Reprinted with permission from Popular Electronics, 5/95, p. 66. (C) Copyright Gernsback Publications, Inc., 1995.

Fig. 76-2. Reprinted with permission from Popular Electronics, 1/95, p. 49. (C) Copyright Gernsback Publications, Inc., 1995.

Fig. 76-3. Radio Receiver Projects You Can Build, McGraw-Hill 4256, pp. 122–129.

Fig. 76-4. Radio-Electronics Experimenters Handbook, p. 65.

Fig. 76-5. Reprinted with permission of National Semiconductor Corporation, National Semiconductor Linear Application Specific IC's Databook 1993, p. 2-38.

Fig. 76-6. Radio Craft, 1993, p. 50.

Fig. 76-7. Analog Devices, Analog Dialogue, Vol. 27, No. 1 (1993), pp. 15-16.

Fig. 76-8. Reprinted with permission from Popular Electronics, 3/94, p. 35. (C) Copyright Gernsback Publications, Inc., 1994.

Fig. 76-9. Radio Receiver Projects You Can Build, McGraw-Hill, pp. 158–165.

Fig. 76-10. Radio Receiver Projects You Can Build, McGraw-Hill, pp. 138–147.

Fig. 76-11. Radio Receiver Projects You Can Build, McGraw-Hill, pp. 102–109.

Fig. 76-12. Radio Receiver Projects You Can Build, McGraw-Hill, pp. 196–203.

Fig. 76-13. Radio Receiver Projects You Can Build, McGraw-Hill, pp. 32–41.

Fig. 76-14. Analog Devices, Analog Dialogue, Vol. 26, No. 2, 1992, p. 19.

Fig. 76-15. Reprinted with permission from Popular Electronics, 5/94, p. 79. (C) Copyright Gernsback Publications, Inc., 1994.

Fig. 76-16. Reprinted with permission from Popular Electronics, 10/94, p. 25. (C) Copyright Gernsback Publications, Inc., 1994.

Fig. 76-17. Reprinted with permission from Electronics Now, 3/94, p. 68. (C) Copyright Gernsback Publications, Inc., 1994.

Fig. 76-18. Reprinted with permission from Electronics Now, 3/94, p. 67. (C) Copyright Gernsback Publications, Inc., 1994.

Fig. 76-19. Reprinted with permission from Popular Electronics, 10/94, p. 62. (C) Copyright Gernsback Publications, Inc., 1994.

Fig. 76-20. Analog Devices, Analog Dialogue, Vol. 27, No. 1 (1993), p. 15.

Fig. 76-21. Radio Craft, 1993, p. 64.

Fig. 76-22. William Sheets.

Fig. 76-23. Reprinted with permission from Electronics Now, 3/94, p. 69. (C) Copyright Gernsback Publications, Inc., 1994.

Fig. 76-24. Reprinted with permission from Electronics Now, 3/94, p. 69. (C) Copyright Gernsback Publications, Inc., 1994.

Fig. 76-25. Reprinted with permission from Electronics Now, 3/94, pp. 70–72. (C) Copyright Gernsback Publications, Inc., 1994.

Fig. 76-26. Reprinted with permission from Popular Electronics, 5/95, pp. 55-56. (C) Copyright Gernsback Publications, Inc., 1995.

Fig. 76-27. William Sheets.

Fig. 76-28. Reprinted with permission from Electronics Now, 3/95, p. 50. (C) Copyright Gernsback Publications, Inc., 1994.

Chapter 77

Fig. 77-1. Reprinted with permission from Electronic Design, 6/94, p. 29. Copyright 1995, Penton Publishing, Inc.

Fig. 77-2. Reprinted with permission of National Semiconductor Corporation, National Semiconductor Linear Applications Handbook, 1991, p. 26.

Fig. 77-3. Reprinted with permission of National Semiconductor Corporation, National Semiconductor Linear Applications Handbook, 1991, p. 26.

Chapter 78

Fig. 78-1. Linear Technology, Design Note #98.

Fig. 78-2. Reprinted with permission from Electronics Now, 7/94, p. 12. (C) Copyright Gernsback Publications, Inc., 1994.

Fig. 78-3. Reprinted with permission from Electronic Design, 2/95, p. 118. Copyright 1995, Penton Publishing, Inc.

Fig. 78-4. Linear Technology, Design Note #100.

Fig. 78-5. Linear Technology, Design Note #98.

Fig. 78-6. NASA Tech Briefs, August 1994, p. 38.

Fig. 78-7. Linear Technology, Advertisement LT/1294.

Fig. 78-8. Reprinted with permission from Electronic Design, 7/94, p. 32. Copyright 1994, Penton Publishing, Inc.

Fig. 78-9. Reprinted with permission from Electronic Design, 1/95, pp. 133–134. Copyright 1995, Penton Publishing, Inc.

Fig. 78-10. Linear Technology, Design Note #98.

Chapter 79

Fig. 79-1. William Sheets.

Fig. 79-2. William Sheets.

Fig. 79-3. William Sheets.

Fig. 79-4. William Sheets.

Fig. 79-5. William Sheets.

Fig. 79-6. William Sheets.

Fig. 79-7. William Sheets.

Chapter 80

Fig. 80-1. Maxim, Vol. III, New Releases Data Book, 1994, p. 3-30.

Fig. 80-2. Reprinted with permission of National Semiconductor Corporation, National Semiconductor Linear Applications Handbook, 1991, p. 1202.

Chapter 81

Fig. 81-1. Reprinted with permission from Popular Electronics, 11/94, pp. 75–76. (C) Copyright Gernsback Publications, Inc., 1994.

Fig. 81-2. Reprinted with permission from Popular Electronics, 11/94, pp. 74–75. (C) Copyright Gernsback Publications, Inc., 1994.

Chapter 82

Fig. 82-1. Reprinted with permission of National Semiconductor Corporation, National Semiconductor Linear Application Specific IC's Databook 1993, p. 1-37.

Fig. 82-2. Reprinted with permission from Popular Electronics, 9/94, p. 73. (C) Copyright Gernsback Publications, Inc., 1994.

Fig. 82-3. 73 Amateur Radio Today, 4/95, pp. 54–58.

Chapter 83

Fig. 83-1. William Sheets.

Fig. 83-2. William Sheets.

Fig. 83-3. Reprinted with permission from Popular Electronics, 5/93, p. 71. (C) Copyright Gernsback Publications, Inc., 1993.

Fig. 83-4. William Sheets.

Fig. 83-5. Sound Light and Music, Delton T. Horn, McGraw-Hill, pp. 105–106.

Chapter 84

Fig. 84-1. Sound Light and Music, Delton T. Horn, McGraw-Hill, pp. 106–110.

Fig. 84-2. Reprinted with permission from Popular Electronics, 1/94, p. 75. (C) Copyright Gernsback Publications, Inc., 1994.

Fig. 84-3. 1994 Electronic Experimenters Handbook, p. 53.

Fig. 84-4. William Sheets.

Fig. 84-5. Sound Light and Music, Delton T. Horn, McGraw-Hill, pp. 98–100.

Fig. 84-6. Sound Light and Music, Delton T. Horn, McGraw-Hill, pp. 114–115.

Fig. 84-7. Amplifiers, Waveform Generators & Other Low-Cost IC Projects, McGraw-Hill, p. 147.

Fig. 84-8. Gordon McComb's Gadgeteer's Goldmine, McGraw-Hill, p. 127.

Chapter 85

Fig. 85-1. Reprinted with permission from Popular Electronics, Fact Card No. 221. (C) Copyright Gernsback Publications, Inc.

Fig. 85-2. William Sheets.

Chapter 86

Fig. 86-1. Reprinted with permission from Popular Electronics, 11/94, pp. 76 and 91. (C) Copyright Gernsback Publications, Inc., 1994.

Fig. 86-2. William Sheets.

Fig. 86-3. Reprinted with permission of National Semiconductor Corporation, National Semiconductor Linear Applications Handbook, 1991, p. 235.

Fig. 86-4. Reprinted with permission of National Semiconductor Corporation, National Semiconductor Linear Applications Handbook, 1991, p. 236.

Chapter 87

Fig. 87-1. Reprinted with permission from Electronic Design, 7/94, p. 94. Copyright 1994, Penton Publishing, Inc.

Fig. 87-2. Reprinted with permission from Electronics Now, 2/94, p. 55. (C) Copyright Gernsback Publications, Inc., 1994.

Fig. 87-3. Reprinted with permission from Electronics Now, 2/94, p. 57. (C) Copyright Gernsback Publications, Inc., 1994.

Chapter 88

Fig. 88-1. Reprinted with permission from Popular Electronics, 6/93, p. 71. (C)

Copyright Gernsback Publications, Inc., 1993.

Fig. 88-2. Analog Devices, The Best of Analog Dialogue, 1967–1991, p. 117.

Fig. 88-3. Analog Devices, The Best of Analog Dialogue, 1967–1991, p. 117.

Fig. 88-4. Reprinted with permission from Electronic Design, 7/94, pp. 96–97. Copyright 1994, Penton Publishing, Inc.

Fig. 88-5. Analog Devices, The Best of Analog Dialogue, 1967–1991, p. 117.

Fig. 88-6. Reprinted with permission from Popular Electronics, 6/93, p. 71. (C) Copyright Gernsback Publications, Inc., 1993.

Fig. 88-7. Reprinted with permission from Electronics Now, 5/95, p. 63. (C) Copyright Gernsback Publications, Inc., 1995.

Fig. 88-8. Reprinted with permission from Electronic Design, 1/94, p. 118. Copyright 1994, Penton Publishing, Inc.

Fig. 88-9. Reprinted with permission from Popular Electronics, 12/94, p. 41. (C) Copyright Gernsback Publications, Inc., 1994.

Fig. 88-10. Reprinted with permission from Popular Electronics, 12/94, p. 42. (C) Copyright Gernsback Publications, Inc., 1994.

Fig. 88-11. Reprinted with permission from Popular Electronics, 12/94, p. 41. (C) Copyright Gernsback Publications, Inc., 1994.

Fig. 88-12. Amplifiers, Waveform Generators & Other Low-Cost IC Projects, McGraw-Hill, p. 72.

Fig. 88-13. Reprinted with permission from Popular Electronics, 12/94, p. 41. (C) Copyright Gernsback Publications, Inc., 1994.

Fig. 88-14. Reprinted with permission from Popular Electronics, 12/94, p. 41. (C) Copyright Gernsback Publications, Inc., 1994.

Fig. 88-15. Reprinted with permission from Popular Electronics, 12/94, p. 41. (C) Copyright Gernsback Publications, Inc., 1994.

Fig. 88-16. Maxim, Vol. III, New Releases Data Book, 1994, p. 3-125.

Fig. 88-17. William Sheets.

Fig. 88-18. Reprinted with permission from Popular Electronics, 6/93, p. 71. (C) Copyright Gernsback Publications, Inc., 1993.

Fig. 88-19. William Sheets.

Fig. 88-20. Reprinted with permission from Popular Electronics, 6/93, p. 72. (C) Copyright Gernsback Publications, Inc., 1993.

Fig. 88-21. Electronics Now, 8/93, p. 12.

Fig. 88-22. William Sheets.

Fig. 88-23. Analog Devices, The Best of Analog Dialogue, 1967–1991, p. 117.

Chapter 89

Fig. 89-1. Reprinted with permission from Electronic Design, 10/88, pp. 126–128. Copyright 1988, Penton Publishing, Inc.

Fig. 89-2. Analog Devices, Analog Dialogue, Vol. 26, No. 1, 1992, p. 12.

Fig. 89-3. Reprinted with permission from Electronic Design, 11/93, p. 99. Copyright 1993, Penton Publishing, Inc.

Fig. 89-4. Analog Devices, Analog Dialogue, Vol. 26, No. 1, 1992, p. 13.

Fig. 89-5. Reprinted with permission from Electronic Design, 10/88, p. 128. Copyright 1988, Penton Publishing, Inc.

Chapter 90

Fig. 90-1. Reprinted with permission from 1987 Radio-Electronics Experimenters Handbook, p. 51. (C) Copyright Gernsback Publications, Inc., 1994.

Fig. 90-2. Reprinted with permission from Electronics Now, 2/94, p. 33. (C) Copyright Gernsback Publications, Inc., 1994.

Fig. 90-3. Reprinted with permission from Electronics Now, 10/94, p. 53. (C) Copyright Gernsback Publications, Inc., 1994.

Fig. 90-4. 1994 Electronics Experimenters Handbook, p. 123.

Fig. 90-5. Reprinted with permission from Electronics Now, 4/95, pp. 39–40. (C) Copyright Gernsback Publications, Inc., 1995.

Fig. 90-6. Reprinted with permission from Electronics Now, 8/94, p. 26. (C) Copyright Gernsback Publications, Inc., 1994.

Fig. 90-7. Analog Devices, The Best of Analog Dialogue, 1967–1991, p. 75.

Fig. 90-8. Reprinted with permission from Popular Electronics, 12/93, p. 62. (C) Copyright Gernsback Publications, Inc., 1993.

Fig. 90-9. Gordon McComb's Gadgeteer's Goldmine, McGraw-Hill, pp. 335–336.

Fig. 90-10. Reprinted with permission from Electronics Now, 2/94, p. 16. (C) Copyright Gernsback Publications, Inc., 1994.

Fig. 90-11. Reprinted with permission from Electronics Now, 8/93, pp. 58–63. (C) Copyright Gernsback Publications, Inc., 1993.

Fig. 90-12. Reprinted with permission from Radio-Electronics, August 1993, pp. 58–63. (C) Copyright Gernsback Publications, Inc., 1993.

Fig. 90-13. Reprinted with permission from Electronic Design, 8/94, p. 116. Copyright 1994, Penton Publishing, Inc.

Fig. 90-14. William Sheets.

Chapter 91

Fig. 91-1. 1994 Electronics Experimenters Handbook, p. 63.

Fig. 91-2. Reprinted with permission from Electronic Design, 7/94, p. 93. Copyright 1994, Penton Publishing, Inc.

Fig. 91-3. Reprinted with permission from Electronic Design, 11/93, pp. 90–92. Copyright 1993, Penton Publishing, Inc.

Fig. 91-4. Reprinted with permission from Electronic Design, 1/95, pp. 80–81. Copyright 1995, Penton Publishing, Inc.

Fig. 91-5. Reprinted with permission from Popular Electronics, 6/95, p. 48. (C) Copyright Gernsback Publications, Inc., 1995.

Fig. 91-6. Reprinted with permission of National Semiconductor Corporation, National Semiconductor Linear Applications Handbook, 1991, p. 524.

Fig. 91-7. Reprinted with permission from Electronic Design, 1/94, pp. 118–119.

Copyright 1994, Penton Publishing, Inc.

Fig. 91-8. Reprinted with permission of National Semiconductor Corporation, National Semiconductor Linear Applications Handbook, 1991, p. 524.

Fig. 91-9. Reprinted with permission of National Semiconductor Corporation, National Semiconductor Linear Applications Handbook, 1991, p. 500.

Fig. 91-10. Reprinted with permission of National Semiconductor Corporation, National Semiconductor Linear Applications Handbook, 1991, p. 1076.

Fig. 91-11. Reprinted with permission of National Semiconductor Corporation, National Semiconductor Linear Applications Handbook, 1991, p. 1076.

Fig. 91-12. Reprinted with permission of National Semiconductor Corporation, National Semiconductor Linear Applications Handbook, 1991, p. 1076.

Fig. 91-13. Reprinted with permission of National Semiconductor Corporation, National Semiconductor Linear Applications Handbook, 1991, p. 1076.

Fig. 91-14. Analog Devices, The Best of Analog Dialogue, 1967–1991, p. 75.

Fig. 91-15. Analog Devices, The Best of Analog Dialogue, 1967–1991, p. 75.

Fig. 91-16. Reprinted with permission of National Semiconductor Corporation, National Semiconductor, NSLAH 1991, p. 1079.

Fig. 91-17. Reprinted with permission of National Semiconductor Corporation, National Semiconductor Linear Applications Handbook, 1991, p. 493.

Fig. 91-18. Reprinted with permission of National Semiconductor Corporation, National Semiconductor Linear Applications Handbook, 1991, p. 1077.

Fig. 91-19. Reprinted with permission from Popular Electronics, 5/93, p. 43. (C) Copyright Gernsback Publications, Inc., 1993.

Fig. 91-20. Reprinted with permission of Burr-Brown Corporation, Burr-Brown Data Sheet INA118, (C) 1989–1995 Burr-Brown Corporation.

Fig. 91-21. Analog Devices, Analog Dialogue, Vol. 27, No. 2, p. 22.

Fig. 91-22. Reprinted with permission of National Semiconductor Corporation, National Semiconductor Linear Applications Handbook, 1991, p. 1076.

Fig. 91-23. QST, 5/95, p. 85.

Fig. 91-24. Reprinted with permission of National Semiconductor Corporation, National Semiconductor Linear Applications Handbook, 1991, p. 523.

Fig. 91-25. Reprinted with permission from Popular Electronics, 1/95, p. 28. (C) Copyright Gernsback Publications, Inc., 1995.

Fig. 91-26. QST, 5/95, p. 85.

Fig. 91-27. Reprinted with permission of National Semiconductor Corporation, National Semiconductor Linear Applications Handbook, 1991, p. 1079.

Fig. 91-28. Reprinted with permission of National Semiconductor Corporation, National Semiconductor Linear Applications Handbook, 1991, p. 1076.

Fig. 91-29. Reprinted with permission from Popular Electronics, 1/95, p. 29. (C) Copyright Gernsback Publications, Inc., 1995.

Fig. 91-30. Power Supplies, Switching Regulators, Inverters, and Converters, McGraw-Hill, pp. 163–164.

Chapter 92

Fig. 92-1. Reprinted with permission from Popular Electronics, 9/94, p. 26. (C) Copyright Gernsback Publications, Inc., 1994.

Fig. 92-2. Reprinted with permission from Popular Electronics, 2/94, p. 25. (C) Copyright Gernsback Publications, Inc., 1994.

Fig. 92-3. Reprinted with permission from Electronic Design, 3/95, pp. 114–116. Copyright 1995, Penton Publishing, Inc.

Chapter 93

Fig. 93-1. William Sheets.
Fig. 93-2. William Sheets.
Fig. 93-3. William Sheets.

Fig. 93-4. William Sheets.

Chapter 94

Fig. 94-1. William Sheets.

Fig. 94-2. Amplifiers, Waveform Generators & Other Low-Cost IC Projects, McGraw-Hill, p. 92.

Fig. 94-3. NASA Tech Briefs, August 1994, pp. 34–35.

Fig. 94-4. William Sheets.

Fig. 94-5. Reprinted with permission from Popular Electronics, Fact Card No. 270. (C) Copyright Gernsback Publications, Inc.

Chapter 95

Fig. 95-1. Spring 1994 Electronics Hobbyist Handbook.

Fig. 95-2. Reprinted with permission from Electronics Now, 5/95, p. 8. (C) Copyright Gernsback Publications, Inc., 1995.

Fig. 95-3. Reprinted with permission from Popular Electronics, Fact Card No. 221. (C) Copyright Gernsback Publications, Inc.

Chapter 96

Fig. 96-1. Reprinted with permission from Electronics Now, 6/93, pp. 41–46. (C) Copyright Gernsback Publications, Inc., 1993.

Fig. 96-2. Reprinted with permission from Electronics Now, 12/93, p. 29. (C) Copyright Gernsback Publications, Inc., 1993.

Fig. 96-3. 1994 Electronics Experimenters Handbook, p. 105.

Fig. 96-4. Reprinted with permission from Electronics Now, 1994 Experimenters Handbook, p. 77. (C) Copyright Gernsback Publications, Inc., 1994.

Fig. 96-5. Gordon McComb's Gadgeteer's Goldmine, McGraw-Hill, pp. 340–341.

Fig. 96-6. 73 Amateur Radio Today, 5/93, p. 56.

Fig. 96-7. Reprinted with permission from Popular Electronics, 8/92, p. 45. (C) Copyright Gernsback Publications, Inc., 1992.

Chapter 97

Fig. 97-1. Reprinted with permission from Popular Electronics, 5/95, p. 29. (C) Copyright Gernsback Publications, Inc., 1995.

Fig. 97-2. Reprinted with permission from Popular Electronics, 1/95, pp. 72–73. (C) Copyright Gernsback Publications, Inc., 1995.

Fig. 97-3. Reprinted with permission from Popular Electronics, 1/95, p. 71. (C) Copyright Gernsback Publications, Inc., 1995.

Fig. 97-4. Reprinted with permission from Popular Electronics, 1/95, p. 73. (C) Copyright Gernsback Publications, Inc., 1995.

Fig. 97-5. Reprinted with permission from Popular Electronics, 12/94, p. 72. (C) Copyright Gernsback Publications, Inc., 1994.

Fig. 97-6. Reprinted with permission from Popular Electronics, 12/94, pp. 74–75. (C) Copyright Gernsback Publications, Inc., 1994.

Chapter 98

Fig. 98-1. Reprinted with permission from Electronics Now, 5/95, pp. 50–51. (C) Copyright Gernsback Publications, Inc., 1995.

Fig. 98-2. Analog Devices, AD8001 Data Sheet.

Fig. 98-3. Reprinted with permission of National Semiconductor Corporation, National Semiconductor Linear Application Specific IC's Databook 1993, p. 3-50.

Fig. 98-4. Reprinted with permission of National Semiconductor Corporation, National Semiconductor Linear Application Specific IC's Databook 1993, p. 3-33.

Fig. 98-5. Reprinted with permission from Radio-Electronics Experimenters Handbook, pp. 83–88, 1994. (C) Copyright Gernsback Publications, Inc., 1994.

Fig. 98-6. Reprinted with permission of National Semiconductor Corporation, National Semiconductor Linear Applications Handbook, p. 1021.

Fig. 98-7. Maxim, Vol. III, New Releases Data Book, 1994, p. 8-16.

Fig. 98-8. 73 Amateur Radio Today, 7/93, p. 74.

Fig. 98-9. Reprinted with permission of National Semiconductor Corporation, National Semiconductor Linear Application Specific IC's Databook 1993, p. 3-32.

Fig. 98-10. 73 Amateur Radio Today, 7/93, p. 74.

Fig. 98-11. Reprinted with permission of National Semiconductor Corporation, National Semiconductor Linear Application Specific IC's Databook 1993, p. 3-13.

Fig. 98-12. Linear Technology, Design Note #92.

Fig. 98-13. Maxim, Vol. III, New Releases Data Book, 1994, p. 8-16.

Fig. 98-14. Reprinted with permission from Popular Electronics, Fact Card No. 268. (C) Copyright Gernsback Publications, Inc.

Fig. 98-15. Maxim, Vol. III, New Releases Data Book, 1994, p. 8-19.

Fig. 98-16. Linear Technology, 2/95.

Fig. 98-17. Analog Dialogue, Analog Devices, Vol. 26, No. 1, 1992, p. 11.

Chapter 99

Fig. 99-1. Analog Devices, The Best of Analog Dialogue, 1967–1991, p.75.

Fig. 99-2. Reprinted with permission of National Semiconductor Corporation, National Semiconductor Linear Application Specific IC's Databook 1993, p. 1-37.

Chapter 100

Fig. 100-1. Reprinted with permission from Popular Electronics, Fact Card No. 269. (C) Copyright Gernsback Publications, Inc.

Fig. 100-2. Analog Devices, The Best of Analog Dialogue, 1967–1991, p. 79.

Chapter 101

Fig. 101-1. Reprinted with permission of National Semiconductor Corporation, National Semiconductor Linear Applications Handbook, 1991, p. 500.

Fig. 101-2. Maxim, Vol. III, New Releases Data Book, 1994, p. 3-126.

Fig. 101-3. Reprinted with permission from Electronic Design, 11/94, pp. 127–128. Copyright 1994, Penton Publishing, Inc.

Fig. 101-4. Reprinted with permission of National Semiconductor Corporation, National Semiconductor Linear Applications Handbook, 1991, p. 207.

Fig. 101-5. QST, 4/95, p. 61.

Fig. 101-6. 73 Amateur Radio Today, 3/95, p 63.

Fig. 101-7. Reprinted with permission from Popular Electronics, 5/94, p. 80. (C) Copyright Gernsback Publications, Inc., 1994.

Fig. 101-8. Reprinted with permission of National Semiconductor Corporation, National Semiconductor Linear Applications Handbook, 1991, p. 499.

Fig. 101-9. Reprinted with permission from Popular Electronics, 5/94, p. 79. (C) Copyright Gernsback Publications, Inc., 1994.

Fig. 101-10. Reprinted with permission of National Semiconductor Corporation, National Semiconductor Linear Applications Handbook, 1991, p. 206.

Chapter 102

Fig. 102-1. Reprinted with permission from Popular Electronics, 11/94, p. 74. (C) Copyright Gernsback Publications, Inc., 1994.

Fig. 102-2. Reprinted with permission from Popular Electronics, 5/94, p. 90. (C) Copyright Gernsback Publications, Inc., 1994.

Fig. 102-3. Reprinted with permission from Electronic Design, 12/94, pp. 118–119. Copyright 1994, Penton Publishing, Inc.

Fig. 102-4. Reprinted with permission from Electronics Now, 8/93, p. 14. (C) Copyright Gernsback Publications, Inc., 1993.

Fig. 102-5. Reprinted with permission of National Semiconductor Corporation, National Semiconductor Linear Applications Handbook, 1991, p. 208.

Fig. 102-6. Gordon McComb's Gadgeteer's Goldmine, McGraw-Hill, p. 197.

Fig. 102-7. Amplifiers, Waveform Generators & Other Low-Cost IC Projects, McGraw-Hill, p. 21.

Fig. 102-8. Reprinted with permission from Popular Electronics, 11/94, pp. 76 and 91. (C) Copyright Gernsback Publications, Inc., 1994.

Fig. 102-9. Reprinted with permission from Electronic Design, 7/94, p. 96. Copyright 1994, Penton Publishing, Inc.

Fig. 102-10. Power Supplies, Switching Regulators, Inverters, and Converters, McGraw-Hill, p. 129.

Fig. 102-11. Reprinted with permission from Electronics Now, 3/95, p. 63. (C) Copyright Gernsback Publications, Inc., 1995.

Chapter 103

Fig. 103-1. 73 Amateur Radio Today, 12/93, p. 70.

Fig. 103-2. 73 Amateur Radio Today, 12/93, p. 72.

Chapter 104

Fig. 104-1. William Sheets.

Fig. 104-2. William Sheets.

Index

Numbers preceded by an "I", "II", "III", "IV", "V", or "VI" are from *Encyclopedia of Electronic Circuits* Vol. I, II, III, IV, V, or VI respectively.

A

absolute-value circuits, I-37, IV-274
  amplifier, I-31
  full wave rectifier, II-528
  Norton amplifier, III-11
  precision, I-37, IV-274
ac amplifier, high input impedance, VI-55
ac line/timer interface, VI-281
ac motors (*see also* motor control circuits)
  control for, II-375
  power brake, II-451
  three-phase driver for, II-383
  two-phase driver for, I-456, II-382
ac power monitor, VI-351
ac/dc indicator, IV-214
ac-to-dc converters, I-165
  fixed power supplies, IV-395
  full-wave, IV-120
  high-impedance precision rectifier, I-164
accelerometer, VI-345
acid rain monitor, II-245, III-361, V-371
acoustic field generator, V-338-341
acoustic sound receiver/transmitter, IV-311
active antennas (*see* antennas, active)
active filters (*see also* filter circuits)
  band reject, II-401
  bandpass, III-190, II-221, II-223
    variable bandwidth, I-286
  digitally tuned low-power, II-218
  five pole, I-279
  fourth-order low-pass, V-184
  high-pass, V-180, V-188
    fourth-order, V-188
    second-order, I-297
  low-pass, V-178, V-181, V-188
    digitally selected break frequency, II-216
    unity-gain, V-187
  low-power
    digitally selectable center frequency, III-186
    digitally tuned, I-279
  programmable, III-185
  RC, up to 150 kHz, I-294
  speech-range filter, V-185
  state-variable, III-189
  ten-band graphic equalizer using, II-684
  three-amplifier, I-289
  tunable, I-289
  universal, II-214
adapters (*see also* conversion and converters)
  dc transceiver, hand-held, III-461
  line-voltage-to-multimeter adapter, V-312

program, second-audio, III-142
  traveller's shaver, I-495
adder circuits, III-327
  binary, fast-action, IV-260-261
AFSK generator, one-chip, VI-23
AGC (*see* automatic gain control (AGC)
air conditioner, auto, smart clutch for, III-46
aircraft receiver, 118- to 136-MHz, VI-542
air motion and pressure
  barometer, VI-338
  electronic anemometer, VI-6
  flow-detector, I-235, II-240-242, III-202-203, IV-82, V-154, VI-4-6, VI-183
  flow-meters (anemometers)
    hot-wire, III-342, V-5, VI-4-6, VI-183
    thermally based, II-241
  pressure change detector, IV-144
  motion detector, I-222, III-364
airplane propeller sound effect, II-592
alarms (*see also* annunciators; sirens), I-4, III-3-9, IV-84-89, V-1-16, VI-7-16
  555-based alarm, V-11
  alarm-tone generator, V-563
  amateur radio on-alarm and timer, VI-32
  audio-sensor alarm, V-8
  auto burglar, I-3, I-7, I-10, II-2, III-4, IV-53
    alarm decoy, VI-13
    automatic-arming, IV-50
    automatic turn-off, 8 minute delay, IV-52
    CMOS low-current, IV-56
    horn as loudspeaker, IV-54
    motion-actuated car/motorcycle, I-9
    security system, I-5, IV-49-56, VI-9, VI-11
    single-IC, III-7, IV-55
  auto-arming automotive alarm, IV-50
  automatic turn-off, IV-54
    8 minute delay, IV-52
  baby-alert transmitter/receiver, V-95-96
  backup battery low alarm, VI-110
  bells, electronic, II-33, I-636
  blown fuse, I-10
  boat, I-9
  body-heat detector, VI-266
  burglar alarms, III-8, III-9, IV-86, VI-8
    burglar chaser, V-16
    latching circuit, I-8, I-12
    NC and NO switches, IV-87
    NC switches, IV-87

one-chip, III-5
  self-latching, IV-85
  timed shutoff, IV-85
camera triggered, III-444
capacitive sensor, III-515
current monitor and, III-338
dark-activated alarm, pulsed tone output, V-13
delayed alarm, V-4
differential voltage or current, II-3
digital clock circuit with, III-84
door-ajar, II-284, III-46, VI-14
  Hall-effect circuit, III-256
door minder, V-5
doorbells (*see* annunciators)
driver, high-power alarm driver, V-2
exit delay for burglar alarms, V-10
fail-safe, semiconductor, III-6
field disturbance, II-507
flasher signal, V-197
flashing brake light for motorcycles, VI-12
flex switch alarm sounder, V-15
flood, I-390, III-206, IV-188, V-374
freezer meltdown, I-13
headlights-on, III-52, V-77
heat-activated alarm, V-9
high/low-limit, I-151
home security system, I-6, IV-87, VI-10-11
ice formation, II-58
infrared wireless system, IV-222-223
latching relay alarm circuit, VI-569
light-activated, V-9, V-273
  high-output, pulsed tone, V-14
  precision design, V-12
  precision with hysteresis, V-14
  self-latch, tone output, V-15
  with latch, V-12
light-beam intruder-detection alarm, V-11, V-13
loop circuit alarms
  closed-loop, V-3
  multi-loop parallel alarm, V-2
  parallel, V-3
  series/parallel, V-3
low-battery disconnect and, III-65
low-battery warning, III-59
low-volts, I-493
motorcycle alarm, VI-13
motorcycle burglar alarm, VI-15
motorcycle horn alarm, VI-14, VI-15
multiple circuit for, II-2
no-doze alarm, V-8
one-chip, III-5
photoelectric, II-4, II-319
piezoelectric, I-12, V-10
power failure, I-581, I-582, III-511
printer error, IV-106
proximity, II-506, III-517, V-485-486, VI-657

current sources, *continued*
  regulator, variable power supply, III-490
  variable power supplies, voltage-programmable, IV-420
  voltage-controlled, III-468, VI-162, VI-163
current-limiting regulator, V-458
current-shunt amplifiers, III-21
current-to-frequency converter, IV-113
  wide range, I-164
current-to-voltage amplifier, high-speed, I-35
current-to-voltage converter, I-162, I-165, V-127, VI-154, VI-155
  grounded bias and sensor in, II-126
  photodiode, II-128
curve tracer, V-300
  diodes, IV-274
  FET, I-397
CW-related circuits
  audio filter, VI-29, VI-405
  CW/SSB receiver, 80- and 40-meter, V-499
  filter, razor sharp, II-219
  identifier, VI-24, VI-408
  keying circuits, IV-244
  offset indicator, IV-213
  SSB/CW product detector, IV-139
  transceiver, 5 W, 80-meter, IV-602
  transmitters
    1-W, III-678
    6-W 40-M, VI-664
    20-M low-power, V-649
    40-M, III-684, V-648
    902-MHz, III-686
    HF low-power, IV-601
    keying circuit, VI-22-23
    one-watt, VI-27
    QRP, III-69
    ultrasonic transceiver, VI-669
cyclic A/D converter, II-30

D

dark-activated (*see* light-controlled circuits)
darkroom equipment (*see* photography-related circuits)
Darlington amplifier, push-pull, V-22
Darlington regulator, variable power supplies, IV-421
Darlington transistor oscillator, VI-455
data-manipulation circuits, IV-129-133
  acquisition circuits, IV-131, VI-378
    CMOS system, II-117
    four-channel, I-421
    high-speed system, II-118
  analog-signal transmission isolator, IV-133
  link, IR type, I-341
  prescaler, low-frequency, IV-132
  read-type circuit, 5 MHz, phase-encoded, II-365
  receiver, carrier-current circuit design, IV-93
  receiver/message demuxer, three-wire, IV-130
  selector, RS-232, III-97
  separator, floppy disk, II-122
  transmission circuits, IV-92
dc adapter/transceiver, hand-held, III-461

dc generators, high-voltage, III-481
dc motors (*see also* motor control circuits)
  direction control, I-452
  driver controls
    fiberoptic control, II-206
    fixed speed, III-387
    servo, bipolar, II-385
    reversible, II-381, III-388
  speed control, I-452, I-454, III-377, III-380, III-388
dc restorer, video, III-723
dc servo drive, bipolar control input, II-385
dc static switch, II-367
dc-to-ac inverter, V-247, V-669
dc-to-dc conversion, IV-118, V-669, VI-164-167
  1-to-5 V, IV-119
  3-to-5 V battery, IV-119
  3-to-25 V, IV-118
  3.3- and 5-V outputs, V-128
  3 A, no heatsink, V-119
  bipolar, no inductor, II-132
  fixed 3- to 15-V supply, IV-400
  isolated, VI-165
  isolated +15V, III-115
  negative step-up converter, VI-166
  push-pull, 400 V/60 W, I-210
  regulating, I-210, I-211, II-125, III-121
  step-up/step-down, III-118
  ultra low-power for personal communications, VI-166
dc-to-dc inverter, VI-285
dc-to-dc SMPS variable power supply, II-480
debouncers, III-592, IV-105, V-316, VI-387, VI-613, VI-614
  auto-repeat, IV-106
  computer applications, IV-105, IV-106, IV-108
  flip-flop, IV-108
debugger, coprocessor sockets, III-104
decibel level detector, audio, with meter driver, III-154
decoders, II-162, III-141-145, VI-168-171
  10.8 MHz FSK, I-214
  24-percent bandwidth tone, I-215
  BCD decoder/driver, multiplexed, VI-189
  direction detector, III-144
  DTMF decoder, VI-169
  dual-tone, I-215
  encoder and, III-144
  FM stereo decoder, VI-170
  frequency division multiplex stereo, II-169
  PAL/NTSC, with RGB input, III-717
  radio control receiver, I-574
  SCA, I-214, III-166, III-170
  second-audio program adapter, III-142
  sound-activated, III-145
  stereo TV, II-167
  time division multiplex stereo, II-168
  tone alert, I-213
  tone dial, I-630, I-631
  tone decoders, I-231, III-143, VI-170
    24% bandwidth, I-215
    dual time constant, II-166
    relay output, I-213
    tone-dial decoder, I-630, I-631

video, NTSC-to-RGB, IV-613
video line decoders, VI-171
weather-alert detector/decoder, IV-140
deglitcher circuit, IV-109, V-336-337
delay circuits/ delay units, III-146-148, V-147-148, VI-172-173
  adjustable, III-148
  analog delay line, echo and reverb effects, IV-21
  door chimes, I-218
  echo and reverb effects, analog delay line, IV-21
  exit delay for burglar alarms, V-10
  headlights, I-107, II-59
  leading-edge, III-147
  long duration time, I-217, I-220
  power-on delay, V-148
  precision solid state, I-664
  pulse, dual-edge trigger, III-147
  pulse generator, II-509
  relay, ultra-precise long time, II-211
  timed delay, I-668, II-220
  time-delay generator, VI-173
  constant-current charging, II-668
  windshield wiper delay, I-97, II-55
demodulators, II-158-160, III-149-150
  5V FM, I-233
  12V FM, I-233
  565 SCA, III-150
  AM demodulator, II-160
  chroma, with RGB matrix, III-716
  FM demodulator, I-544, II-161, V-151, V-155
    narrow-band, carrier detect, II-159
  linear variable differential transformer driver, I-403
  LVDT demodulators, II-337, III-323-324
  stereo, II-159
  telemetry, I-229
demonstration comparator circuit, II-109
demultiplexers (*see also* multiplexers), III-394
  differential, I-425
  eight-channel, I-426, II-115
descramblers, II-162
  gated pulse, II-165
  outband, II-164
  sine wave, II-163
derived center-channel stereo system, IV-23
detect-and-hold circuit, peak, I-585
detectors (*see* fluid and moisture; light-controlled circuits; motion and proximity; motor control circuits; peak detectors; smoke detectors; speed controllers; temperature-related circuits; tone controls; zero-crossing)
deviation meter, IV-303
dial pulse indicator, telephone, III-613
dialers, telephone
  pulse-dialing telephone, III-610
  pulse/tone, single-chip, III-603
  telephone-line powered repertory, I-633
  tone-dialing telephone, III-607
dice, electronic, I-325, III-245, IV-207
differential amplifiers, I-38, III-14, V-18, V-21, VI-185-287
  high-impedance, I-27, I-354
  high-input high-impedance, II-19

speaker amplifier, IV-555
speakerphone, II-632, III-608
speakerphone adapter, V-606-607
speech activity detector, II-617, III-615
speech network, II-633
status monitor using optoisolator, I-626
switch, solid-state, line-activated, III-617
tap, III-622
tape-recorder starter controlled by, I-632
telecom converter -48 to +5 V at 1 A, V-472
timer, tele-timer, V-623
toll-totalizer, IV-551
tone-dialing, III-607
tone ringers, I-627, I-628, II-630, II-631
Touchtone generator, III-609
touch-tone decoder, IV-555
vocalizer, dialed-phone number, III-731
voice-mail alert, V-607
teleprinter loop supply, VI-497
television (*see* amateur television; video circuits)
temperature-related circuits (*see also* thermometers), I-641-643, I-648, I-657, II-645, III-629-631, IV-565-572, V-616-620, VI-629-647
0-50 C, four-channel temperature, I-648
A/D converter for temperature measurement, VI-234-235, VI-634
absolute temperature log with RS-232, VI-636
alarms, II-4, II-643, II-644, V-9
amplifier, precision RTD, for +5 V, VI-643
automotive water-temperature gauge, II-56, IV-44, IV-48
body-heat detector, VI-266
boiler temperature control, I-638
compensation adjuster, V-617
control circuits, I-641-643, II-636-644, III-623-628, IV-567, VI-631, VI-641, VI-646
defrost cycle, IV-566
heater element, II-642
heater protector, servo-sensed, III-624
heat sniffer, electronic, III-627
liquid-level monitor, II-643
low-power, zero-voltage switch, II-640
piezoelectric fan-based, III-627
proportional, III-626
signal conditioners, II-639
single setpoint, I-641
thermocoupled, IV-567
zero-point switching, III-624
converters
logarithmic, V-127
temperature-to-digital, V-123
temperature-to-frequency, I-168, I-646, I-656, II-651-653, V-121
temperature-to-time, III-632-633
cool-down circuit for amplifiers, V-354, V-357
defrost cycle and control, IV-566
differential temperature, I-654, I-655, VI-645
flame temperature, III-313
freeze-up sensor, VI-647

furnace fuel miser, V-328-329
heater control, I-639, I-640, II-642, III-624, VI-632
heat sniffer, III-627
hi/lo sensor, II-650
hook sensor on 4- to 20-mA loop, V-618
IC temperature, I-649
indicator, II-56, IV-570
isolated temperature, I-651
LCD contrast temperature compensator, VI-195
logarithmic converter, V-127
low-temperature sensor, V-619
measuring circuit/sensors, II-653, IV-572
meters/monitors, I-647, III-206, IV-569
op amp, temp-compensated breakpoint, V-401
oscillators, temperature-controlled, I-187, II-427, III-137
over-temperature switch, IV-571
over/under sensor, dual output, II-646
proportional temperature controller, V-633
pyrometer, optoelectronic, VI-640
regulator, zero-voltage switching, VI-647
remote sensors, I-649, I-654, V-619
room temperature display, bar graph, VI-641
sensors, I-648, I-657, II-645-650, III-629-631, IV-568-572, V-619
-50 to 300 F, single supply, VI-638
0-50-degree C four channel, I-648
0-63 degrees C, III-631
5 V powered linearized platinum RTD signal conditioner, II-650
automotive-temperature indicator, PTC thermistor, II-56
Centigrade thermometer, II-648
coefficient resistor, positive, I-657
differential, I-654, I-655
full-range Fahrenheit, VI-643
output referenced to ground, two-wire, VI-638
over/under, dual output, II-646
DVM interface, II-647
hi/lo, II-650
integrated circuit, I-649
isolated, I-651, III-631
low-temperature, V-619
remote, I-649, I-654, V-619, VI-638
soil heater for plants, V-333
soldering iron control, V-327
thermal monitor, IV-569
thermocouple amplifier, cold junction compensation, II-649
thermocouple multiplex system, III-630
zero-crossing detector, I-733
signal conditioners, II-639
single-setpoint, temperature, I-641
temperature-to-digital converter, V-123, VI-646
temperature-to-frequency converter, I-168, I-646, I-656, II-651-653, V-121, VI-639
temperature-to-time converters, III-632-633
thermocouples
amplifier, cold junction compensation, II-649, VI-635, VI-

642, VI-644
control, IV-567
multiplex system, III-630
thermometers (*see* thermometers)
thermostat (*see* thermostats)
thermostatic fan switch, V-68
thermostatic relay circuit, VI-643, VI-645
transconducer, I-646, I-649
under-temperature switch, IV-570
zero-crossing detector, I-733
temperature-to-frequency converter, I-168, I-656, II-651-653, VI-639
temperature-to-frequency transconducer, linear, I-646
temperature-to-time converters, III-632-633
ten-band graphic equalizer, active filter, II-684
Tesla coils, III-634-636
test bench amplifier, V-26
test circuits (*see* measurement/test circuits)
text adder, composite-video signal, III-716
theremins, II-654-656
digital, II-656
electronic, II-655
thermal flowmeter, low-rate flow, III-203
thermocouples, II-649, VI-635, VI-642, VI-644
amplifiers, I-355, I-654, II-14, II-649
digital thermometer using, II-658
multiplex, temperature sensor system, III-630
pre-amp using, III-283
thermometers (*see also* temperature-related circuits), II-657-662, III-637-643, IV-573-577
0-50 degree F, I-656
0-100 degree C, I-656
1.5-V, VI-637
5-V operation, V-617
adapter, III-642
add-on for DMM digital voltmeter, III-640
centigrade, I-655, II-648, II-662
calibrated, I-650
ground-referred, I-657
differential, I-652, II-661, III-638, VI-640
digital, I-651, I-658, V-618, VI-637
temperature-reporting, III-638
thermocouple, II-658
μP controlled, I-650
electronic, II-660, III-639, IV-575, IV-576
Fahrenheit, I-658
ground-referred, I-656
high-accuracy design, IV-577
implantable/ingestible, III-641
Kelvin, I-653, I-655, II-661
linear, III-642, IV-574
low-power, I-655
meter, trimmed output, I-655
remote, II-659
single-dc supply, IV-575
variable offset, I-652
thermostats, I-639, I-640, V-60, VI-630
third-overtone oscillator, I-186, IV-123
three-in-one test set, III-330
three-minute timer, III-654
three-rail power supply, III-466

781

wireless microphones (*see* microphones)
wireless speaker system, IR, III-272
wiring
  ac outlet tester, V-318
  ac wiring locator, V-317
  two-way switch, V-591
write amplifiers, III-18
WWV converter, VI-147
WWV receiver, VI-538-539, VI-558

X

xenon flash trigger, slave, III-447
XOR gates, IV-107
  complementary signals generator, III-226

oscillator, III-429
up/down counter, III-105

Y

yelp oscillator/siren, II-577, III-562

Z

Z80 clock, II-121
Z-Dice game 248-249, VI-248
zappers, battery, II-64, II-66, II-68
zener diodes
  clipper, fast and symmetrical, IV-329
  increasing power rating, I-496, II-485

limiter using one-zener design, IV-257
test set, V-321
tester, I-400
variable, I-507
voltage regulator, programmable, IV-470
zero crossing detector, I-732, I-733, II-173
zero crossing switch, VI-606
zero meter, suppressed, I-716
zero-point switches
  temperature control, III-624
  triac, II-311
zero-voltage switches
  closed contact half-wave, III-412
  solid-state, III-410, III-416

ABOUT THE AUTHORS

Rudolf F. Graf has 45 years of engineering, sales, and marketing experience in the electronics field. He has written more than 30 books (about three million copies printed) and well over 100 articles. He is a senior member of the IEEE, a licensed amateur radio operator (KA2CWL), and has a BSEE degree from Polytechnic Institute of Brooklyn and an MBA from NYU. He is self-employed.

William Sheets is a self-employed circuit design engineer. He has more than 25 years of experience in RF, analog, and digital electronics. He has written numerous articles in electronics publications and co-authored five books with Graf. His interests include amateur radio (K2MQJ), photography, and travel. He has designed and built numerous items, including a satellite TV system, many transmitters and receivers, and a computer. He has an MEE degree from NYU, is married, and lives in upstate New York.